VIDEO CODING

The Second Generation Approach

VIDEO CODING

The Second Generation Approach

EDITED BY

Luis Torres
Department of Signal Theory and Communications,
Universitat Politècnica de Catalunya
Barcelona, Spain

and

Murat Kunt
Signal Processing Laboratory,
Swiss Federal Institute of Technology
Lausanne, Switzerland

KLUWER ACADEMIC PUBLISHERS
Boston/London/Dordrecht

A C.I.P. Catalogue record for this book is available from the Library of Congress

ISBN 0-7923-9680-4

Published by Kluwer Academic Publishers,
P.O. Box 17, 3300 AA Dordrecht, The Netherlands.

Kluwer Academic Publishers incorporates
the publishing programmes of
D. Reidel, Martinus Nijhoff, Dr W. Junk and MTP Press.

Sold and distributed in the U.S.A. and Canada
by Kluwer Academic Publishers,
101 Philip Drive, Norwell, MA 02061, U.S.A.

In all other countries, sold and distributed
by Kluwer Academic Publishers Group,
P.O. Box 322, 3300 AH Dordrecht, The Netherlands.

Printed on acid-free paper

Printed in the Netherlands

CONTENTS

PREFACE

In recent years there has been an increasing interest in Second Generation image and video coding techniques. These techniques, introduce new concepts from image analysis that greatly improve the performance of the coding schemes for very high compression. This interest has been further emphasized by the future MPEG-4 standard.

Second Generation image and video coding techniques are the ensemble of approaches proposing new and more efficient image representations than the conventional canonical form. As a consequence, the human visual system becomes a fundamental part of the encoding/decoding chain. More insight to distinguish between First and Second Generation can be gained if it is noticed that image and video coding is basically carried out in two steps. First, image data are converted into a sequence of messages and, second, code words are assigned to the messages. Methods of the First Generation put the emphasis on the second step, whereas methods of the Second Generation put it on the first step and use available results for the second step. As a result of including the human visual system, Second Generation can be also seen as an approach of seeing the image composed by different entities called objects. This implies that the image or sequence of images have first to be analyzed and/or segmented in order to find the entities.

It is in this context that we have selected in this book three main approaches as Second Generation video coding techniques:

- Segmentation-based schemes
- Model-based schemes
- Fractal-based schemes

We have compiled the necessary information to further facilitate developments in this exciting field. The first chapter provides an introduction to the subsequent technical chapters. Furthermore, a self-contained chapter is included which covers the basic concepts of image and video coding and describes the existing standards. This will assist the reader in understanding the evolution from First Generation to Second Generation schemes.

We hope you enjoy reading this book as much as we did writing and preparing it.

Luis Torres, Barcelona
Murat Kunt, Lausanne
November 1995

ACKNOWLEDGEMENTS

We are indebted to many people who have contributed one way or the other to this book. In the first place we want to thank all the authors of the chapters for their outstanding work, and for suffering the red pencil of the often too tough and picky editors. We also believe that the ideas contained in this book are the fruits of the work of many people, too many to mention here, who in recent years have contributed to many journal papers and conferences to disseminate new image and video coding techniques.

We are also thankful to the Directorate General XIII of the European Union for their support through the RACE and ACTS projects that have greatly contributed to the advances in this new field. Within this Directorate, special thanks go to Eric Badique for his continual belief in what some people occasionally refer to as, "exotic video coding techniques". We are also indebted to Mike Casey, acquisitions editor of Kluwer Academic Publishers, for the help provided while editing and compiling this book. Carlos Nistal from UPC, has provided all the computer support needed to handle and solve the variety of problems that have arisen in the electronic editing of this book.

Our final debt of gratitude is to ours wives, Mercè and Jocelyne, for their support during the last twelve months we have spent editing and writing this book. This book now appears, thanks also to these Second Generation wives.

CONTRIBUTORS

Kiyoharu Aizawa
Department of Electrical Engineering
University of Tokyo
Tokyo 113, Japan
E-mail: aizawa@ee.t.u-tokyo.ac.jp

Frédéric Dufaux
The Media Laboratory
Massachusetts Institute of Technology
Cambridge, Massachusetts 02139, USA
E-mail: dufaux@media.mit.edu

Nariman Farvardin
Electrical Engineering Department
and Institute for Systems Research
University of Maryland
College Park, Maryland 20742, USA
E-mail: farvar@eng.umd.edu

Robert Forchheimer
Image Coding Group
Department of Electrical Engineering
Linköping University
S-581 83 Linköping, Sweden
E-mail: robert@isy.liu.se

Narciso García
Escuela Técnica Superior
de Ingenieros de Telecomunicación
Universidad Politécnica de Madrid
28040 Madrid, Spain
E-mail: narciso@gti.upm.es

Antoni Gasull
Department of Signal Theory
and Communications
Universitat Politècnica de Catalunya
08034 Barcelona, Spain
E-mail: gasull@tsc.upc.es

Mohammad Gharavi-Alkhansari
Beckman Institute
for Advanced Science and Technology
University of Illinois at Urbana-Champaign
Urbana, Illinois 61801, USA
E-mail: mohammad@vision.ai.uiuc.edu

Michael Gilge
Philips Communication Systems
90411 Nuremberg, Germany
E-mail: dtvf_gil@pki-nbg.philips.de

Thomas S. Huang
Beckman Institute
for Advanced Science and Technology
University of Illinois at Urbana-Champaign
Urbana, Illinois 61801, USA
E-mail: huang@ifp.uiuc.edu

Fernando Jaureguizar
Escuela Técnica Superior
de Ingenieros de Telecomunicación
Universidad Politécnica de Madrid
28040 Madrid, Spain
E-mail: fjn@gti.upm.es

Murat Kunt
Signal Processing Laboratory
Swiss Federal Institute of Technology
1015 Lausanne, Switzerland
E-mail: kunt@epfl.ch

Haibo Li
Image Coding Group
Department of Electrical Engineering
Linköping University
S-581 83 Linköping, Sweden
E-mail: haibo@isy.liu.se

Ferran Marqués
Department of Signal Theory
and Communications
Universitat Politècnica de Catalunya
08034 Barcelona, Spain
E-mail: ferran@tsc.upc.es

Fabrice Moscheni
Signal Processing Laboratory
Swiss Federal Institute of Technology
1015 Lausanne, Switzerland
E-mail: moscheni@ltssg3.epfl.ch

Montse Pardàs
Department of Signal Theory
and Communications
Universitat Politècnica de Catalunya
08034 Barcelona, Spain
E-mail: pardas@tsc.upc.es

Xiaonong Ran
Systems Technology
Corporate Technology Group
National Semiconductor Corporation
2900 Semiconductor Drive, Santa Clara
California, USA
E-mail: xran@berlioz.nsc.com

José Ignacio Ronda
Escuela Técnica Superior
de Ingenieros de Telecomunicación
Universidad Politécnica de Madrid
28040 Madrid, Spain
E-mail: jir@gti.upm.es

Philippe Salembier
Department of Signal Theory
and Communications
Universitat Politècnica de Catalunya
08034 Barcelona, Spain
E-mail: philippe@tsc.upc.es

Luis Torres
Department of Signal Theory
and Communications
Universitat Politècnica de Catalunya
08034 Barcelona, Spain
E-mail: torres@tsc.upc.es

1

SECOND GENERATION VIDEO CODING TECHNIQUES

Luis Torres* and Murat Kunt**

** Department of Signal Theory and Communications,*
Universitat Politècnica de Catalunya,
Barcelona, Spain

*** Signal Processing Laboratory,*
Swiss Federal Institute of Technology,
Lausanne, Switzerland

ABSTRACT

The main objective of this chapter is to present the underlying concepts in second generation video coding. It is also intended as a broad introduction to the rest of the chapters of this book. Video coding results of two approaches in the context of segmentation-based video coding schemes are also presented. The chapter ends with conclusions and new trends in the field. A basic introduction to the main characteristics and limitations of first generation coding will be also presented in order to have a self-contained chapter. This chapter is mainly intended for newcomers to the field of second generation video coding. More experienced readers may wish to go directly to specific chapters.

1 INTRODUCTION

The digital representation of an image or of a sequence of images requires a very large number of bits. The goal of image coding is to reduce this number as much as possible, and to reconstruct a faithful duplicate of the original picture. For many years waveform-based image coding approaches have been the only approaches utilized in image compression. One of the the main problems of the now so-called *first generation* coding techniques, is that they did not question the image representation structure imposed by the canonical representation of the images. As a consequence, they share the concept of pixel or block of pixels as the basic entities that are coded. In addition, they also share in common the absence of consideration for the human visual system (HVS) in the design of the coder.

At the beginning of the eighties was quite clear that first generation techniques had reached a saturation [23] and new and innovative techniques were needed. Although at that time video coding schemes were in their infancy, especially approaches dealing with digital television, the techniques developed some years later in the framework of the standards JPEG [16], H.261 [11] and MPEG [21, 22], confirmed that the codecs based in the so called first generation approaches presented severe limitations for very low bit-rate video coding applications (bit-rates below 64 kbits/s). In 1995 the new standard for very low bit-rate video coding H.263 [12] was defined, although it is only intended for conversational services and it is not generic.

In order to overcome the limitations imposed by first generation image coding techniques, *second generation* image coding was formally introduced in 1985 [18], although some schemes following the same basic approach had been introduced earlier [32]. Earlier precursors of what is today known as model-based video coding dates back to 1851 [26]. G. R. Smith demonstrated at the Great Exhibition of London a *comic electric telegraph* consisting of iron bars attached to a flexible model of a face. In 1981 some authors reported novel schemes for the transmission of video at very low bit-rates [36, 20]. In 1983 a new compression scheme based in analysis/synthesis was proposed [6]. This scheme used computer animation to transmit videophone scenes.

We will call second generation image and video coding techniques the ensemble of approaches proposing new and more efficient image representations than the conventional canonical form. As a consequence, the human visual system becomes a fundamental part of the encoding/decoding chain. More insight to distinguish between first and second generation can be gained if it is noticed that image and video coding is basically carried out in two steps. First, image data are converted into a sequence of messages and, second, code words are assigned to the messages. Methods of the first generation put the emphasis on the second step, whereas methods of the second generation put it on the first step and use available results for the second step. As a result of including the human visual system, second generation can be also seen as an approach of seeing the image composed by different entities called objects. This implies that the image or sequence of images have first to be analyzed and/or segmented in order to find the entities.

It is in this context that we have selected in this book three main approaches as second generation video coding techniques:

- Segmentation-based schemes
- Model-based schemes
- Fractal-based schemes

We note that although the wavelet decomposition has also been used to exploit self-similarities of images at different scales, as fractal coding [30], it will not be considered a second generation coding technique in this book. We also note that although second generation image coding techniques were developed in the context of coding and compression, their use is finding other important applications such as content-based multimedia data access or content-based scalability as defined in the proposal package description of the future standard MPEG-4 [17].

In order to present the segmentation-based, model-based and fractal-based schemes we have divided the book into eight chapters related to these topics. Chapter 3 deals with the spatial image sequence segmentation problems in the context of video coding. Chapter 4 presents the main approaches that have been proposed to encode the resulting partitions of the segmentation stage. Chapter 5 gives a unified description of the transform techniques mainly used to encode arbitrarily shaped regions resulting from the segmentation stage. Chapter 6 provides techniques to estimate the motion of image regions and further extends the segmentation into the temporal domain. Chapter 7 reviews fractal coding as a new promising approach for video coding and discusses the fractal concepts in the context of the second generation. Chapters 8 and 9 present the underlying ideas in model-based image and video coding. In particular, Chapter 8 gives a detailed review of the main model-based approaches presented up to date while Chapter 9 presents a specific implementation of a model-based scheme for still images that can be extended to video sequences. Chapter 10 ends with new proposals for very low bit-rate video coding based on extensions of known signal-theoretic techniques. In addition, Chapter 2 provides a self-contained summary of the main approaches followed in first generation schemes with particular emphasis in their limitations and possible improvements.

This chapter is organized as follows. Section 2 presents a basic overview of the main limitations of the canonical representation of the image. Section 3 gives some remarks on first generation that help to understand the evolution towards second generation. Section 4 presents the underlying general concepts of second generation techniques and makes a brief overview of the topics covered in the following chapters concerning segmentation-based, model-based and fractal approches. Section 5 shows results of two complete segmentation-based video schemes relying on the segmentation of the spatial and temporal domains. Section 6 presents the new set of functionalities that are being pursued in the context of MPEG-4 and describes how second generation techniques can also be applied in this context. Finally, Section 7 presents some conclusions and gives new trends in the field.

2 THE PROBLEMS OF THE CANONICAL REPRESENTATION OF IMAGES

The first problem one faces in compressing visual information is that of representation. How do we represent visual information that each one of us has around us in the four dimensional (4-D) space we are living (3 space and 1 time dimensions)?

The first answer came in the late forties, thanks to the television. The 4-D space is projected on the image plane of a sensor at a regular pace producing a series of single images. An image is scanned line by line and an electrical signal representing the brightness of the scanned area is generated. This analogue signal is then formatted and modulated for transmission. Although this system has worked quite successfully since more than a half century, it has a number of drawbacks. The analogue 4-D space is projected on a plane. Even if this projection could be made continuous in time, one dimension (and hence information) is lost by projection. In practice, the projection cannot be made continous in time for technical reasons (the image plane need to be scanned). This introduces a first sampling, the sampling in time of the 4-D space. Then, the image plane is sampled vertically to produce the video signal. This is a second sampling in space. We all know, since the early days of the kindergarten, that whenever a sampling operation is performed, attention must be payed to the conditions of the sampling theorem. None of these samplings do that. As a consequence, we define two sets of input information. One set is that of these signals which respect the sampling theorem and can go through the system. The other set, complementary, contains signals that do not respect the theorem. These are necessarily distorted when processed. Typical examples are the wheels of a car going to the right, turning in the aberrant direction as if the car was going to the left, or the material of dresses some TV speakers wear inducing moire patterns and motion though staying still.

In the fully digital world in which we are living today, a third sampling is introduced along the scanning direction of the video signal. It is as careless as the first two with regards to the sampling theorem. Notice that all these three samplings are implemented at a constant sampling rate or period. To be fully digital, all these samples are quantized using a given number of bits per sample. Eventually, we have an enormous set of bits or bit rate that represents our 4-D world. Ironically, we call this representation canonical, implying that we cannot do any better. In fact, we cannot do any worse! The criticism we can formulate may be listed as follows. The 4-D information is processed in the same way we process stationary 1-D signals. Visual information is everything but stationary. Not only do we use sampling without respect for its theorem, but we use practical values for sampling rates that have nothing to do with the input signal. The famous 25 or 30 frames per second rates are dictated not by the input signals but by the frequency of the voltage power line. Almost nothing or very little is done, noticing that the last element of the processing chain is not the display device but the human visual system, with its marvelously rich and poor features. Because we are unable to do adaptive sampling, because we need easily implementable systems and we are still doing our best, we obtain a canonical representation which may be usable for a subset of visual information. Even within this subset, the system may have the right parameters for some part of the information but, due to the nonstationarities, has the wrong parameters for the other parts. As a consequence, the canonical form is not canonical but dramatically redundant. One last comment that can be made is that the structure imposed by the system is not data driven. It is a fully arbitrary structure, independent from the data. Assuming a binary representation of the canonical form, if we were able to generate images representing all the possible combinations of these bits, most of the results would look like anything but visual information. Natural images are only a very small subset. The problem at hand is compressing the canonical form. Viewed from the previous angle it looks like Sysiphus and his rock. It keeps us busy and happy.

The set of compression methods developed till the mid eighties did not question the structure imposed by the canonical representation and tried to combine picture samples in various ways to obtain a compressed bit stream. Natural images contain man made or nature made objects defined not by a set of pixels regularly spaced in all dimensions but by their shape, texture, and color. Even the pointillism was a way for some painters to express their view, a painter never paints by pixels. Shape, texture and color are the most frequently used features. So why not to imitate painters by technical means? As Sysiphus, let's push our rock up hill and try to extract these features from the available data. Computer vision has been trying to implement these operations for quite some time with more or less success. However the constraint is different. In image compression we need to recover a faithfull replica of the original scene, whereas computer vision aims at extracting semantical information or descriptions.

Efforts directed in these new directions developed the so-called second generation techniques for image and video compression.

3 FIRST GENERATION VIDEO CODING TECHNIQUES

In order to better understand the new concepts introduced in second generation video coding schemes, it is important to review the basics in which the coding techniques of the first generation are established. To that end Chapter 2 provides a self contained summary of the main approaches followed in first generation schemes with particular emphasis in their limitations and possible improvements. In addition it has been thought interesting to provide in Chapter 2 a review of the current image and video coding standards with two objectives in mind. First to present what will be normal use in video communications in the following years and second to have a complete book in actual and future trends in video coding. We present in this section a very brief summary of what is introduced in Chapter 2 with special emphasis on concepts of first generation schemes that provide insight in the second generation.

The basic schemes that are mainly considered first generation are pulse code modulation, predictive coding, transform coding, vector quantization and the myriad of schemes including combinations or particular cases of those such us subband and wavelets. For a more complete information and additional references on these coding schemes, the reader is referred to [29, 25, 14, 37, 8, 2].

First generation image and video coding techniques achieve compression by reducing the statistical redundancy of the image data. To that end, the image is considered as a set of pixels that have to be uncorrelated. This objective is accomplished using waveform coding. The problem is stated in the well known field of rate-distortion theory to achieve the minimum possible waveform distortion for a given coding rate or, equivalently, to achieve a given acceptable level of waveform distortion at the least possible encoding rate [33, 1]. This approach requires the knowledge of the source distribution function and an adequate definition of the distortion measure. A mean square distortion measure has generally been used. This approach does not pay any attention to the semantic meaning of the selected set or block of pixels.

This model has been the basis of all first generation schemes and has been very popular in the last twenty years or so. But the last years have seen some sort of disappointment concerning this model, due mainly to the following general reasons:

- First generation video coding techniques have not proven so far useful to achieve very low bit-rates with good visual quality for generic sequences.
- All the attempts to provide a true knowledge of the source distribution function have had a very limited success and only simple models are used [14].
- It has been quite difficult to incorporate the most important part of a video coding system, the human visual system, into the model. The widely used mean square distortion measure keeps little relation with the human visual system.

One of the main achievements coming from the first generation is the hybrid scheme formed by combining motion compensated prediction in the temporal domain and a decorrelation technique in the spatial domain (Chapter 2, Figure 9). This scheme may serve as a starting point to introduce some of the basic concepts of the second generation approach. In this scheme the input image is divided into square blocks of usually 8x8 or 16x16 pixels. Three main problems arise when considering this structure:

- The blind division, without taking into account the semantic content of the image, results in block effect when high compression ratios are desired.
- Motion models are applied to square blocks of pixels that may have little resemblance to the true motion of the objects that form the image.
- The properties of the human visual system are not taken into account.

Chapter 2 gives a detailed study of the hybrid scheme analyzing its performance in a variety of bit-rates and different video sequences. For an additional study concerning the theoretical performance limits of motion compensation prediction the reader is referred to [9] and for a general study on motion analysis in image sequence coding to [34].

The main conclusions of the actual video coding schemes of the first generation in the context of very low bit-rate video coding can be stated as follows:

- The image deteriorates rapidly at high compression ratios, mainly due to the block effect appearance in the decoded image.
- The achievable bit rates range from 8 to 36 kbits/s depending on the difficulty of the video sequence and the desired quality. 8 kbits/s is only achievable for very steady head and shoulders sequences while it seems that bit rates around 24 kbits/s may be reached for more complicated sequences.
- The fact that these techniques are still being applied in a block-based approach will still establish an upper bound, at least in some other fields related to image and video coding, such as content-based multimedia data access or content-based scalability, as defined in the proposal package description of the future standard MPEG-4 [17].

One may wonder if this hybrid scheme is capable of improvements that may give further compression and/or achieve new functionalities. Two main stages can be improved: the motion analysis and the intra-frame spatial decorrelation stages. The reader is referred to Chapter 2 for additional comments on this and also to [10] for a good discussion on the fact that very accurate motion compensation may not be the key to a better picture quality due to the severe rate-constriction of very low bit-rate coders. Further discussions on the limits and improvements of the hybrid scheme of first generation techniques are presented in [19].

As an important point with respect to first generation techniques, the new concept of functionalities has to be emphasized. Functionalities have been described as the set of tools and algorithms that allow manipulation of the content of the image [17]. It seems clear that due to the inherent block based nature of the first generation schemes the amount and quality of the functionalities that can be implemented is limited. More details on the topic of functionalities is presented in Section 6.

We would like to conclude this section by noting that first generation schemes are excellent *texture coding schemes*. They have proven to provide very good image quality for moderate compressions rates, but their use for generic sequences at very low bit-rates remains to be seen. In this context it seems appropriate to explore new compression methods that may overcome the limitations of first generation techniques. The rest of the chapter is dedicated to the so called second generation image and video coding techniques.

4 SECOND GENERATION VIDEO CODING TECHNIQUES

In 1985 when the first original paper was published on second generation image coding techniques [18], it was clear what this concept was. Now, 10 years later, the field of image and video coding has seen a tremendous explosion of new ideas, techniques, implementations and standards. It then seems adequate to briefly review the main concepts in which the techniques belonging to second generation are founded.

We have seen in Section 3 that first generation techniques are based in the framework of waveform coding. We have already seen how this framework has been implemented and the problems that arise for very low bit-rate video coding applications. The basic idea underlying second generation is to overcome these limitations. How can it be done? One possible answer is to incorporate the properties of the human visual system, that, at the end, is the most important part of the coder/decoder chain. How these properties can be incorporated is not an easy task. The first thing to do is to have a detailed knowledge of the HVS. Then, the second thing is to find an adequate model for the image that fulfills this knowledge. If we know that the human visual system is able to recognize an image or a sequence of images using only a very few information points, then the task reduces to finding the points that convey most of the information. If we want not only to recognize but also to appreciate more detailed information, then we should also be able to incorporate this characteristic into the model. It is adequate in this context to review very briefly the main characteristics of the HVS that have led to the introduction of new image and video coding techniques.

4.1 The human visual system

The human visual system (HVS) plays a very important role in second generation video coding techniques. For this reason there have been many reports, articles and books in the last years explaining the behaviour and influence of the HVS in this type of coding approach. We think then, that it is not necessary to repeat in this book the main topics concerning the HVS considered extensively elsewhere [18, 2, 28, 9]. Instead, our approach has been to present in Chapter 9 and 10 some properties of the HVS in the context of video coding that had not been presented extensively before.

In particular, Chapter 9 presents some experiments that show the importance of the HVS in the design of model-based coding systems. It is concluded there that the edge information is of paramount importance and the HVS can be satisfied with an image model based in a decomposition in primary, smooth and texture components.

Chapter 10 considers also the so called perception-based coding where it is shown that the HVS highly influences the design of future video codecs. In particular the effect of eye-movements on the spatio-temporal modulation transfer function of the HVS is studied and applied in the design of video coding schemes. It is also shown there that for the visual system, spatial acuity is only a function of retinal velocity and not of the image velocity itself. Taking into account this fact, some hints are provided to improve the performance of actual first generation hybrid coding schemes and to propose new paradigms for future codecs.

For additional information on the HVS and its influence on the design of image and video coders the reader is referred to [18], to the general references provided in [2, 28] and the book chapter by Girod for topics on the HVS related to motion compensation in the context of video coding [9].

We summarize below the main properties concerning the human visual system that are of interest to understand the proposal of second generation image and video coding techniques made in this book.

- The human visual system can be modelled through the eye, the retina and the cortex.
- The lens of the eye can be modelled as a two-dimensional low pass filter.
- The retina presents a high-pass behaviour due to the so called lateral inhibition of the HVS.
- The ganglion cells and the lateral geniculate nucleus of the human visual system are responsible for the simultaneous contrast effect and edge detection. These properties can be modelled by a two-dimensional high pass filter.
- The visual cortex is sensitive to the orientation of the excitation.
- To the human eye, many fractal curves and surfaces look very similar to natural curves and surfaces.

These properties lead to the following very basic but important conclusions:

- The edge and contour information are very much appreciated by the human visual system and are responsible for our perception.
- The texture information has relative importance. Texture is associated with *additional* information. It influences our perception when taken together with the contour information.
- The images of many natural objects can be approximated by members of a class of deterministically self-similar sets.

These conclusions explain the reasons that have led to the proposition of different alternatives concerning the integration of the human visual system into the coder design. First, is the proposal of a contour-texture approach to image coding. This has been considered in this book under the scope of segmentation-based video coding schemes (Chapters 3, 4, 5 and 6). Second, is the proposal of fractal-based coding schemes (Chapter 7). Third, is the proposal of a model-based approach that takes into consideration the human visual system in the analysis/synthesis parts of the video coding stages. This has been considered in this book when designing a 3-D scene model of a person's face and in defining a three component model for an image (Chapters 8 and 9). Finally, the proposal of new paradigms in video coding taking into account the human visual system (Chapter 10). A summary of the current image and video coding schemes of the first and second generation is presented in Table 1 of Chapter 8. In this table, however, the reader will notice that segmentation-based schemes are considered a particular case of two-dimensional (2-D) model approaches.

As a last comment regarding the influence of the human visual system, there is the important issue of defining distortion measures in the context of second generation. If the human visual system is so important in second generation video coding schemes, one may wonder which is the best measure to be used to evaluate the performance of the video coding system. If the mean square distortion measure used in rate-distortion theory is used, then we may incur the error of evaluating a human visual system-based coder with tools that have nothing to do with the HVS. A possible answer is suggested in Chapter 10, where the Kantorovich metric is proposed to evaluate the performance of compressed images. Although results are only available for fractal image coding, this metric is a promising measure to be used in second generation video coding schemes.

4.2 The color information in second generation techniques

Although practically all video communication systems provide color information, the topic of color has not been intensively considered in second generation video coding techniques. One of the main reasons is the relatively minor importance of the color information in the fine details of an image. This consideration has led to the concentration of almost all efforts in the development of new techniques to the consideration of only the luminance information.

The color information is usually subsampled and treated in a YIQ or YUV space (where Y represents luminance and IQ and UV represents the chrominance). We notice in passing that the process of subsampling, usually done on the chrominance components, can also be seen as a coarse introduction of the properties of the human visual system in first generation techniques. It is well known that the human visual system is much less sensitive to the details of the color information than to the details contained in the luminance information.

It can be generally stated that in second generation video coding the same process that is applied to the luminance signal is applied to the color signal. For instance, in motion estimation only the motion corresponding to the luminance information is found and it is assumed that the color has the same motion. To see one of the few schemes in which the motion estimation takes into account the color information the reader is referred to [3]. In texture coding the color information is always assigned much less bits than the luminance, although the encoding process and the bit assignment approach is generally the same. In the rest of the book, comments on the processing of color information will be done only when needed.

4.3 Segmentation-based video coding

The first still image coding approach that was presented as a result of the properties of the human visual system studied in Section 4.1 was based on a contour/texture representation model of the image [18]. The decomposition and posterior coding of a still image in contours and textures does not represent, in principle, a theoretical problem. A good segmentation technique able to decompose the image in homogeneous entities called regions or objects has to be found. Then, the resulting contours and textures are coded. The basic idea behind this approach is that the contours correspond, as much as possible, to those of the entities defining the image, that is, the objects. The situation in video sequences is very different. To start with, the introduction of image motion poses the following questions:

- Must the segmentation be done in the spatial domain (spatially-based segmentation) or in the temporal domain (motion-based segmentation) or in both simultaneously?
- How is the concept of contour defined in still images extended to video sequences?
- Is it as efficient to use motion compensation as in first generation techniques?

To obtain a contour/texture representation the image sequence has also to be segmented. The result of the segmentation process gives a set of connected regions, called the partition sequence. We note in passing, that the partition sequence has been sometimes called the contour sequence as in still images. The partition is represented by the shape of the regions and their evolution in the time domain. These two informations have to be coded, as well as the interior of the regions, called also sometimes texture. Different performance, image quality and compression ratios are obtained as a function of the ability of the segmentation technique to provide homogeneous regions and the compression capabilities of the partition and texture coding techniques used.

We recall here that Chapter 3 explains in detail possible ways of segmenting an image sequence in the context of video coding. Chapter 4 explains different methods of coding the partition sequence while Chapter 5 presents a review of the texture coding techniques most used in this context, concentrating on generalized orthogonal transforms. Chapter 6 deals with the problem of motion estimation and segmentation in region-based schemes. In the following, a very brief introduction to the contents of these chapters is provided.

Segmentation of video sequences

The segmentation can be done in the spatial domain, in the temporal domain or in both simultaneously. In the first case different criteria can be used to define homogeneous spatial entities such as contrast, size, etc. In the second case the motion information is what is used as homogeneity criterion. Some schemes segment both domains simultaneously. The election of the homogeneity criteria is conceptually difficult, as different criteria give different segementations, which affect the performance of the overall coding system.

In the case of spatial segmentation three different approaches have been proposed. The first one is a pure intra-frame procedure where the temporal information is not taken into account. The second considers the image sequence as a 3-D signal (space plus time) and the third one is based on a recursive segmentation of the image sequence. After a detailed description of each approach, it is concluded in Chapter 3 that the recursive segmentation is the most appropriate due to requirements of time coherence and to avoid random fluctuations of the partition sequence. Of special importance in video coding applications are the considerations of bit rate regulation and the time delay introduced in the segmentation process. Following the election of the time recursive approach, a general structure for segmentation is proposed. In order to improve the quality of the segmentation, the partition and texture coding stages are included in the segmentation process. The structure is discussed in both intra-frame and inter-frames modes. An interesting side result of this proposal is a hierarchical representation of the image very well suited for coding applications. Two implementations, one based on Compound Random Fields and the other on Mathematical Morphology are presented. Segmentation results of both implementations are shown.

In many situations what gives coherence to moving objects is precisely their motion. It does make sense then, to consider a temporal segmentation. The temporal segmentation is very intimately linked to the problem of motion estimation. Chapter 6 explains in detail these two problems. The main results are briefly summarized here.

In the context of motion estimation, it is concluded that an affine model is more convenient in segmentation-based video coding schemes. Several procedures are outlined and discussed to find the model parameters. Object tracking is also introduced in Chapter 6 as a way to solve the problems of the inconsistency of the regions in the temporal domain posed by motion estimation.

Concerning motion segmentation, Chapter 6 presents three main approaches to the problem. The first one consists of the segmentation of the optical flow. The second one estimates simultaneously the motion and the segmentation and the last one relies on a combination of spatial segmentation and motion information. In the context of video coding the last approach seems to provide the most promising results.

It is very important to remark that a lot of effort has been made in motion estimation in the field of image analysis and there exists many methods to estimate motion. But in that field the amount of overhead information that has to be sent to the receiver it is not of concern. However, in video coding this information is fundamental and puts constraints and restrictions that may limit the quality of the motion estimation. Very few results have been reported on motion estimation and its use in complete segmentation-based video coding schemes. Along with those provided in Chapters 3, 4 and 6, an additional reference is [4], where bit rates of 9.6 kbits/s are claimed for videoconference sequences in a completely region-based video scheme using a combination of spatial and temporal segmentation.

As a last comment on motion estimation it is fair to say that although segmentation-based video coding techniques rely on the concept of arbitrarily shaped regions, the motion estimation approaches presented in the literature up to date rely mainly at one stage of the process or the other on block matching techniques. One of the main reasons for this is the easy implementation of block matching algorithms and the very simple motion model assumed. If good motion estimation algorithms are designed in the future based totally in arbitrarily shaped regions, a big improvement in segmentation-based video coding schemes may be expected. Some video coding schemes based in a hybrid approach of the first generation and an affine motion model based in regions have already been presented [15, 35].

Coding of the partition sequence

The resulting partition sequence of the segmentation stage has to be coded. Chapter 4 gives a complete description of the main approaches followed so far in the coding of the partition sequence. Two modes of operation have to be distinguished: intra-frame and inter-frame. In the intra-frame mode only the shape and the positions of the regions have to be coded. Lossless and lossy approaches have been taken. The main lossless techniques may be divided into contour-oriented and shape-oriented. Among the first, chain code techniques have been widely used. Among the second, the morphological skeleton and the quadtree techniques have also been proposed. It is concluded in Chapter 4 that the best technique for intra-frame lossless partition coding is the chain code. Lossy techniques for the intra-frame mode have also been proposed. The amount of losses is very critical as we have already mentioned that the human visual system is very sensitive to the contours of the image. The so-called multigrid chain code has shown very promising results in a lossy scheme.

In the inter-frame mode, in addition to the coding of the shape and the position of the regions, the labels of each region have also to be transmitted. Chapter 4 presents a general motion compensation strategy for the partition and it is concluded that with this scheme it is possible to divide by two the cost of the inter-frame mode with respect to the intra-frame mode.

Texture Coding Techniques

We have already explained in Section 4.1 that texture contributes to improve the quality of second generation video coding techniques (as it does also for first generation). It is important then, to provide texture representations that may help the coding of video sequences in second generation schemes. It is fair to say that practically all texture coding approaches in the context of segmentation-based video schemes are based in first generation image coding, that is, waveform coding techniques. It is clear that the main effort has been dedicated to adapting the already existing coding techniques to the arbitrary shape of the regions of the images. Very few works report the specific coding of the prediction error image, although most of them assume that the same technique used in intra-frame coding can be applied in inter-frame coding.

Chapter 5 presents a review of the main texture coding techniques that have been used in segmentation-based image and video coding schemes. Special emphasis is on a generalized transform coding algorithm that allows for coding the texture inside image regions of arbitrary shape. This approach has been generally used in most of the schemes presented so far in the context of segmentation-based image and video coding. It has the additional advantage of being able to be used in the coding of the prediction error in motion compensation video schemes. Approaches relying upon vector quantization and wavelets are also presented as well as modifications of the generalized transform coding algorithm. Chapter 5 concludes presenting a complete coding scheme valid for very low bit-rate video coding.

As an additional reference, worth mentioning are the results presented in [27] where it is claimed that if the mean value of the region is used to code the region, the number of gray levels can be decreased by a factor of twelve after a requantization process. It would have been interesting to pursue this work for the coding of the region with other approximations different than the mean value, but to the best of our knowledge this work has not been extended.

4.4 Fractal-based image and video coding

Fractal coding is a new and promising technique that has recently attracted the interest of the researchers in the field of image and video coding. It is based on the similarities between different scales of the same image. Fractal coding can be considered a pure second generation technique as it tries to reconstruct an image by representing the regions as geometrically transformed versions of other regions in the same image [5].

Fractal coding has been also used in segmentation-based schemes as a measure of homogeneity in image segmentation and for combined contour and texture coding. In addition, fractals have been used to model the motion of video sequences in a block-based approach although it could be extended to regions easily. Chapter 7 presents a complete self-description of these topics along with coding results for still images and video sequences. For still images good visual results are obtained in the range of 0.43 and 0.15 bits/pixel. For video sequences, bit rates of 80 kbits/s are presented for videoconference applications. The use of fractal coding for very low bit-rate video coding as an independent technique remains to be seen, but its combination with other approaches such as segmentation-based video coding are worthwhile to follow.

4.5 Model-based video coding

Model-based schemes define three dimensional space structural models of the scene. In this case, the coder and decoder share the same object model. The coder analyzes the input image and the decoder generates images using the model. These schemes have been mainly used for videoconference and videotelephony applications as all intend to model the human head.

Model-based approaches to video coding assume a parameterized model for each object of the image. The parameters associated to each object are coded and transmitted. To extract the set of parameters, the image is first analyzed using tools from the field of image analysis and computer vision. This analysis of the image gives information on the size, location and motion of the objects that compose the image. This extracted information is then parameterized. It is interesting to notice that the 3-D model-based coding methods introduced in Chapter 8 may be considered as a particular case of segmentation-based methods. In the first case the shape of the contours is fixed previously (for instance, a wire frame), while in the second it can have any shape. Chapter 8 presents the main techniques that have been developed in model-based video coding. Good visual results for video conference sequences are presented at rates of 16 kbits/s. When using a hybrid approach of 3-D model-based and 2-D triangle based, acceptable visual results are presented at 5 kbits/s.

It is interesting to notice that this approach is more distant from first generation techniques than the contour/texture approach. In this last scheme the contour and texture information is still coded using one of the common approaches of waveform schemes. For this reason, some researchers in the field define model-based as belonging to a third generation approach to video coding [13].

It is extremely difficult to model every generic object that may form an image. For this reason, almost all the efforts so far in model-based video coding have been concentrated in modeling images of the type head and shoulders. Although current models are concentrated mainly in human faces, their interest is enormous, in particular for conversational communication services such as videoconference and videotelephony. However, other interesting applications have been proposed such as speech driven image animation for talking heads and virtual space teleconferencing.

Chapter 9 presents a model-based still image coding scheme highly dependent on the importance of the human visual system. An image is modelled as being composed of primary, smooth and texture components. These components are defined as a function of their importance with respect to the human visual systems. Specific coding techniques for coding each component are developed. Very good visual results are presented for bit-rates of 0.12 bits/pixel. The scheme is, at the moment of printing, being extended to the coding of video sequences.

5 SEGMENTATION-BASED VIDEO CODING RESULTS

The other chapters of this book dealing with segmentation-based video coding schemes, present techniques to segment a video sequence and propose a variety of texture and partition coding approaches. This section is dedicated to show results of two complete segmentation-based video coding schemes. The objective is not to make comparisons among both schemes, but to show the quality that can be provided in this type of approaches when used in a very low bit-rate context. The first scheme relies on spatial segmentation using a Mathematical Morphology approach and the second scheme relies on object-based analysis-synthesis and temporal segmentation. Both schemes will be very briefly introduced. For complete details of the first scheme the reader is referred to [31] and to [24, 7] for the second.

5.1 A video coding scheme based in spatial segmentation

The general structure of the video coding scheme based in spatial segmentation is presented in Figure 1. The scheme involves a time-recursive segmentation relying on the pixels homogeneity, a region-based motion estimation, and motion compensated contour and texture coding. The segmentation step follows the implementation based in Mathematical Morphology explained in Chapter 3. The partition coding is based in the motion compensation of partitions presented in Chapter 4. The texture is coded using texture compensation followed by the coding of the prediction error using the generalized orthogonal transform technique explained in Chapter 5. A translational motion model is used to find the motion associated to each region. For more details the reader is referred to [31]. One of the important features of the approach is that no assumption is made about the sequence content. Moreover, the algorithm structure leads to a scalable coding process giving various levels of quality and bit rates. The coding as well as the segmentation are controlled to regulate the bit stream.

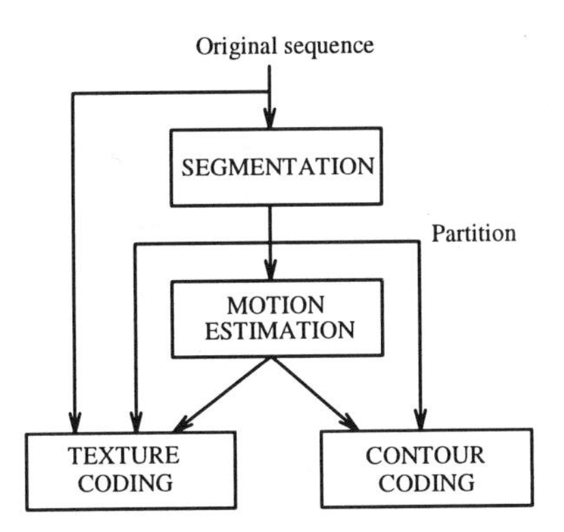

Figure 1 General structure of the spatial segmentation-based coder

The algorithm has been designed to meet the following basic system requirements:

- The codec is mainly devoted to very low video bit-rates, that is below 64 kbits/s.
- It should be generic with respect to the input sequences. In particular, no assumption about the scene content should be made.

- It should be flexible with respect to the coded sequence quality. The coding strategy should provide an easy way to define various levels of quality.
- The system is principally devoted to fixed rate transmission. As a consequence, the bit stream should be regulated.
- It should be suitable for interactive applications. Therefore, large processing delays should be avoided.

Figure 2 and Figure 3 present the sequences *Carphone* (frames 10 and 110) and *Foreman* (frames 10 and 110) coded at 26 kbits/s and 32 kbits/s respectively at a frame rate of 5 Hz. These sequences are representative of the Class B sequences proposed in MPEG-4.

Figure 2 Above: original sequence *Carphone*. Below: coded frames at 26 kbits/s.

Figure 3 Above: original sequence *Foreman*. Below: coded frames at 32 kbits/s.

5.2 Object-based analysis-synthesis coder[1]

The principle of object-based analysis-synthesis coding is shown in Figure 4. In this scheme images of the incoming sequence are decomposed into arbitrarily shaped moving objects by an object-based image analysis, which relies on a certain source model. Each moving object is described by three sets of parameters defining its shape, motion, and color. The color parameters consist of the luminance and chrominance values of the object surface. While coding of color parameters requires a relatively high amount of bits, coding of shape and motion parameters does not. Therefore, in a first step the current image is synthesized using only the shape and motion parameters of the current image and the motion compensated color parameters of the previous image. Then, for each object those areas are detected where this synthesis does not lead to a sufficient quality. For these so called model failure (MF) areas, the frame to frame changes can not be described by the underlying source model. On the contrary, all other areas of the object are denoted as areas of model compliance (MC). For all objects, motion and shape parameters are coded and transmitted to the

[1]The text and results of this section have been kindly provided by M. Wollborn and P. Gerken of the University of Hannover, Germany.

receiver. In order to reach a high coding efficiency, color parameters are coded and transmitted only for the MF areas of the object. Since the MF areas are arbitrarily shaped, in addition to the color parameters also shape parameters are transmitted for these areas.

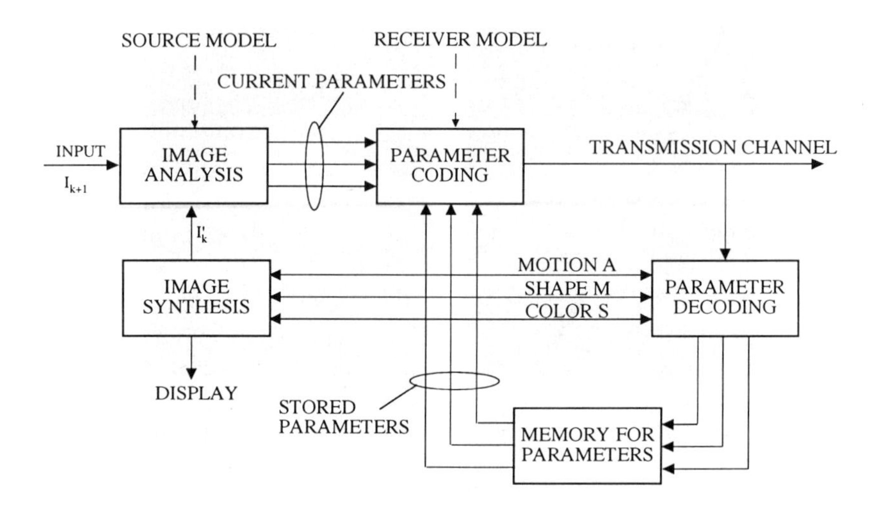

Figure 4 General structure of the object-based analysis-synthesis coder

The contents of the parameter sets highly depend on the source model upon which the image analysis is based. In an implementation specialized for very low bit-rate (8 to 64 kbit/s) applications, as proposed in [7], the source model of flexible two-dimensional objects with translational motion is applied. When using this source model, shape parameters describe the approximated silhouettes of the moving objects. Motion parameters are described by a field of displacement vectors, for the estimation of which the arbitrary shapes of the objects are taken into account. The shape parameters are predictively coded by transmitting the vertices of a polygon/spline approximation of the object silhouettes. The motion parameters are coded by a common DPCM technique. The color parameters are coded using a vector quantization algorithm which takes into account the statistical dependencies between the luminance and chrominance values of the respective picture elements to be coded. In this implementation, the area for which color parameters have to be coded is restricted. For example at a transmission rate of 8 kbit/s the area of the color parameters is about 2 to 4 % of the total image area. By this, even for very low transmission rates, the bit-rate available for the color parameters is around 1 bit/pixel, thus guaranteeing a relatively high quality of the reconstruction image compared to conventional hybrid block-based coding where only around 0.2 bit/pixel is available. Figure 5 and Figure 6 present the sequences *Akiyo* (frame 25) and *Hall monitor* (frame 38) coded at 10, 24 and 48 kbits/s. These sequences are representative of the Class A sequences proposed in MPEG-4.

Figure 5 Above left: original sequence *Akiyo*. Above right: coded at 10 kbit/s. Below left: coded at 24 kbits/s. Below right: coded at 48 kbits/s.

6 MPEG-4 FUNCTIONALITIES

The last couple of years have seen tremendous activity in the standardization of image and video coding systems. As a result of that effort, the new standards JPEG [16], H.261 [11], MPEG [21, 22] and H.263 [12] have fulfilled the needs of different applications for a variety of compression ratios. In the last two years rejuvenated interest was put into the definition of a new standard for very low bit-rate video coding in the range of 8-64 kbits/s. For some time the unique interest was in achieving more compression, but it was realized that besides compression other innovative applications deserved further attention. After some MPEG meetings, the concept of functionalities was introduced and finally materialized in the MPEG-4 Proposal Package Description (PPD) [17].

Figure 6 Above left: original sequence *Hall Monitor*. Above right: coded at 10 kbit/s. Below left: coded at 24 kbits/s. Below right: coded at 48 kbits/s.

We review briefly in this section the main objectives of MPEG-4 with special emphasis in these new functionalities.

MPEG-4 is an emerging coding standard that supports new ways (notably content-based) of communication, access and manipulation of digital audio-visual data. Recognizing the opportunities offered by low-cost, high performance technology, and the challenge of rapidly expanding multimedia databases, MPEG-4 will offer a flexible framework and an open set of tools supporting a range of both novel and conventional functionalities. This approach will be particularly attractive because rapidly progressing technology will facilitate the downloading of tools in a practical way.

MPEG-4 will address the new expectations and requirements, by providing an audio-visual coding standard allowing for interactivity, high compression and/or universal accessibility. In order to take advantage of rapidly evolving relevant technologies, the standard will provide for a high degree of flexibility and extensibility. Content-based interactivity involves the ability to interact with meaningful objects in an audio-visual scene. Providing interaction with natural, synthetic and hybrid synthetic/natural audio-visual objects is very important to enable new interactive audio-visual applications.

Eight key functionalities have been identified which support the MPEG-4 focus and are thought to be not well supported by existing or other emerging standards. These functionalities are to be supported by the collection of coding tools and the MSDL (MPEG-4 Syntatic Description Language). This flexible MSDL should allow the use of different coding tools to provide different combinations of these functionalities as needed by specific applications. A basic description of the eight functionalities with special emphasis in the ones related to content-based multimedia data access and content-based scalability are described in the following:

1. Content-based multimedia data access tools. MPEG-4 shall provide data access based on the audio-visual content.
2. Content-based manipulation and bitstream editing. MPEG-4 shall provide an MPEG-4 Syntactic Description Language and coding schemes to support content-based manipulation and bitstream editing withouth the need for transcoding.
3. Hybrid natural and synthetic data coding. MPEG-4 shall support efficient methods for combining synthetic scenes or objects with natural scenes or objects, the ability to code and manipulate natural and synthetic audio and video data, and decoder-controllable methods of compositing synthetic data with ordinary video and audio, allowing for interactivity.
4. Improved temporal random access. MPEG-4 shall provide efficient methods to randomly access, within a limited time and with fine resolution, parts (e.g. frames or objects) from an audio-visual sequence.
5. Improved coding efficiency. For specific applications targeted by MPEG-4, MPEG-4 shall provide subjectively better audio-visual quality at comparable bit-rates compared to existing or other standards.
6. Coding of multiple concurrent data streams.
7. Robustness in error-prone environments.
8. Content based scalability. MPEG-4 shall provide ability to achieve scalability with a fine granularity in content, quality, (e.g. spatial resolution, temporal resolution), and complexity. In MPEG-4, these scalabilities are especially intended to result in content-based scaling of audio-visual information.

These functionalities may provide applications such as: content-based retrieval of information from on-line libraries, used or automated selection of decoded quality of objects in the scene, database browsing at different content levels, scales, resolutions and qualities. One of the main objectives is the ability to manipulate the image in the spatial and temporal domains. It is clear from this description that besides the issue of compression, new image and video sequence representations are needed to fulfill these requirements. This implies that the segmentation and model-based techniques explained in the context of video coding can be used and/or modified to cope with the new requirements. Applications such as object tracking or object manipulation are easy to implement with some of the techniques explained in this book.

7 CONCLUSIONS AND NEW TRENDS

This chapter has reviewed the main concepts in which second generation image and video coding techniques are founded. The main techniques selected as second generation have been introduced as a way to overcome the main conceptual problems posed by first generation. The approach relies on an understanding of the image as a composition of objects which have a semantic meaning. Segmentation, model and fractal-based schemes have been presented as representatives of a new generation of codecs that, taking into account the properties of the human visual system, pay attention to the content of the image. New functionalities have also been shown as a very new and exciting field of applications.

Now it is time to ask which is the next step. As usual this question has a very difficult answer. It is clear that the next step in segmentation-based coding is not to segment objects but to recognize them. If we were able to recognize objects, then everything, motion compensation, object tracking, etc., would be easier. It seems also evident that for model-based the step is to model automatically any kind of objects. Fractal schemes may also show their potentiality for very low bit-rate video coding, either by themselves or in combination with some other advanced schemes. In a more conservative context we will see an explosion of hybrid first and second generation schemes taking into account the good properties of both. In addition, combined second generation source and channel video coding should provide a better efficiency.

Speech may come to the rescue of video (or vice versa) by combining in an intelligent way both fields. In addition, as presented in Chapter 10, promising avenues to pursue are the further involvement of the human visual system in the designing of the coding scheme, new approaches to the estimation and employment of motion along with signal-dependent coding, the re-use of the transmitted information and the designing of new coding architectures.

We also would like to draw the attention to the fact that a multitude of research efforts are being dedicated to the efficient coding of videoconference and videotelephony applications. There is a great danger of designing very highly evolved schemes focused only upon these applications. It is clear that videoconference and videotelephony are very attractive communication systems, but we should not forget to develop more exciting schemes able to code any generic sequence.

When second generation schemes were proposed in 1985, they showed an increase in the compression ratio of about 10 times with respect to what was available at that time. How far are we in increasing the compression ratio of what we have today by factors of 10? Can we encode video sequences at 500 bits/s with good quality? It seems that if we are able to, we are still far from even knowing how to do it. For this reason some people now seem to be happy with what we have achieved in video coding and are turning their attention towards functionalities. In addition, to improve the efficiency of video coding schemes, it is sure that the techniques explained in this book will have a bright future in the new world of functionalities that is appearing.

REFERENCES

[1] T. Berger. *Rate distortion theory.* Prentice-Hall, Englewood Cliffs, 1971.

[2] R. J. Clarke. *Digital compression of still images and video.* Academic Press, 1995.

[3] E. Dubois and J. Konrad. Estimation of 2-D motion fields from image sequences with application to motion-compensated processing. In M. I. Sezan and R. L. Lagendijk, editors, *Motion analysis and image sequence processing*, pages 53–87. Kluwer Academic Publishers, 1993.

[4] T. Ebrahimi, H. Chen, and B. G. Haskell. A region based motion compensated video codec for very low bit-rate applications. In *IEEE International Symposium on Circuits and Systems - ISCAS'95*, pages 457–461, Seattle, USA, April 1995.

[5] T. Ebrahimi, E. Reusens, and W. Li. New trends in very low bitrate video coding. *Proceedings of the IEEE*, 83(6):877–891, June 1995.

[6] R. Forchheimer and O. Fahlander. Low bit-rate coding through animation. In *Picture Coding Symposium*, pages 113–114, Davis, May 1983.

[7] P. Gerken. Object-based analysis-synthesis coding of image sequences at very low bit rates. *IEEE Transactions on Circuits and Systems for Video Technology*, 4(3):228–235, June 1994.

[8] A. Gersho and R. Gray. *Vector quantization and signal compression.* Kluwer Academic Publishers, 1992.

[9] B. Girod. Motion compensation: visual aspects, accuracy and fundamental limits. In M. I. Sezan and R. L. Lagendijk, editors, *Motion analysis and image sequence processing*, pages 125–152. Kluwer Academic Publishers, 1993.

[10] B. Girod. Rate-constrained motion estimation. In *Proc. SPIE Visual Communications and Image Processing VCIP-94*, volume 2308, pages 1026–1034, Chicago, USA, September 1994.

[11] ITU-T Recommendation H.261. Video codec for audiovisual services at px64 kbit/s. Technical report, ITU, 1993.

[12] Draft ITU-T Recommendation H.263. Video coding for narrow telecommunication channels at $<$ 64 kbit/s. Technical report, ITU, July 1995.

[13] H. Harashima, K. Aizawa, and T. Saito. Model-based analysis-synthesis coding of videotelephone images- conception and basic study of intelligence image coding. *IEICE Transactions*, E72(5):452–458, 1989.

[14] N. S. Jayant and P. Noll. *Digital coding of waveforms, principles and aplications.* Prentice-Hall, Englewood Cliffs, 1984.

[15] H. Jozawa. Segment-based video coding using an affine motion model. In *Proc. SPIE Visual Communications and Signal Processing VCIP-94*, volume 2308, pages 1605–1614, Chicago, USA, October 1994.

[16] ISO/IEC IS 10918-1 (JPEG). Digital compression and coding of continuous-tone still images: requirements and guidelines. Technical report, ISO, 1994.

[17] ISO/IEC JTC1/SC29/WG11. MPEG-4 Proposal Package Description (PPD). July 1995.

[18] M. Kunt, A. Ikonomopoulos, and M. Kocher. Second generation image coding techniques. *Proceedings of the IEEE*, 73(4):549–575, April 1985.

[19] C. Labit and J. P. Leduc. Very low bit-rate (VLBR) coding schemes: a new algorithmic challenge? In *Proc. SPIE Visual Communication and Image Processing VCIP-94*, volume 2308, pages 25–37, Chicago, USA, September 1994.

[20] A. Lippman. Semantic bandwidth compression: Speechmaker. In *Picture Coding Symposium*, pages 29–30, Montreal, June 1981.

[21] ISO-IEC IS 11172 (MPEG-1). Coding of moving pictures and associated audio for digital storage media up to about 1.5 Mbit/s. Technical report, Motion Picture Experts Group, 1993.

[22] ISO-IEC DIS 13818 (MPEG-2). Generic coding of moving pictures and associated audio. ITU-T recommendation H.262. Technical report, Motion Picture Experts Group, March 1994.

[23] H. Musmann, P. Pirsch, and H. J. Grallert. Advances in picture coding. *Proceedings of the IEEE*, 73(4):523–549, April 1985.

[24] H.G. Musmann, M. Hotter, and J. Ostermann. Object-oriented analysis-synthesis coding of moving images. *Signal Processing: Image Communication*, 1(2):117–138, October 1989.

[25] A. N. Netravali and B.G. Haskell. *Digital pictures: representation and compression.* Plenum Press, New York, 1988.

[26] D. Pearson. Developments in model-based video coding. *Proceedings of the IEEE*, 83(6):892–906, June 1995.

[27] H. Peterson, S. A. Rajala, and E. J. Delp. Image segmentation using human visual system properties with applications in image compression. In *SPSE/SPIE Symposium on Electronic Imaging*, volume 1077-20, Los Angeles, USA, January 1989.

[28] H. A. Peterson and E. J. Delp. An overview of digital image bandwidth compression. *Journal of Data and Computer Communications*, 2(3):39–49, Winter 1990.

[29] M. Rabbani and P. Jones. *Digital image compression techniques.* SPIE Optical Engineering Press, Bellingham, 1991.

[30] R. Rinaldo and G. Calvagno. An image coding scheme using block prediction of the pyramid subband decomposition. In *IEEE International Conference on Image Processing*, volume II, pages 878–882, Austin, Texas, November 1994.

[31] P. Salembier, L. Torres, F. Meyer, and C. Gu. Region-based video coding using mathematical morphology. *Proceedings of the IEEE*, 83(6):843–857, June 1995.

[32] W. F. Schreiber, C. F. Knapp, and N. D. Kay. Synthetic highs, an experimental TV bandwidth reduction system. *Journal of SMPTE*, 68:525–537, August 1959.

[33] C. E. Shannon. A mathematical theory of communication. *Bell System Technical Journal*, 27; Part I and II:379–423; 623–656, 1948.

[34] G. Tziritas and C. Labit. *Motion analysis for image sequence coding.* Elsevier Science B.V., The Netherlands, 1994.

[35] H. Wagner and B. Girod. Region-based motion field estimation. In *Picture Coding Symposium*, pages 4.5.1–4.5.2, Lausanne, Switzerland, March 1993.

[36] R. K. Wallis, W. K. Pratt, and M. Plotkin. Video conferencing at 9600 bps. In *Picture Coding Symposium*, pages 104–105, Montreal, June 1981.

[37] J. W. Woods. *Subband image coding.* Kluwer Academic Publishers, 1991.

2

PIXEL-BASED VIDEO COMPRESSION SCHEMES

Narciso García, Fernando Jaureguizar, and José Ignacio Ronda

*Escuela Técnica Superior de Ingenieros de Telecomunicación,
Universidad Politécnica de Madrid,
Madrid, Spain*

ABSTRACT

Classical image and video compression techniques pursue the fulfillment of pixel-based fidelity requirements, which imposes the use of waveform-based compression techniques. Here we review the fundamentals of these techniques with the target of identifying their advantages and limitations. Focusing on the most consolidated approaches, a section is devoted to an overview of the international standards for video compression and the limits in the compression rate that they can provide. Finally, we indicate the main tendencies born within the pixel-based approach towards overcoming the limitations of current-day schemes.

1 INTRODUCTION

Recently, the application of image and video compression techniques has been responsible for a significant improvement in the human capability to transmit, store, and process visual information. Among the main achievements we can mention a huge increase in the storage capacity of image data bases, and the possibility to transmit High Definition Television (HDTV) signals through an (analog) terrestrial TV channel (or, alternatively, the quadruplication of the number of standard TV programs that the terrestrial or cable TV channel can provide). Furthermore, the integration of video in the computer environment has been made possible, setting the starting point for multimedia and virtual reality technologies. These developments have been made possible by the convergence of VLSI technology with the image and video compression theory developed within the last three decades, with the support of international standardization activities.

The need for video compression can be evaluated from the data in Tables 1 and 2. Table 1 summarizes the characteristics of the main application environments for digital video, which differ from one another in the type of the signal to encode and the minimal quality requirements. The uncompressed data rate corresponding to these applications is shown in the second column of Table 2, which also indicates the average number of bits available for the transmission of each color pixel as a function of the capacity of the channel for the most typical types of transmission/storage facilities. Some of the compression targets shown in Table 2 are achievable, at the required quality, with current standardized schemes (boldface figures and below), while others (above boldface figures) are waiting further developments.

Application	Motion	Resolution	Quality
Cont. HDTV	Unrestricted	$1920 \times 1152 \times 50$	Excellent
Dist. HDTV	Unrestricted	$1920 \times 1152 \times 50$	Very good
Cont. TV	Unrestricted	$720 \times 576 \times 25$	Excellent
Dist. TV	Unrestricted	$720 \times 576 \times 25$	Very good
VR	Unrestricted	$352 \times 288 \times 25$	Good
VT	Restricted	$352 \times 288 \times 15$	Moderate
VLBR-VT	Restricted	$176 \times 144 \times 15$	Very moderate

Table 1 Typical parameters associated to different visual communication applications. The resolution of the signal is given as *horizontal* $\times$ *vertical* $\times$ *temporal*, and corresponds to a typical (not exclusive) choice of parameters for Europe. More details on signal formats can be found in Section 3. *Contribution* applications usually imply transmission between production centers, which requires enough quality to allow any form of video processing, while *distribution* refers to the delivery of the signal to the final viewer. *Restricted* motion indicates typical motion of a sitting speaking person. Abbreviations: **Cont:** Contribution, **Dist:** Distribution, **VR:** Video recording, **VT:** Visual telephony, **VLBR:** Very low bit-rate.

Current applications are based on a quite homogeneous technology which is usually considered to define **first generation** of image and video compression techniques. A coarse characterization of this technology can be given in the following terms:

- These techniques strive to obtain images at the receiving side which are similar to the original pixel for pixel.
- They decompose the signal into *fixed elements*, independent of the image content.

Application	Pixel Rate (Mpel/s)	Bits per pixel				
		Broadcast 20 Mb/s	Broadcast 5 Mb/s	CD-ROM 1 Mb/s	ISDN phone 64 kb/s	Analog phone 20 kb/s
HDTV	110.5	**0.18**	0.045	0.009	–	–
TV	10.4	1.92	**0.48**	0.096	0.006	–
VR	2.5	8	2	**0.40**	0.026	0.008
VT	1.5	–	3.32	0.66	**0.042**	0.014
VLBR-VT	0.37	–	–	2.70	0.172	**0.054**

Table 2 Pixel rates and average number of bits per color pixel (bpp) required for different applications and transmission/storage channels. Figures above 8 or below 0.005 bpp are not shown. Boldface figures delimit the possibilities of current standardized schemes.

- They eliminate spatial redundancy by means of general signal analysis techniques.
- They eliminate temporal redundancy through motion compensation techniques which do not pay attention to the structure of the image content.
- They make little use of the properties of the human visual system (HVS).

It is apparent from this characterization why these techniques are also termed *pixel-based* or *waveform-based* [29]. The main criticism to this approach applies to the first and last points: for applications oriented to human observers, a pixel-based fidelity criterion is too demanding according to the characteristics of human perception, which tends to overlook many details and appreciate no semantic content in many others. Taking advantage of HVS properties is one of the principles which guide the **second generation** of image and video compression techniques to which this book is devoted. Other applications, such as the compression of images for their transmission to remote automatic analysis centers, will still demand the preservation of the pixel-by-pixel quality of the original image, maintaining the need for pixel-based compression.

In this chapter we review the most important pixel-based techniques for image and video compression. The chapter is organized as follows. In Section 2 we present the basic paradigms on which pixel-based image and video coding schemes are based. Section 3 describes the main standards based on these paradigms. An assessment of the good properties and limitations of these technical approaches is included in Section 4 along with a review of the most significant efforts made so far to extend the capabilities of classical schemes beyond their current limitations. Section 5, finally, summarizes the main conclusions of the chapter.

2 FUNDAMENTALS OF PIXEL-BASED TECHNIQUES

2.1 Introduction

This section presents the main concepts on which pixel-based image and video compression techniques are based. A system for video compression and decompression is depicted in Figure 1, where the signal undergoes a 6-stage process for its encoding, regulated at times by a bit-rate control module. Corresponding counterparts for each stage constitute the decoder. We describe briefly these elements:

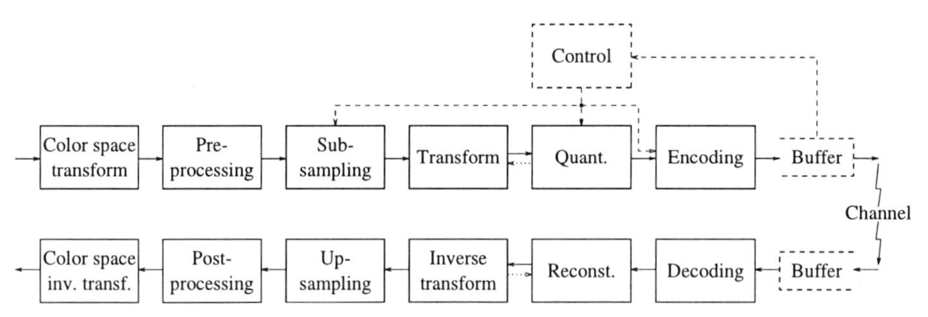

Figure 1 Block diagram of real time image and video coder (above) and decoder (below). The dashed elements are required when the output bit-rate must be adjusted.

- **Color space transformation:** Redundancy within the color representation of images is usually reduced in a first step. In most cases the operation consists in the mapping of the original representation in terms of the fundamental colors (red, green, blue – RGB) onto another in which the color information (chrominance – C_R, C_B components) appears separated from intensity information (luminance – Y component) [17] and each can be processed according to its own perceptual relevance.

- **Preprocessing:** The efficiency of the compression process can be improved if some features of the input image are suppressed or, on the contrary, reinforced beforehand. For example, some techniques will work better with low-pass filtered images. The degree of preprocessing of any type should be optimized together with the parameters of the rest of the compression system in order to optimize the overall quality.

At the receiving side the decoded signal can be improved at the **postprocessing** stage, on which great interest has lately developed. Approaches to this module range from aiming at removing the main compression artifacts (such as the blocking effect in block-transformed compression [40]) to estimating the original image from the compressed data [39].

- **Subsampling:** Subsampling reduces the amount of data to compress by changing the sampling structure of the discrete image or video signal into another sampling structure with a reduced sample rate (either in space, time or both) [6].

- **Transformation:** This block performs the description of the signal in a more efficient vocabulary than that of pixel values. In pixel-based schemes, this process is oriented to the removal of statistical (and, to some extent, psychovisual[1]) redundancy. This stage constitutes the heart of the compression system in most cases, and we dedicate the largest part of this section to this matter.

  Besides elimination of redundancy, the transform is also responsible to a large extent for the organization of the information content of the image for transmission. For instance, some applications require that this information adopts a *hierarchical representation* [4][48][49], consisting of a set of ranked units, so that each unit can be processed according to its relevance. The hierarchical approach introduces flexibility in the selection of the rate-distortion compromise from the receiver side, so that a receiver can decide among a menu of qualities and associated bit-rates, allowing the development of multi-terminal compatible services. Another feature is the possibility to optimize channel encoding by providing more protection to the more relevant data, thus achieving graceful degradation characteristic in lossy transmission environments such as terrestrial broadcasting or ATM networks. Some international image and video coding standards, as JPEG [13] and MPEG-2 [15], include different forms of hierarchical representations.

- **Quantization:** In this stage the number of possible values for the transformed signal parameters is reduced, introducing irreversible degradation in the signal, by associating each incoming value with a member of a finite set of output symbols (labels). Quantization can be independent for each sample or can be made blockwise *(vector quantization)*. In some compression schemes quantization is bound to the transformation stage through a feedback loop, indicated in Figure 1 as a dotted arrow. At the decoder

[1] Thus anticipating an important feature of second generation schemes.

side the module **Reconstruction**[2] maps the incoming labels onto the corresponding *reconstruction values.*

- **Encoding:** This module assigns a sequence of bits to the previously transformed and quantized signal making use of lossless source coding techniques.

- **Buffer** and **control:** Since a common feature of video compression systems is that their compression efficiency varies depending on the local characteristics of the video signal (image content, type of motion in the scene), when considering real-time transmission applications, the use of a smoothing buffer between the coder and the channel is necessary in order to adapt the variable bit-rate output of the first to input requirements of the second. To avoid the overflow of this buffer, the coder operation has to be regulated by a control system which becomes another module to take into account in the compression scheme.

Once the block diagram for image and video compression has been introduced, we are ready to go into the details of its main building blocks: subsampling, transform, quantization, encoding, and control. Regarding the transform stage, we will consider first, for the sake of clarity of presentation, the concepts applied for still image compression, and then we will see how they extend to the case of the video signal.

2.2 Subsampling

Subsampling implies, in general, multidimensional filtering followed by decimation [6] and can be motivated by the desire to adjust the sample rate to the minimum required for the alias-free representation of the perceptually meaningful spectrum of the signal, or simply by the need to increase the compression rate of the system drastically[3]. A third possible purpose is the reduction of the pixel rate to be processed by the rest of the system due to hardware limitations. It is a matter of optimization what degree of compression must be obtained through subsampling and what degree must be left to the more sophisticated

[2] Some authors use the term *inverse quantization* for this module. Others include it within the quantizer, so that the quantizer output becomes the reconstruction value for each sample. We prefer to consider *reconstruction* as an independent module, following the tendency of the texts of international standards.

[3] As an example of a compression system almost exclusively based on subsampling we cannot help mentioning the HD-MAC system [10], which had expensive consequences for the European tax-payer.

transform+quantization stages. Subsampling of chrominance components is usually different considering their lower perceptual relevance. A 2:1 relation of resolutions between luminance and chrominance in both spatial dimensions is considered a good choice.

Sampling structures for video signals can be classified into **progressive** and **interlaced**. In *progressive* video format (like the one resulting from the orthogonal sampling of a film), a single instance in time is enough to cover the complete spatial area. In *interlaced* video (like the one resulting from sampling the analog TV signal), several (generally 2) line interlaced instances in time are needed to cover the complete image area. The complete image is called *frame* and each partial image receives the name of *field*. Figure 2 shows the spatio-temporal sampling structure employed in *interlaced* formats (such as ITU-R 601 [17]).

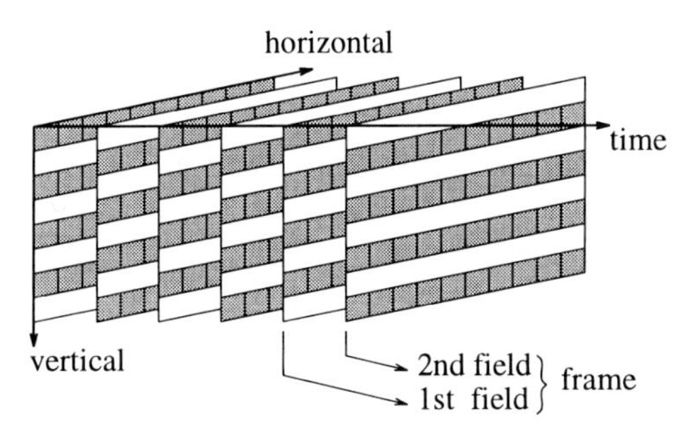

Figure 2 Interlaced video format.

2.3 Spatial transform

Two main approaches exist to eliminate statistical redundancy in a signal: the **predictive** approach and the **transformational** approach. The first one is the most straightforward, as it consists in directly *subtracting* from each signal element what can be estimated about it from the context. The second is more elaborate, as it consists of the translation of the signal to a different space in which the information that it contains appears to some extent separated from the redundancy. The transformational approach takes the main forms of **block transformation**, **pyramidal schemes** and **subband/wavelet analysis.** The lack of a spatial transform stage in the compression chain results in the so-called *pulse code modulation* (PCM) encoding system, which we cite for reference.

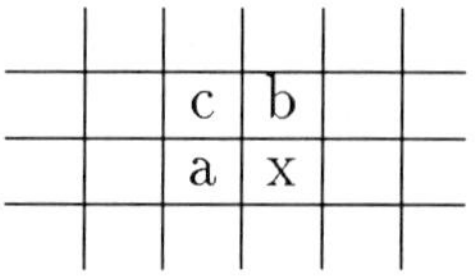

		c	b		
		a	x		

0	P = 0
1	P = a
2	P = b
3	P = c

4	P = $a + b - c$
5	P = $a + (b - c)/2$
6	P = $b + (a - c)/2$
7	P = $(a + b)/2$

Figure 4 Spatial predictions employed, for the pixel **x**, in lossless JPEG.

non-overlapping rectangular blocks of samples[4] and the application of a linear invertible transform to each block, as first suggested in [12]. Reordering the block as a column vector $\mathbf{x}$, the transformation corresponds to the matrix multiplication

$$\mathbf{y} = \mathbf{A}\mathbf{x}. \tag{2.3}$$

The transformation can also be considered as a change of base in the linear space of blocks, so that the transformed vector $\mathbf{y}$ (vector of *transform coefficients*) corresponds to the coordinates of vector $\mathbf{x}$ in a new base. Particularly important are orthonormal transforms (those for which the transform matrix $\mathbf{A}$ is orthonormal). In this case, the rows of $\mathbf{A}$ are the coordinates of the (unitary) vectors of the new base. An important property of orthonormal transforms is that the encoding error in the transformed domain has the same mean square value as the corresponding error in the original domain.

A transform for 1-D signals can be extended to 2-D rectangular image blocks through the definition of a separable linear transform operating independently in each dimension:

$$\mathbf{Y} = \mathbf{A}\mathbf{X}\mathbf{A}^T, \tag{2.4}$$

where $\mathbf{X}$ and $\mathbf{Y}$ are the matrices representing, respectively, the original and transformed rectangular blocks of pixels. If one of the coefficients is proportional to the average value of the pixels in the block, it is termed *DC coefficient*, while the others are referred to as *AC coefficients*.

The rationale behind the transformation approach is that a more efficient encoding can be achieved if the energy of a block of pixels can be concentrated in a few transform coefficients. More precisely, assuming that the signal is stationary and that the rate-distortion characteristic of the quantizer for the transform coefficient k is given by [26]

$$\sigma_{q_k}^2 = \epsilon^2 2^{-2R_k} \sigma_k^2, \tag{2.5}$$

[4] These blocks are usually equal-sized, with the result of a segmentation which does not take into account the image content.

where $\sigma_{q_k}^2$ is the quantization error energy of the coefficient, R_k is the number of bits allocated to the coefficient, σ_k^2 is its energy, and ϵ is a parameter dependent on the statistical distribution of the coefficients, we can quantify the gain in SNR resulting from the application of transform coding (TC) to a block of N pixels as

$$G_{TC} = \frac{\frac{1}{N}\sum_{k=0}^{N-1}\sigma_k^2}{(\Pi_{k=0}^{N-1}\sigma_k^2)^{1/N}} \tag{2.6}$$

for an optimal fixed bit allocation among the coefficients.

The transform which maximizes this gain is known as the Karhunen-Loeve transform (KLT) [12], which is an orthonormal transform whose base vectors are the eigenvectors of the autocorrelation matrix of the signal. The KLT is also optimal for a compression scheme based on discarding the coefficients of less energy.

The direct use of KLT for decorrelation is difficult because of its dependency on the statistics of the signal and the lack of efficient algorithms for its computation. Both problems are partially solved by the discrete cosine transform (DCT) [42], which is asymptotically optimal for first order autoregressive processes as the correlation coefficient tends to one [1]. Moreover, fast FFT-like algorithms for the computation of the transform exist [42]. The extension to higher dimensions of the DCT can be achieved by defining the corresponding multidimensional separable transform. Some degree of HVS adaptation in transform coding is achieved, on the ground of the modelization of the HVS as a linear space-invariant system, by quantizing each transform coefficient differently according to its visibility [36]. In this way the system participates from one of the defining features of second generation systems.

The encoding of the coefficients in practical schemes also diverges from the assumptions of the basic theory, as the individual coefficient processing can be easily improved by introducing run-length encoding of null coefficients and different forms of adaptivity [14][21].

The asymptotic gain over PCM for transform coding (as the block size and the number of quantization levels tend to infinity) shares with prediction the expression 2.1. For finite size blocks KLT is more efficient than optimal prediction of the same order, especially at low bit-rates, and is less sensitive to image statistics [22]. Furthermore, some additional advantages allow transform coding to outperform predictive coding: in the transformed domain it is easy to make a dynamical allocation of bits to the different transformed coefficients, depending on the signal content; transform coding is less sensitive to channel errors, and has better characteristics for HVS adaptation. Finally, in the case of video compression, spatial block transform coding integrates easily, as we will see, with block-based temporal motion compensated prediction.

Block transform coding is agreed to provide very good quality for color images at rates above 0.75 bpp, and moderately good quality at rates between 0.25 and 0.75 bpp [56]. The typical encoding artifact for transform coding is the *blocking effect*, which consists in the visibility of the boundaries of the blocks. This phenomenon is a key factor in the selection of the block size for the transform, since very large blocks, though more efficient, may become too visible under blocking effect. In order to reduce this effect, the basic transform approach has been extended to overlapping blocks. The lapped orthogonal transform (LOT) [31] makes use of overlapped blocks without increasing the total number of transform coefficients for the image.

Other problems of block transform, specially at low bit-rates, are its limitations in the description of edges, which can appear blurred or ringed, and the *checkerboard* effect, resulting from the encoding of a block with very few DCT coefficients. In spite of these limitations, the fact that block-transform provides a good performance-complexity trade-off for current hardware technology has determined the adoption of the DCT-based international standard JPEG [13] for lossy image compression. The most important video compression standards, which we later describe, also rely on DCT for spatial decorrelation.

Pyramidal schemes

Pyramidal schemes are a particular case of *hierarchical representation* of images, consisting of a graded set of images of different resolutions. One of its most important versions for compression purposes, the **Laplacian pyramid** [4], exploits the intuition that *a large amount of the information of an image can be contained in a subsampled version of the same.* In this approach a two-level pyramid (see Figure 5) is obtained by subsampling the original image (I_0) in both dimensions after low-pass filtering (yielding image L_1). The pyramid will be made up of L_1 as the upper level and $L_0 = I_0 - I_1$ as the lower, where I_1 is obtained from L_1 by interpolation. As the complementary image L_0 has lower entropy, a net bit-rate reduction results in spite of the increase in the total number of samples. The process can be iterated to create pyramids of any number of levels. Decoding starts with the image at the top level and continues downwards.

The transform process can be made independent of the quantization stage or coupled with it; both situations are also depicted in Figure 5. In the first one (switch in position 1) the subsampled image is upsampled by zero insertion, filtered and subtracted from the original to obtain the complementary image. In the second case the subsampled image is quantized and reconstructed before upsampling, so that the complementary image compensates to some extent the effect of the quantization in the lower-resolution level and the only encoding error is due to the quantization of the complementary image L_0. Both solutions are compared experimentally in [41], with the result that the second case is superior.

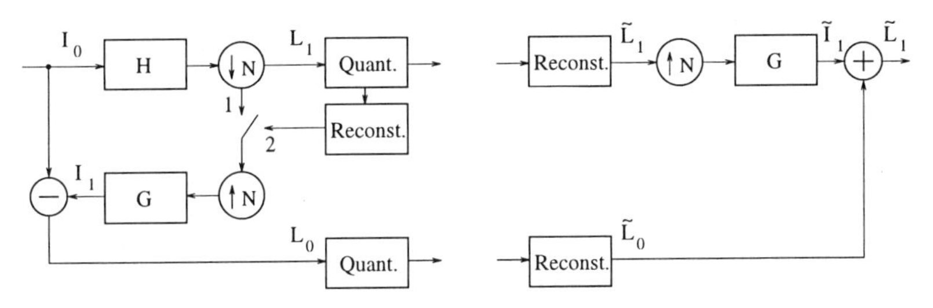

Figure 5 Two-stage pyramidal decomposition. The upper level image $L_1$is a reduced-resolution version of the original image I_0. L_1 is upsampled by insertion of zeros followed by filtering, and subtracted from I_0 to produce the lower level image L_0. The two positions of the switch correspond to the decoupling of transform and quantization (position 1) and their coupling (position 2).

Subband/wavelet analysis is a different approach which holds relations with pyramidal schemes and can also be applied to obtain hierarchical representations. We present it in the next section.

Subband/wavelet analysis

In **subband analysis** [5][50][55] the signal is decomposed into a set of elementary signals with an overall sample rate equal to that of the original in such a way that each subsignal carries the information associated to a part of the original spectrum. This is achieved, as depicted in Figure 6, by filtering the signal through a filter bank and decimating the results. At the receiving side, subsignals are upsampled by zero insertion, filtered and added together to recover the original signal. Several variations of this scheme result from iterating the decomposition in all or some of the branches.

The basic requirement for a subband filter bank is the capability to provide an alias free signal in the receiver. A stronger further requirement is that

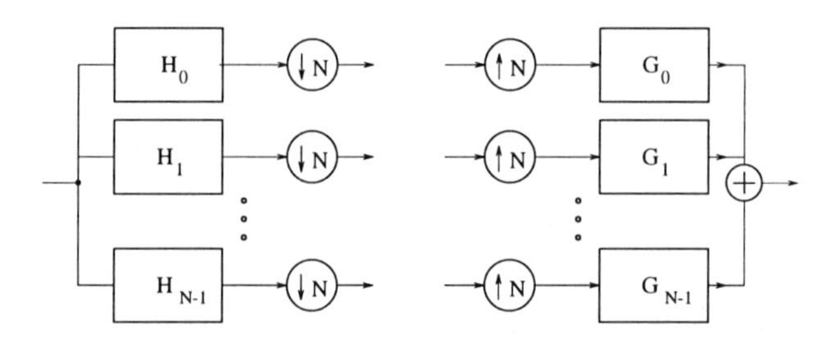

Figure 6 Subband analysis and synthesis filter banks. The total sample rate of the subsignals obtained from the analysis filter banks equals that of the original signal. Under certain conditions, the overall operation of the analysis and synthesis filter banks is equivalent to a delay.

the overall system does not introduce linear distortion *(perfect reconstruction filter banks)*. Another desirable property in image processing is phase-linearity of the filters. Finally, in order to obtain uncorrelated subbands, the pass-bands of the analysis filters should not overlap. Filter structures have been provided which enjoy strictly the first three properties, while the fourth one can only be approximated. For the application of subband analysis to images, which are defined on a finite domain, they must be conceptually converted into infinite signals employing, for example, periodical repetition. The theoretical asymptotical performance of subband coding in terms of gain over PCM is the same as that of predictive and transform coding [26].

Relations exist between subband analysis and block transform, either overlapped or not [31], since a block-based image transformation can be seen as the application of a filter bank based on FIR filters. Similitude with the Laplacian pyramid is also easy to check [43] by comparing Figures 5 and 6 for $N = 2$: while the upper branches are equal, in the subband case the lower produces a signal of half the sampling rate.

For multidimensional signals, apart from the obvious extension of the scheme by operating in one dimension at each time (separable subband decomposition), specific non-separable filtering+subsampling structures have been found [55]. Subband schemes are also interesting as a way to decompose the signal in a hierarchical way. Subband decomposition can be applied in a pyramidal way by decomposing at each stage the low frequency signal into two subsignals (see Figure 7).

The interaction between subband decomposition and wavelet theory [43] has given an important momentum to this signal analysis approach. Wavelets pro-

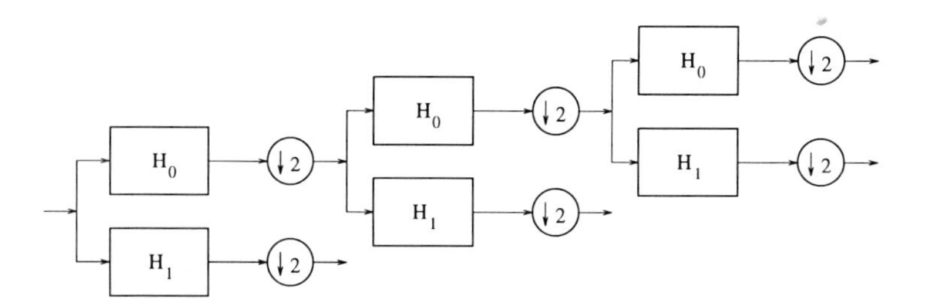

Figure 7 Subband analysis applied in a pyramidal way.

vide a powerful tool for the analysis of the time or space-varying content of a signal at different scales. A *wavelet basis* is a basis of the space of finite energy signals on the real domain whose elements $\{\psi_{m,n}\}$ are derived from a single function ψ called *mother wavelet* through translations and dilations (typically, $\{\psi_{m,n}(x) = 2^{-m/2}\psi(2^{-m}x - n)\}$). Parallelism with subband processing of discrete signals stems from the properties of some FIR filter banks, for which the structure of Figure 7 implements a decomposition of the input in terms of discrete signals which are asymptotically scaled versions of each other. For these filter banks there exist associated wavelets, and the operation (whose computational complexity is, interestingly, linear in the number of samples, in contrast with other transforms as DCT) is called **discrete wavelet transform** (DWT).

Recent work on wavelet transform [2] points towards the superiority of this signal decomposition scheme over non-overlapped block-based transforms, especially for high compression rates (with reports of acceptable quality near 0.1 bpp [11]). This can be attributed to the greater ability of wavelets to capture fine details of the image and to the lower visibility of the quantization errors due to the overlapped nature of the basis functions. Visual masking properties can also be incorporated in the system, as in the case of block transform. Finally, regular filter banks compress better than those which do not have an associated wavelet, with increasing compression efficiency as the number of vanishing moments of the wavelet increases. This seems to be a consequence of the suitability of these wavelets for the description of smooth functions.

2.4 Spatio-temporal transform

Extension of image analysis techniques

The generalization of the summarized analysis schemes to the three dimensional case of video signals, can be carried out in different ways. The direct approach, consisting in an equal consideration of the three dimensions of the signal and the application of the same decorrelation technique in every dimension, has been proposed with moderate success. In the case of predictive schemes, the prediction of a pixel can be obtained either from pixels of the same image or from a previously transmitted image. In the first case, we say that we use *spatial prediction*, while the second scheme is referred to as *temporal prediction* [35]. In the case of block-transform coding, the use of three-dimensional blocks [35][44] was discarded early because of the delay it introduced, the presence of time artifacts, and its relatively low performance. Three-dimensional hierarchical [54], subband [27] and wavelet processing [28] have also been suggested.

The efficiency of these schemes is limited due to the little correlation that exists in the presence of motion between samples occupying the same position in consecutive images. **Motion compensation** tries to overcome this problem.

It must be pointed out that there are practical limitations in the use that can be made with the transformational approach in the third dimension, as the *depth* of the temporal operation has serious implications in delay and coder and decoder memory requirements. The predicted approach, on the other hand, only suffers from the second inconvenience.

Motion compensation

Motion compensation (MC) establishes a correspondence between elements of nearby images in the video sequence. The main application of MC is providing a useful prediction for a given image from a reference image.

Motion compensation in first generation video coding makes use of a simple motion model and tries to find for each image the parameters of the model which best allow the prediction of the image from the reference [23]. This motion model assumes that purely translational motions occur in different areas of the image, and, consequently, the prediction for a rectangular block of pixels in an image will consist of a block of the same size in the reference image, related to it by a displacement or **motion vector** (see Figure 8). Motion vectors are transmitted as side information, and the associated overhead becomes important in very low bit-rate (VLBR) applications. Within the framework of this simple motion model, the main issue is the computation of the motion vectors. The most employed methods for this task are different forms of the *block matching* [7][24][25][33] approach, which consists in the search for a block within the previous image which, according to some distance criterion such as the mean square error or the mean absolute error, resembles the block to encode. Current VLSI technology allows the application of an exhaustive search for this task, although other methods are sometimes employed, such as *phase correlation* [59], which operates in the discrete Fourier transform domain.

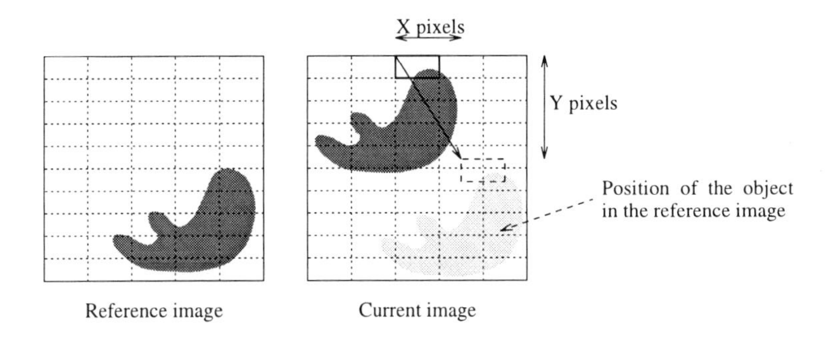

Figure 8 Block-based motion compensation. It is assumed that each block in the current image corresponds to the rigid translation of a block of the same size in the reference image, described by a motion vector (X, Y).

In section 4.1 we point out the main weaknesses of this form of MC within the analysis of the limitations of standard first generation techniques. Evolutions of this scheme will be presented in section 4.2 as some of the main directions to overcome these problems.

Hybrid coding

Hybrid predictive-transform coding or, for short, **hybrid coding** consists in the cohabitation of a prediction loop in the temporal dimension with a suitable decorrelation technique in the spatial domain. Including MC, this approach has been incorporated by the most important video compression international standards. An easy integration between MC and spatial processing occurs when both work on a non-overlapped-block basis, as it is the case of the MC that we have introduced, and the non-overlapped-block spatial transform. In this case, it can be decided for each block whether it must be encoded non-predictively (intraframe), if no suitable prediction has been found, or predictively (inter-frame), in which case the prediction error block is spatially decorrelated and encoded for transmission together with the motion vector. In the last case, spatial decorrelation is very often ineffective, as the prediction error signal is very little correlated and its information content is of high frequency [45]. However, block-transform is usually applied in the same way as it is applied to intraframe blocks, in order to make use of perceptually optimized quantization.

The hybrid video compression scheme operates by structuring the video sequence in three different levels: **picture**, **macroblock** and **block.** A *block* is the set of pixels to which the DCT transform is applied. Blocks consist of squares of pixels of an image component (Y, C_R, or C_B) and constitute a non-overlapping tesselation of that component. For convenience, they are grouped to form an intermediate level structure called *macroblock* which contains all the blocks of the different image components included in a given rectangular spatial region. The highest level processing unit can be the *field* or the *frame*; we will use the term *picture* to refer to this highest level unit, regardless its nature.

Figure 9 shows the **picture-level processing**. The first module, *Picture reordering*, is necessary in those schemes in which the prediction of a picture is obtained from future pictures as well as from past ones, with the result that the order in which pictures are processed is different from that of their arrival. The *Prediction memory* block in the diagram is a buffer containing the decoded versions of some of the previously processed pictures; in the simplest case it will contain exclusively the previous one, but more complex schemes make use of additional pictures. The processing of each picture consists of an operation over each of its macroblocks; as in any predictive scheme, processing includes the computation of the decoded version of the picture, which is stored in a buffer so that it can be used as a source of predictions for forthcoming pictures.

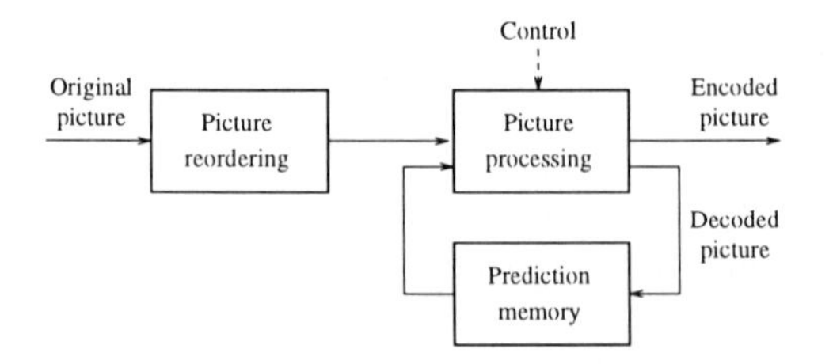

Figure 9 Hybrid coding. Processing at the picture level.

Macroblock-level processing is outlined in Figure 10. First, a choice of predictions for the macroblock is obtained from the pictures stored in the prediction memory. Each prediction can consist of a macroblock obtained from a single picture (either directly extracting the pixels from the picture or computing them by interpolation) or in the linear combination of macroblocks obtained from two different pictures. Subsequently, the best prediction is selected and subtracted from the original macroblock to obtain the prediction error macroblock, which is encoded and transmitted; the prediction error macroblock can consist of the macroblock itself if no prediction has been found suitable (intraframe mode). Side information for the macroblock is the mode of prediction and the motion vector. The macroblock is then reconstructed in the same way the decoder will do and stored in the prediction memory.

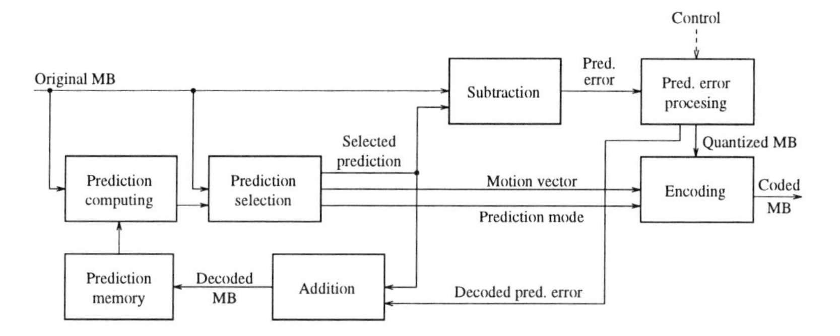

Figure 10 Hybrid coding. Processing at the macroblock (MB) level.

Block-level processing, as shown in Figure 11, starts with the 2-D DCT transformation of the prediction error block. After the coefficients are quantized, they are encoded. The quantized block is also reconstructed and inverse-DCT transformed.

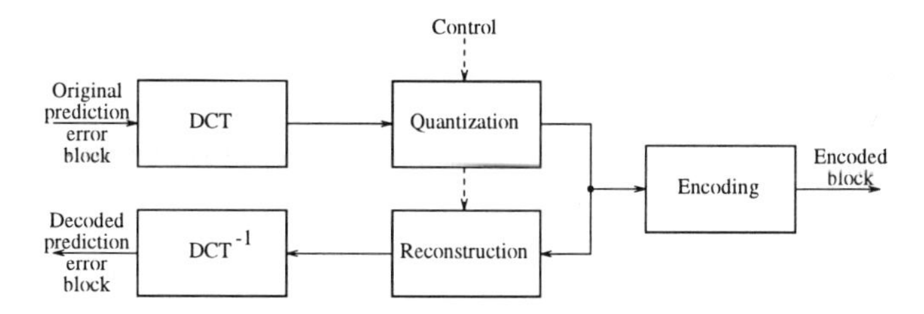

Figure 11 Hybrid coding. Processing at the block level.

Prediction schemes

While early hybrid schemes were based on *time-stationary* prediction schemes, such as using the previous picture as a reference for each picture, *time-varying* predictive schemes were introduced later in order to improve compressing efficiency while endowing the encoded bit-stream with properties which are important for some applications.

Time varying prediction schemes operate by defining three types of pictures and assigning the picture type to different pictures in either a predefined or adaptive order. These picture types are: **Intra** (I), which are encoded independently (without making use of any temporal prediction), **Predictive** (P), which are processed using the previous I or P picture as the reference, and **Bidirectionally predicted** (B), which benefit from predictions obtained from the closest previous and future I or P pictures. Figure 12 illustrates the relationship between the three different picture types.

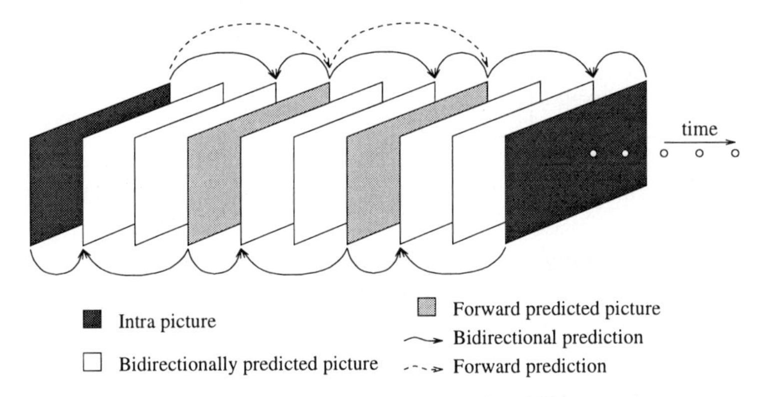

Figure 12 Relationship among the three different picture types in MPEG.

Intra pictures provide *random access points* to start the decoding of the sequence, and, if no B pictures are included between contiguous P and I pictures, the coded video sequence is divided into independent groups of pictures, which simplifies editing operations. Besides, I pictures provide fast forward and backward reproduction and limit the propagation of transmission errors associated with predictive schemes. This forced introduction of non-predictively coded areas in the picture is called *refreshment*, and can also be made at lower levels (forcing the intra coding of several macroblocks in each picture).

The disadvantage of I pictures is that they typically require several times more bits than predictive pictures, which, apart from reducing the overall compression rate, introduces noticeable peaks in the bit-rate which make necessary a larger buffer for the adaptation of the coder to the network. For this reason, macroblock-based refreshment is preferred in transmission applications.

Bidirectionally predicted pictures benefit from the use of predictions from both the previous and the following I or P pictures. Each prediction can consist of a macroblock from one of the reference pictures (obtained, if necessary, with interpolation in the case of fractional-pixel motion vector accuracy) or of an average between macroblocks from both reference pictures (see Figure 13). In storing applications, B pictures allow fast forward operation of the decoder by skipping B pictures and decoding only I and P pictures.

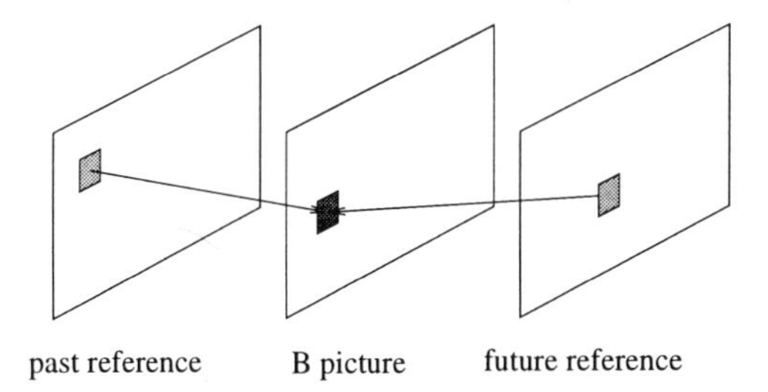

Figure 13 Bidirectional prediction of a macroblock in a B picture.

Bidirectional compression is more efficient for three reasons. First, prediction from a future picture makes possible the differential encoding of uncovering areas. Second, bidirectional prediction averages two different macroblocks with a similar content of information but with independently added quantization noise, so that, statistically, the average reinforces the information content while reducing the error energy. Third, as B pictures are not used for further prediction, their encoding with large distortion has no consequence on longer-term sequence quality.

Drawbacks for the compression efficiency of B pictures are the increase in overhead in bidirectionally predicted macroblocks (two different motion vectors have to be transmitted for interpolated macroblocks) and the larger range of motion vectors required for the prediction of P pictures (which get further apart from each other). Finally, B pictures increase end-to-end delay as a result of the reordering operations in the coder and the decoder (see Figure 14) and increase hardware complexity (more complex motion estimation) and memory size (prediction memory must contain two pictures).

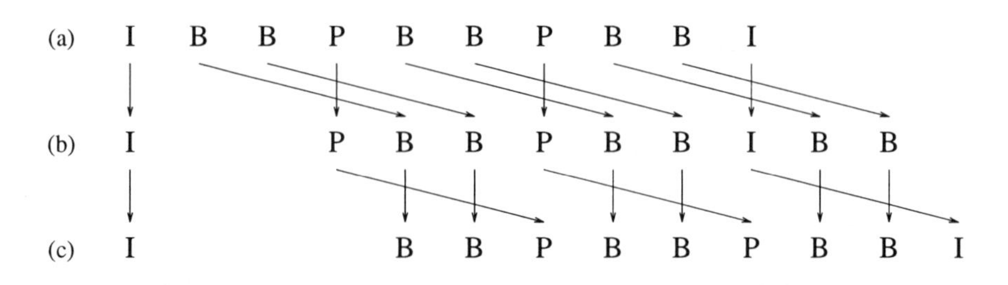

Figure 14 End-to-end delay with B pictures: (a) temporal order; (b) coding, transmission and decoding order; (c) displaying order. There is a delay of $M+1$ pictures, where M is the number of B pictures between consecutive references (I or P pictures).

Another issue in the definition of a prediction scheme is the selection of the picture level processing unit. Encoders operating with interlaced video sequences can identify the picture with either the video frame or the video field. Frame-based pictures contain pixels which are more closely located in space, so that in the absence of motion a higher spatial correlation can be expected than in the case of two separate pictures obtained from splitting the frame into two fields. On the other hand, in the presence of motion, consecutive lines in the frame picture will be uncorrelated, and it becomes more appropriate to split the frame into its two fields and employ MC for redundancy reduction.

2.5 Quantization

Scalar quantization

The basic algorithms for optimal scalar quantizer design in the mean square error sense were independently established by Lloyd and Max [30][32]. Rate-distortion optimality of the quantizer provided by Max-Lloyd algorithms is restricted to the case in which it is followed by fixed-length coding. If this is not the case for the encoding stage, different techniques must be used. An important result is that the optimal quantizer under the minimum entropy restriction for the quantized signal for Laplacian-distributed coefficients or for a large number of quantization levels is the uniform quantizer [26].

This is one of the reasons for preferring uniform quantization in standardized coding schemes. Another reason is that the optimization of this stage is compromised by the need to employ variable quantizers as a means for bit-rate control.

Vector quantization

The block quantization of samples, known as **vector quantization** (VQ) [9], is able to take advantage itself of the spatial redundancy in the image. Even if the quantized source produces symbols independently, vector quantization is, according to rate-distortion theory, the way to approach the theoretical rate-distortion limit for the source. In video coding applications VQ can be intraframe (if each vector is obtained from a single image) or interframe (if each vector contains pixels from adjacent images).

The operation of a vector quantizer is based on the selection of a finite set of vectors as representatives of the whole space. This set of vectors, known as the *codebook* determines a segmentation of the input space into cells, according to a minimum distance criterion. For each input block the quantizer finds the closest vector in the codebook and transmits its label. In order to obtain the optimal codebook, the true distribution of the blocks is necessary. If it is not known, which is generally the case, the LBG algorithms [9] finds a suboptimal codebook based on a set of training samples. If a model of the image is known, then a synthetic generation of the codebook has been proposed [53].

Quantization is in general a very consuming task, and different ways to structure the codebook have been suggested which simplify the search, such as hierarchical vector quantization [9]. On the other hand, the operation of the decoder is very simple, as it simply accesses a memory which stores the codebook.

Non-overlapped block transformation followed by scalar quantization can be seen as a particular case of VQ with quantization bins indirectly determined by the transform basis vectors and the scalar quantizer. If we compare these two schemes, we can appreciate that, as in block transform coding, straightforward VQ systems show blocking effect at low bit-rates as well as a poor edge representation, although in this case can be improved by assigning the edges a separate codebook *(classified VQ)* [34].

Despite the good properties of VQ, its high coder computational complexity has so far biassed the standardization decisions towards scalar quantization. Some proposals for future standards as MPEG-4, however, apply the VQ scheme to the encoding of textures.

2.6 Encoding

The output from the quantization stage can be seen as a fixed-rate sequence of symbols. The design of the encoding stage usually assumes that the transform stage has been successful in making these symbols independent so that any technique for the encoding of sources of independent symbols is, in principle, suitable. The simplest approach is the use of *variable length codes* (VLC), such as the optimal ones provided by the Huffman algorithm [26].

Modifications to this basic approach are necessary because of the difficulty VLCs have in reaching the first-order entropy limit. For example, in standard systems based on DCT coding run-length is applied in the encoding of null coefficients [13][18][21], thus, taking advantage of the high probability of zero-valued coefficients, specially as we move towards coefficients representing higher spatial frequencies; in order to accumulate as many zeroes as possible together, a zig-zag scan of the block in decreasing order of energy is employed. This different treatment of null coefficient suggests in some cases the application of further optimization in the quantizer, in order to increase the zero frequency and decrease the bit-rate. This results in a joint quantizer-encoder optimization.

Arithmetic coding [58], though more computationally consuming, does not suffer from the limitations of VLCs to approach the first order entropy limit, and is able, with little modifications, to reach higher-order entropy limits (which might be lower than the first-order entropy limit if the encoded coefficients are not independent, as in the case of a not totally successful transform stage). Recently, standardization bodies have appreciated the advantages of arithmetic coding and included it in the corresponding standards [13][19].

2.7 Rate control

The rate control module in video compression systems [38][46] is responsible for the adaptation of the coder output to the input requirements of the network. The most important property that a buffer control system must enjoy is the capability to prevent the overflow of the buffer, as it implies a loss of video information and synchronism which will be perceived as highly annoying at the receiving side. The control algorithm is also responsible for efficiently allocating the available bandwidth to optimize the global quality.

As the buffer inserts a delay in the video transmission which is proportional to its size, this is a parameter which has to be kept within the limits imposed by the service requirements. The control policy is responsible for making an efficient use of this limited buffering capability by selecting for every segment of the video signal the mode of operation of the encoder which best suits the buffer occupancy and the available information on the instantaneous characteristics of the video source.

Control is usually achieved by varying the quantization step, although other coding parameters can be used as well, as the picture rate [18][19] or the picture resolution. Variable coding parameters used for control have to be transmitted to the decoder unless the control algorithm is known by both sides and operates on the basis of previously transmitted samples.

Although most control algorithms, including those proposed in the reference models for standards, are basically heuristic, work has also been conducted to obtain optimal control policies, either assuming a complete (deterministic) knowledge of the source [38] or its stochastic characterization [46][47].

3 VIDEO CODING STANDARDS

3.1 Introduction

As international standards reflect practical trade-offs between efficiency and implementation feasibility, their analysis becomes a solid approach for the assessment of a technology. Besides, in their conception different requirements are taken into account (such as delay, cost, robustness, etc.), which makes one discard non-realistic alternatives, and an extensive evaluation of the capabilities of the system is conducted. For this reason, apart from its intrinsic practical interest, we dedicate in our review of pixel-based techniques a section to the discussion of the main standard video compression algorithms. After a description of their technical details, results will be provided on the performance of some of them within the application environment in which rivality with new approaches is likely to be harder, namely, VLBR coding.

Some of the features which make different standards diverge is the different structure of the video signal they deal with. Table 3 summarizes the main standardized digital video formats. The *Chroma Format* column in the table makes reference to the relation of spatial sampling ratios between luminance and chrominance, with the notation 4:4:4 for equal sampling, 4:2:2 for half horizontal resolution of the chrominance and 4:2:0 for half horizontal and vertical chrominance resolution.

Video Format	Luminance Frame Size	Frame Rate (Hz)	Sampling	Frame Aspect	Chroma Format	Pixel Rate (Mpel/s)
HDTV	1920 × 1152	50	Prog.	16:9	4:2:2	110.5
	1920 × 1080	60	Prog.	16:9	4:2:2	124.3
ITU-R 601	720 × 576	25	Inter.	4:3	4:2:2	10.4
	720 × 480	29.97	Inter.	4:3	4:2:2	10.4
SIF-625	352 × 288	25	Prog.	4:3	4:2:0	2.5
SIF-525	352 × 240	29.97	Prog.	4:3	4:2:0	2.5
CIF	352 × 288	29.97	Prog.	≈4:3	4:2:0	3.0
QCIF	172 × 144	29.97	Prog.	≈4:3	4:2:0	0.74
Sub-QCIF	128 × 94	29.97	Prog.	≈4:3	4:2:0	0.36

Table 3 A selection of digital video formats defined in different international recommendations. The two selected HDTV formats correspond to progressive scanning, and, respectively, to American [52] and European proposals [3]. ITU-R 601 formats correspond to 625 and 525 line TV systems [17]. SIF-625 and SIF-525 derive from the previous and are specified in MPEG-1 [14]. CIF and QCIF are introduced in ITU-T Rec. H.261 [18] and Sub-QCIF appears in ITU-T Draft Rec. H.263 [19]. Abbreviations: **Prog**: Progressive, **Inter:** Interlaced.

3.2 Video standards

As we can see in Table 4, which summarizes the main video coding standards, all of them, with only one exception, implement particular approximations to the hybrid coding scheme introduced in section 2.4, with common features including the use of non-overlapped blocks of 8 × 8 pixels and the employment of the DCT for spatial transformation. This form of hybrid coding has been found adequate for HDTV and TV transmission with contribution and distribution quality, VHS-quality TV recording, and visual telephony. Differences in the predictive scheme are summarized in Table 5. A brief description of the most important video standards follows.

- **ITU-T H.261** *"Video codec for audiovisual services at p × 64 kbit/s"* [18] (1990) (previously for $n \times 384$ kb/s (1988)). This standard was developed for compressing the video component of visual telephony. Initially, the aim of this standard was a video data rate of $n \times 384$ kb/s ($n = 1,2,...5$), but a later demand for standardization for narrow band ISDN and the advances

Standard	Application	Technique	Format	Bit-rate	Bpp
H.263	VT	HPT	QCIF [1]	20 kb/s	0.05
H.261	VT	HPT	CIF [1]	320 kb/s	0.21
MPEG-1	VR	HPT	SIF [2]	1.15 Mb/s	0.45
MPEG-2	Dist. TV	HPT	ITU-R 601	4-9 Mb/s	0.39-0.87
J.81	Cont. TV	HPT	ITU-R 601	30 Mb/s	2.89
J.80	Cont. TV	S-DPCM	ITU-R 601	140 Mb/s	13.5

Table 4 Summary of the main video coding standards. *Application* and *format* fields should be taken as indicative, as standards enjoy different degrees of generality. Abbreviations: **VT:** Visual telephony, **VR:** Video recording, **Dist**: Distribution, **Cont**: Contribution, **HPT:** Hybrid predictive-transform, **S-DPCM:** spatial DPCM.
(1) Skipping a number of consecutive frames is allowed. For the computation of the number of bits per pixel we consider that every other frame is skipped, which would constitute a common practice at the indicated bit-rate.
(2) The coded area of the picture includes only 352 of its 360 columns.

Standard	Format	Main unit	Picture types
H.263	Progressive	Frame	I,P,B
H.261	Progressive	Frame	P (1)
MPEG-1	Progressive	Frame	I,P,B
MPEG-2	Interlaced	Frame/field	I,P,B
J.81	Interlaced	Field	P (1)

Table 5 Hybrid coding standards: prediction scheme.
(1) I frames, thought, not considered as a special case by the standards, can also be employed.

in video coding technology imposed a redefinition of this data rate to the new rate of $p \times 64$ kb/s ($p = 1,2,\dots,30$) which implies a net video bit-rate from approximately 40 kb/s to less than 2 Mb/s. Accepted signal formats by this are CIF (progressive $352 \times 288 \times 29.97$ with 4:2:0 chrominance subsampling) and QCIF (a quarter of samples of CIF), with the possibility to skip up to 3 consecutive frames. With values of $p = 1$ or 2, QCIF format and about 10 frames per second are normally employed for videotelephony applications. With values of $p \geq 6$, CIF format and 15 or more frames per second are intended for videoconference applications. This standard can be considered to provide from very moderate to good quality, depending on the bit-rate.

The compression approach is the canonical hybrid scheme with block-based MC and only P frames. The possibilities for the encoding of each macroblock are *intra* and *predictive* with or without MC. Motion vectors have integer coordinates (pixel precision) and are limited to a narrow range.

- **ITU-T H.263** *"Video coding for narrow telecommunication channels at < 64 kbit/s"* [19] (expected for 1996). Also intended for compressing the video component of visual telephony, this standard constitutes an important evolution of H.261 intended for VLBR operation, as required by the use with the ITU-T V.34 modem on analog telephone lines (up to 28.8 kb/s, which leaves about 20 kb/s for video). The encoded signal can be sub-QCIF (progressive 128 × 96 × 29.97 with 4:2:0 chrominance subsampling), QCIF, CIF or other larger (4CIF, 16CIF) input formats with a minimum number of consecutive skipped frames. This coder provides the same quality as H.261 with less than half the number of bits.

  To this goal contribute the introduction of efficiently encoded B frames, fractional-pixel motion vector accuracy, motion vectors pointing outside the picture, MC on 8 × 8 blocks instead of on 16 × 16 macroblocks, overlapped-block MC (see section 4.2), and arithmetic coding.

- **ISO/IEC MPEG-1** *"Coding of moving pictures and associated audio for digital storage media up to about 1.5 Mbit/s"* [14]. Video part: ISO 11172-2 (1992). This standard aims at the compressed representation of progressive video and its audio on various digital storage media (such as CDs, DATs, Winchester disks and optical drives) offering a continuous transfer rate of about 1.5 Mb/s. It can be used, nevertheless, at a wide range of rates and picture resolutions, although the system parameters are optimized for the main application. This is defined by a video bit-rate of 1.15 Mb/s and the dimensions of the SIF signal (progressive 352 × 240 × 29.97 or 352 × 288 × 25, with 4:2:0 chrominance subsampling). VHS-quality playback is expected from these system parameters.

  The main difference from H.261 is the introduction of I and B frames, which are very convenient for storage applications. Other improvement is the use of half-pixel motion vector accuracy. In MPEG-1 the type of frame determines the set of processing modes for the macroblocks in the picture. While in I frames all macroblocks are intra encoded, in P frames two prediction modes are available: *intra* and *motion-compensated forward-predicted* (obtaining the prediction from the previous I or P frame). For B frames, the repertoire of predictive modes enriches itself with the inclusion of *backward-predicted* (with the prediction from next I or P frame) and interpolated (constructing the prediction as the weighted average between a forward and a backward prediction). The range of the motion vectors is made dependent on the distance between current and reference pictures (with the corresponding enlargement of the search area for motion vectors).

- **ISO/IEC MPEG-2** *"Generic coding of moving pictures and associated audio"* [15]. Video part: ISO 13818-2 or ITU-T H.262 (1994). Initially targeted for digital transmission of broadcast TV-quality video at bit-rates between 4 and 9 Mb/s, it absorbed during its development the compression of HDTV (initially covered by the never-born MPEG-3 standard). The standard allows compression of both progressive and interlaced video at various data bit-rates from about 1.5 Mb/s to more than 60 Mb/s, enabling applications ranging from home entertainment quality video up to HDTV. Subsets of the standard are known as *profiles* (which delimit syntax, i.e., algorithms) and *levels* (delimiting signal parameters). Different profile-level pairs indicate the main applications of MPEG-2 coders, which will range from competing with MPEG-1 to compressing HDTV for distribution (at 18-36 Mb/s), including TV transmission with PAL broadcast quality (at 4 Mb/s) or *near transparent* [16] quality (at 9 Mb/s).

MPEG-2 can deal either with progressive or interlaced formats. It is actually this capability which constitutes the main progress over MPEG-1, but it is also responsible for an important increase in the syntax complexity. MPEG-2 inherits the set of picture types from MPEG-1. When compressing interlaced signals, it can operate on a field or a frame-basis. When operating with frame-pictures, the processing mode of a macroblock now depends on more parameters: the encoder can choose between obtaining a single prediction for the whole macroblock or splitting it (according either to spatial or temporal criteria) into two submacroblocks and computing different motion vectors for each of them. It is also possible to split the macroblock on a field basis in order to obtain the blocks to DCT-transform.

- **ITU-T J.81** *"Transmission of component-coded digital television signals for contribution-quality applications at the third hierarchical level of ITU-T Recommendation G.702".* Formerly CCIR Rec. 723 (1991 as ETSI standard ETS 300 174) [21]. This standard specifies the coding and transmission of digital television signals at bit-rates of 34-45 Mb/s (video and audio) in the format specified by recommendation ITU-R 601 (interlaced $720 \times 480 \times 29.97$ or $720 \times 576 \times 25$ with 4:2:2 chrominance subsampling). Nct video capacity depends on the number of optional channels (audio, teletext), ranging from 26 to 31 Mb/s for Europe. This encoding can be considered to provide very high quality, corresponding to the transparent compression necessary for contribution applications.

  ITU-T J.81 operates on a field basis, with three different processing modes for each macroblock: *intra*, *interfield predicted* (obtaining the prediction from the previous field without MC) and *interframe predicted* (with the prediction obtained from the corresponding field in the previous frame and half-pixel MC). An extension of this algorithm for HDTV signals made possible the world first all-digital transmission of HDTV (through satellite and fiber optic links), which took place in Europe in 1990 [3].

3.3 Low bit-rate performance of the main standard coders

While the performance of standard compression coders at high bit-rates is considered in general satisfactory enough, some important VLBR applications are, as indicated in the introduction, on the verge of the possibilities of current techniques. In order to provide concrete data on these possibilities, both concerning objective and subjective quality, we present in this section the results of a comparison between the three of the above mentioned coders which include VLBR operation among their applications.

The original sequences have QCIF spatial resolution at 25 Hz. They all have 150 frames (6 seconds) and the rate of coded frames has been $25/3 = 8.3$ Hz (2 frames out of three are skipped). Sequences differ in complexity: *Claire* shows a talking head typical of visual telephony. It can be coded at very low bit-rate because of its uniform and fixed background. Sequence *Carphone* also shows a talking head but with more motion in the speaker, and a non uniform changing background. Finally, sequence *Foreman*, which is the most complex, shows a talking person with much more motion both in the person and in the (not uniform) background. Figure 15 shows some original frames from these sequences.

To compare the coders under the same conditions, only the first frame was allowed to be intra coded and no buffer regulation was used. Resulting bit-rates were achieved coding the sequences with different quantization steps. Not considering the first frame, the results in terms of objective quality are presented in Figure 16, which shows the peak SNR (PSNR)[5] vs the resulting bit-rate for the three sequences encoded with standard coders: H.261, MPEG-1 with only predicted frames, (IPPPPPP... structure), MPEG-1 with two B frames between every two P frames (IBBPBBPBB... structure), and H.263 including unrestricted motion vector mode, syntax-based arithmetic coding, advanced prediction mode and PB frames (IBPBP... structure). The first point of each curve marks the minimum achievable bit-rate for the coder. For the considered range of bit-rates, a fixed ordering in the capabilities of the four schemes can be observed with H.263 at the top (maximum PSNR) and H.261 at the bottom, with an average difference between these two coders of about 3 dB which increases as the bit-rate gets lower. These results are an indication of the improvement in objective quality produced by the different enhancements on the basic hybrid scheme.

---

[5] $PSNR(\text{dB}) = 20\log[255/rmse]$, were $rmse$ is the root mean squared error for the image.

(a) (b) (c) (d)

Figure 15 Original frames from sequences *Claire* (a), *Carphone* (b) and *Foreman* (c) and (d).

Figures 17 and 18 show the same frame for each sequence coded with H.261, IBBPBBP... MPEG-1 and H.263 with all its enhancements. To make meaningful comparisons, each coded sequence is shown at a bit-rate, selected as the minimum bit-rate which can be achieved in the encoding of the sequence with every coder. It can be seen that H.261 shows more artifacts (blocking effect) than the others, being H.263 the one to provide the best objective and subjective quality, this last due mainly to the reduction of the blocking effect achieved with overlapped-block MC (see section 4.2).

To see the performance of the H.263 coder at low bit-rates, each sequence has been coded at the minimum achievable bit-rate and at two and three times this

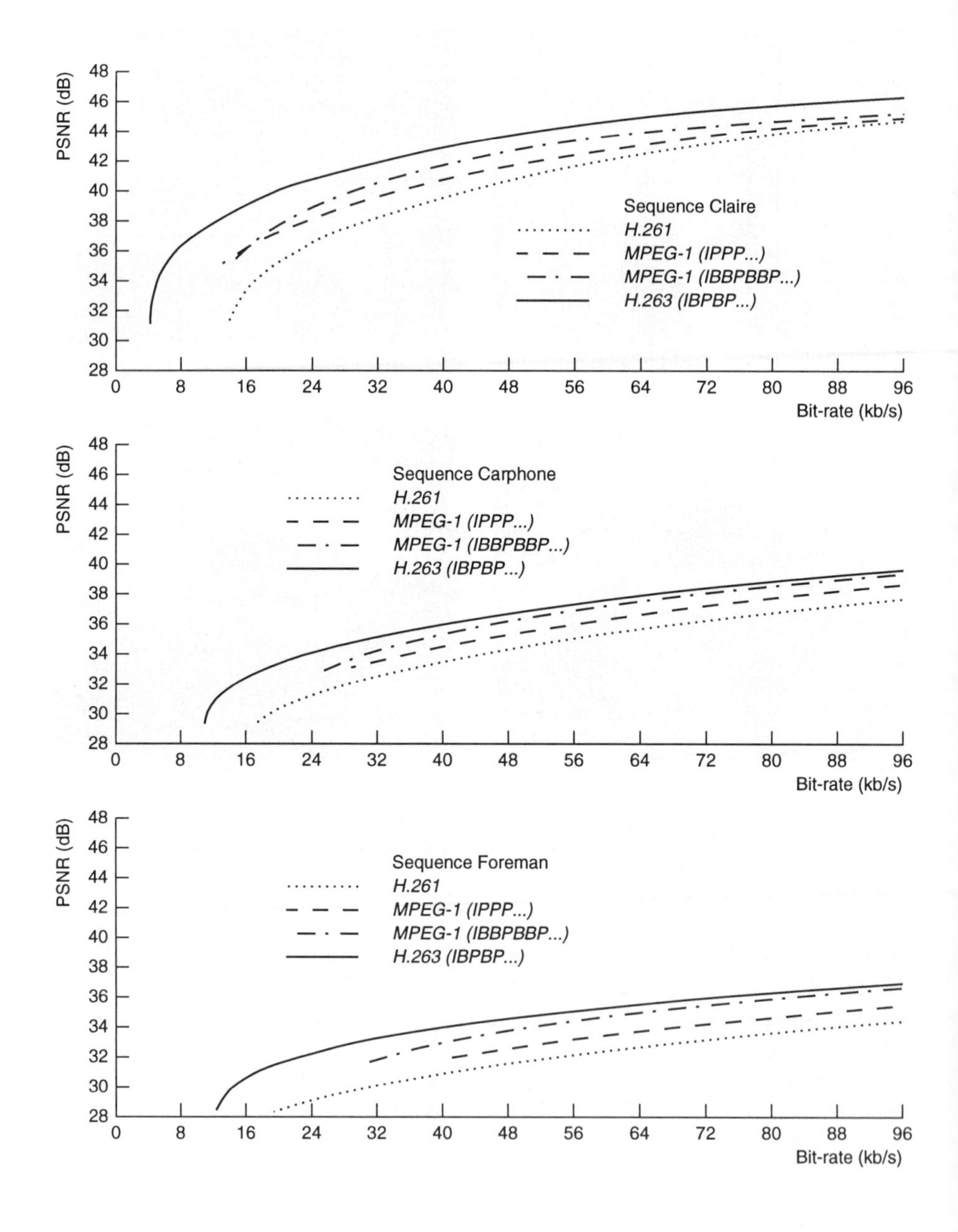

Figure 16 Peak signal-to-noise-ratio vs bit-rate curves for three different sequences encoded with H.261, MPEG-1 and H.263 coders.

Figure 17 Frames from sequences *Claire* coded at 14 kb/s with H.261 (a), MPEG-1 (b) and H.263 (c); and *Carphone* coded at 24 kb/s with H.261 (d), MPEG-1 (e) and H.263 (f).

Figure 18 Two frames from sequence *Foreman* coded at 30 kb/s with H.261 (a)(d), MPEG-1 (b)(e) and H.263 (c)(f).

rate. Figure 19 shows the results for the most simple (*Claire*) and the most complex (*Foreman*) sequences.

Besides the previous comparison, conducted on well-known sequences commonly used for the evaluation of pixel-based algorithms, additional results on new sequences used in the testing and evaluation procedures for the future MPEG-4 standard are presented in Figures 20, 21 and 22. These sequences have a quarter of the SIF spatial resolution (progressive 176 $\times$ 128 $\times$ 29.97, with 4:2:0 chrominance subsampling). They all have 300 frames and the rate for the coded frames has been 10 Hz (as in the previous sequences, 2 frames out of three are skipped). Each sequence belongs to a different class as defined in MPEG-4 tests: *Akiyo* (class A) shows low spatial detail and low amount of motion, *News* (class B) presents medium spatial detail and low amount of motion or vice versa, and *Children* (class E) is hybrid of natural and synthetic sequences.

As can be seen from Figures 19, 20, 21 and 22, there appears a quality reduction as the bit-rate decreases in the H.263 coder. The lowest bound in bit-rate for an acceptable quality ranges from 8 kb/s to 48 kb/s, depending on the complexity of the sequence.

4 PIXEL-BASED SCHEMES: PROPERTIES AND POSSIBLE IMPROVEMENTS

4.1 Properties of pixel-based schemes

Before discussing the foreseeable directions in which current pixel-based techniques for video compression can evolve, we concentrate now on their good and bad properties. While some of these properties stem from the pixel-based paradigm itself, others are associated with the limitations of the current standardized approaches and might be overcome within the pixel-based approach.

This approach is in principle *the most general*, as the only assumptions made on the contents of the signal are the existence of statistical dependence between neighboring pixels and those deriving from the motion model. But these assumptions are enough to impose some limitations: for example, predictive and block transform coding work better with smooth images but are unsuitable for sharp edges representation.

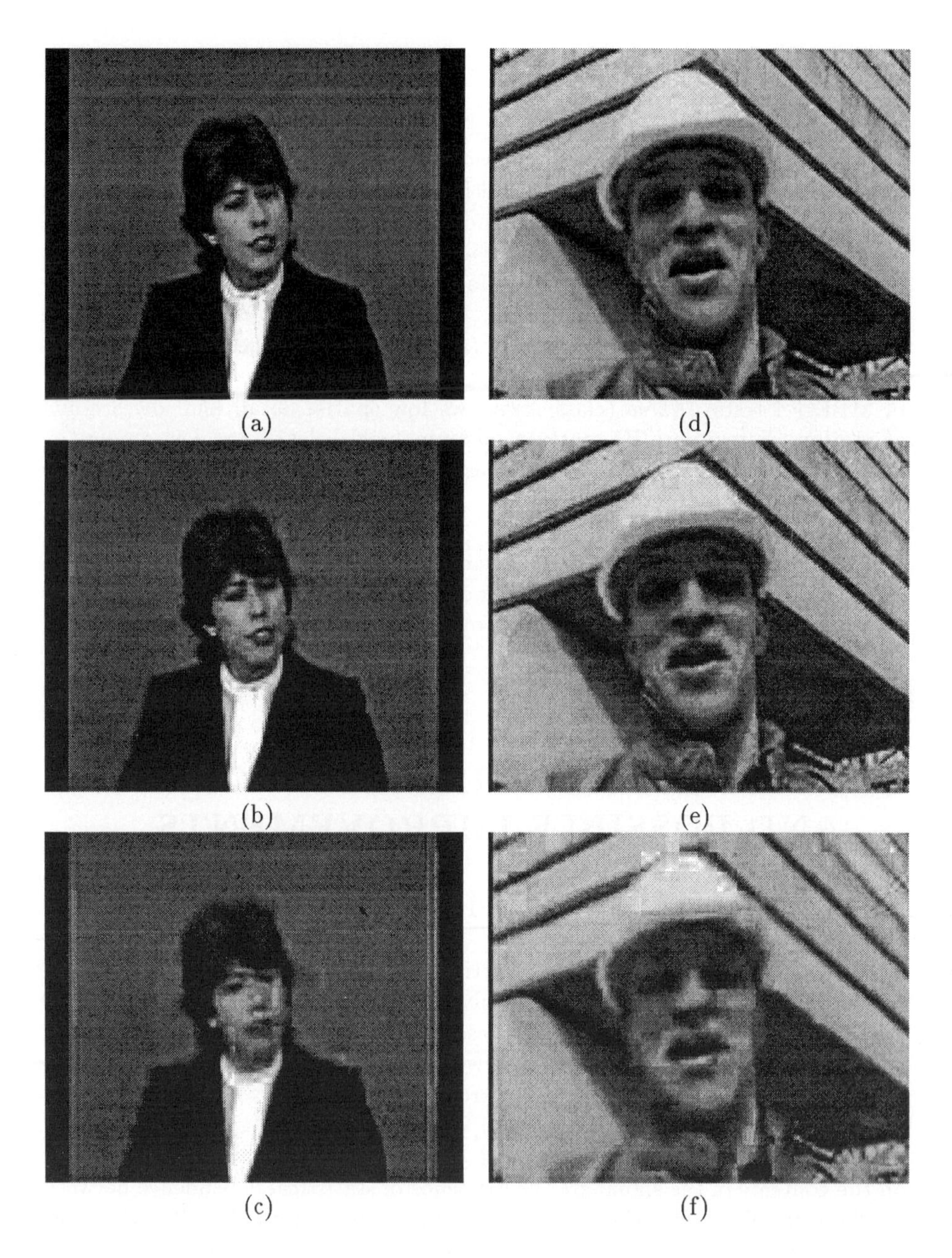

(a) (d)

(b) (e)

(c) (f)

Figure 19 Frame from sequence *Claire* coded with H.263 at 12 kb/s (a), 8 kb/s (b) and 4 kb/s (c), and frame from sequence *Foreman* coded with H.263 at 36 kb/s (d), 24 kb/s (e) and 12 kb/s (f).

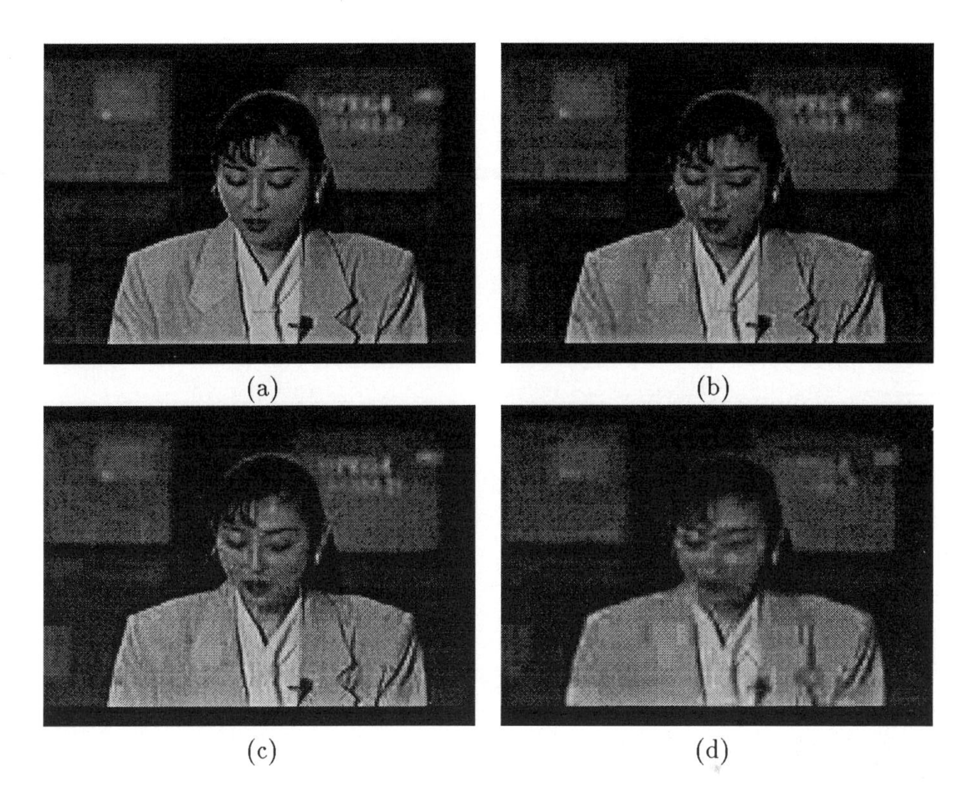

Figure 20 Frame from sequence *Akiyo*: original (a) and coded with H.263 at 12 kb/s (b), 8 kb/s (c) and 4 kb/s (d).

Current pixel-based techniques behave quite properly from the *controllability* point of view. We can define this property intuitively as the possibility to regulate softly the output bit-rate of the coder. Easy control will be possible if the compression-quality curve for the encoder is smooth. In pixel-based techniques quantization is the main control parameter; modifying the quantization step, quality can be smoothly traded with compression. Controllability of transform coders is better than that of predictive coders because quantization, in those coders is linked with redundancy reduction through a feedback loop, there being a steep quality fall as quantization gets coarser.

Non-overlapped block-based techniques suffer from specific limitations. The decomposition of the picture in a predetermined way has obvious disadvantages with respect to any ad-hoc decomposition of each picture according to its content: predefined blocks are bound to contain edges, which are bit-consuming el-

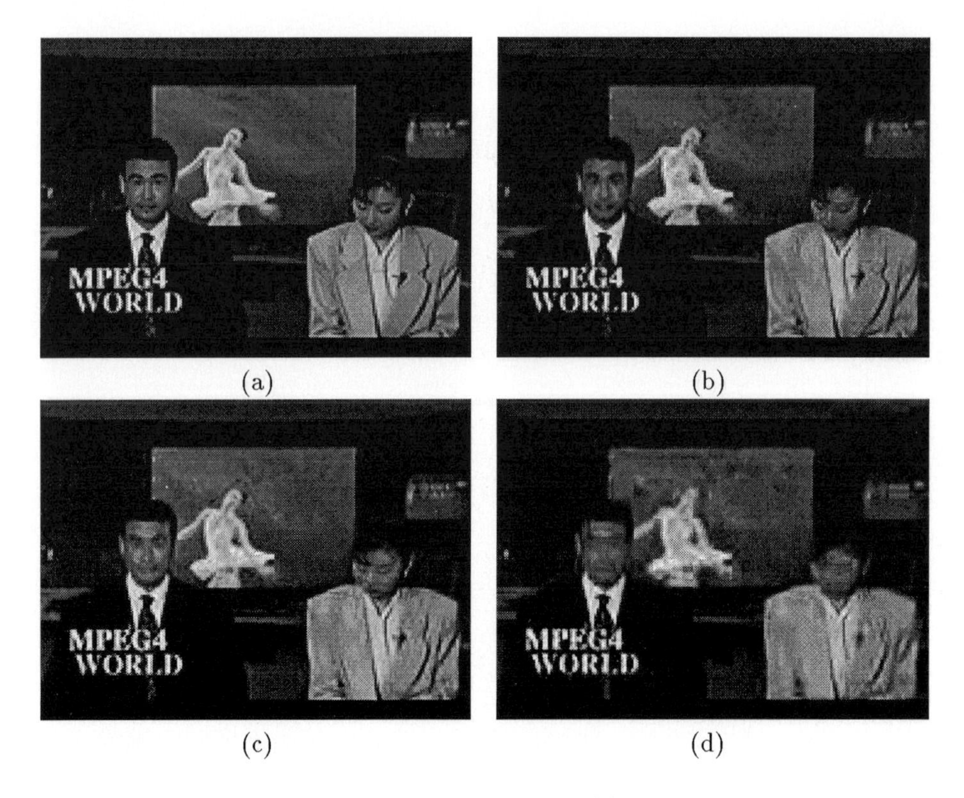

Figure 21 Frame from sequence *News*: original (a) and coded with H.263 at 30 kb/s (b), 20 kb/s (c) and 10 kb/s (d).

ements, while a free segmentation will likely consist of more uniform regions. In the case of video signals, non-overlapped block-based MC is limited by another version of the same problem, as a block shape is imposed as a segmentation of the motion field. Blocks containing the border of a *true* motion-uniform region will be predicted only partially, with the result that some pixels of the block can be well predicted while others cannot, contributing greatly to the energy of the prediction error (see Figure 23).

Another problem is that the motion vector field resulting from block matching does not in general correspond to the real motion. This implies that adjacent blocks belonging to the same solid object can be predicted from blocks situated far apart in the reference field, with the result that the reconstructed image shows block discontinuities which will only be partially compensated by the transmission of the block prediction errors. Moreover, the correlation between

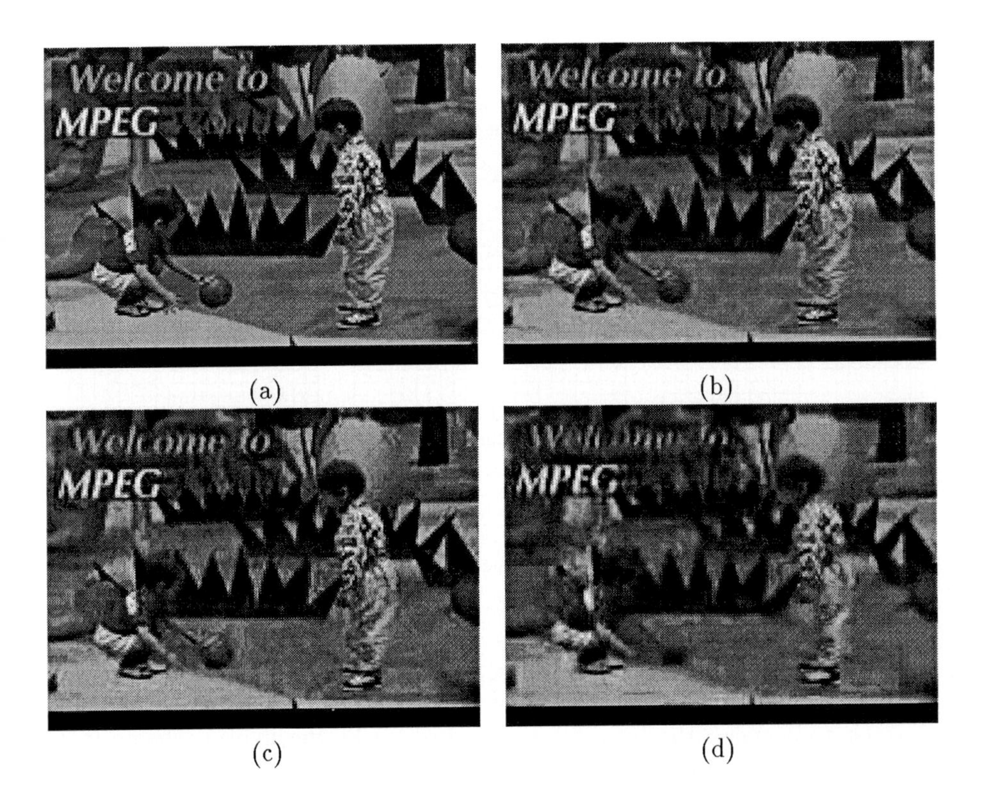

Figure 22 Frame from sequence *Children*: original (a) and coded with H.263 at 72 kb/s (b), 48 kb/s (c) and 24 kb/s (d).

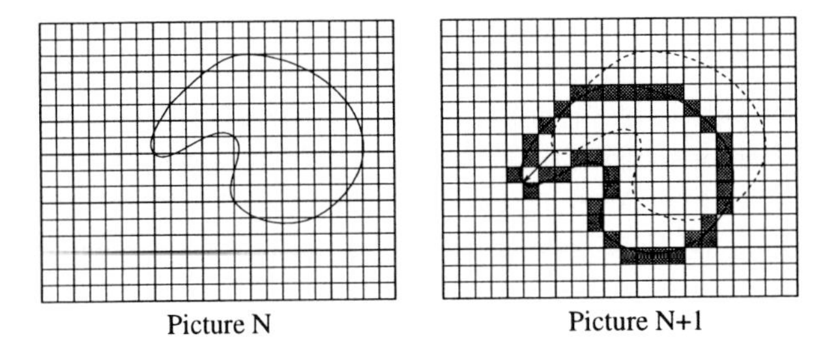

Figure 23 Block MC. Shaded blocks cannot be predicted from the previous picture because they include the border of a moving rigid object.

motion vectors obtained by block-matching is lower than that of real motion vectors, which makes their encoding more costly.

Finally, as block-based motion estimation assumes a purely translational motion model, the accuracy of MC is compromised (see Figure 24). At low bit-rate this problem gets more serious, as, it becomes necessary that a part as large as possible of the image can be transmitted without prediction error information.

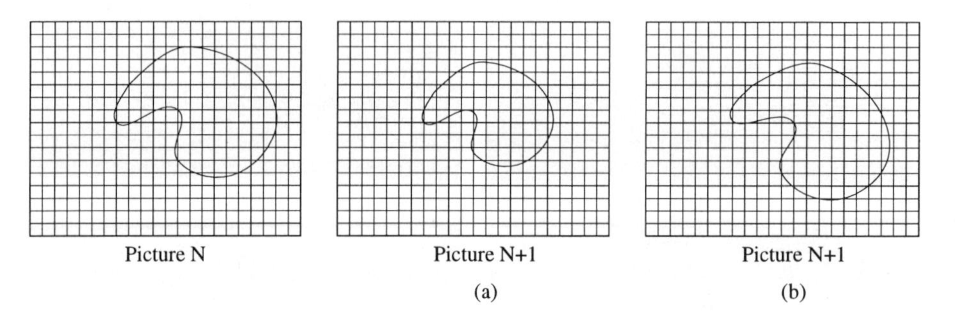

Figure 24 Two examples of motion which cannot be compensated by assuming a purely translational motion model: zooming (a) and rotation (b).

4.2 Improvements in classical schemes

In order to decrease the bit-rates attained so far by the hybrid scheme more efficient techniques must be applied for spatio-temporal decorrelation. As discussed before, overcoming the limitations of non-overlapped block transforms implies the use of other schemes which do not integrate so straightforwardly with block-based MC. Efforts have consequently been conducted toward the integration of MC with overlapped block transforms and with DWT [8].

In any case, an analysis of the bit-rate budget for very low-bit-rate applications shows that *the transmission of the motion field requires an important part of the bit-rate* [29]. Thus, most research is oriented to improving the prediction loop so that it becomes possible to dedicate very few bits to the encoding of the prediction error. We will review some of these approaches below.

Advanced motion compensation

The improvement of the classical MC scheme in order to cover motions other than purely translational and reduce the blocking effect, has been object of diverse efforts in the last years [29]. The difficulty of these developments lies in the need to keep reasonable the computational complexity of coder and decoder and not to increase the amount of motion parameters to transmit.

Two main approaches summarize the recent evolution of pixel-based MC: **control grid interpolation** and **overlapped block MC**. These methods have in common that they can preserve the sampled description of the motion vector field characteristic of the classical schemes.

Control grid interpolation (CGI) [51][57] assumes that the current picture can be approximated by a warping of the reference picture. The correspondence between pixels in both pictures is given explicitly for the pixels which belong to a predefined grid as a set of motion vectors (see Figure 25), while for the other pixels it is obtained by interpolation from the available vectors.

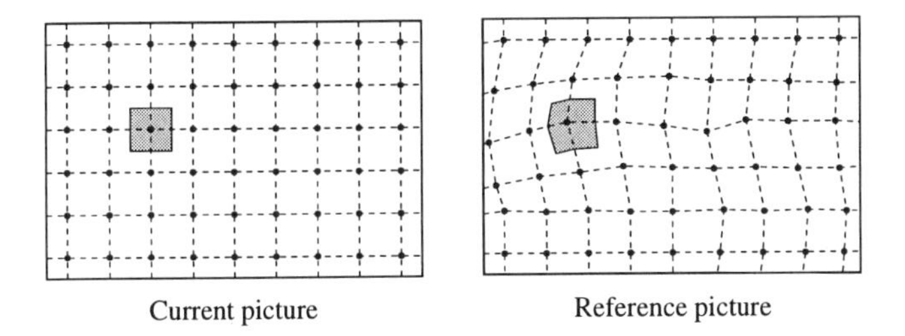

Figure 25 Control grid interpolation. Motion vectors corresponding to the grid nodes are computed and sent to the decoder, which estimates the motion vectors corresponding to all pixel locations by interpolation. The prediction for a square block (shaded) is no longer a square block in the reference picture, unless zero-order motion vector interpolation is applied.

Typical selections for the grid are rectangular [51] or triangular [29] sampling structures, and can be fixed or image dependent [57]. To obtain the motion parameters (i.e. the motion vectors for the grid nodes) the energy of the *displaced frame difference* (DFD) has to be minimized:

$$\mathrm{DFD}(x, y) = I(x, y) - I_r(x - v_x, y - v_y), \tag{2.7}$$

where I is the current picture, I_r the reference picture, and $(v_x(x, y), v_y(x, y))$ represents the motion field, defined by its values at the grid nodes plus the interpolation algorithm (a trivial interpolation scheme would result in classical block-based MC). The computation of the optimal motion vectors now becomes, in general, more complicated than in the block-based case because the optimal motion vector in a grid node depends on the surrounding motion vectors. An algorithm for approximated optimization is suggested in [51].

As CGI does not impose any block-based decomposition on the image, no particular advantage results from the encoding of the prediction error image through non-overlapped block transform, except to indicate that some areas are better encoded in intra mode. Assuming that there is no need for this, nothing prevents the use of more advanced transforms for spatial decorrelation.

Overlapped block MC (OBMC) [37] constitutes a direct attempt to avoid blocking artifacts in the predicted picture with moderate increase in computational complexity with respect to classical block-based MC and is compatible with block-based transform. This is achieved by modifying the block-based MC so that a different prediction is obtained for each pixel by averaging predictions obtained from different motion vectors associated to close spatial locations on a regular grid. The resulting DFD has the form

$$\mathrm{DFD}(x, y) = I(x, y) - \sum_{i \in C(x,y)} \alpha_i(x, y) I_r(x - v_{xi}, y - v_{yi}), \tag{2.8}$$

where $C(x, y)$ is the set of indices of the grid nodes in the neighborhood of pixel (x, y) and α_i is a weighting function. Figure 26 illustrates the differences between the CGI and OBMC approaches.

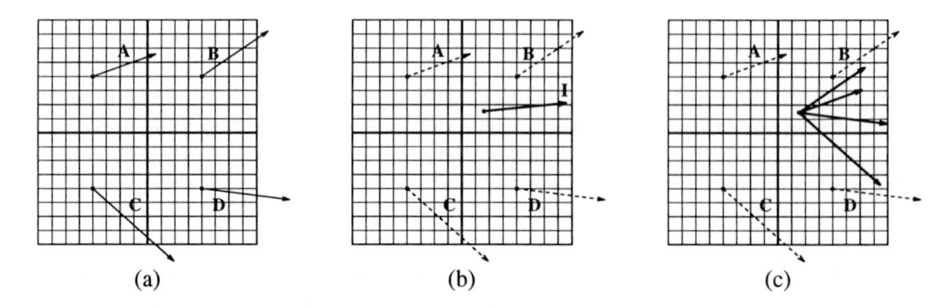

Figure 26 Advanced pixel-based MC techniques. Both CGI and OBMC approaches rely on a description of the motion field given by its value on the nodes of a grid (vectors A, B, C y D in (a)). In order to obtain the prediction for an arbitrary pixel, CGI interpolates the motion field in order to assign a single vector to the pixel (vector *I* in (b)). OBMC, in its term, obtains for the pixel several predictions, based on the motions vectors defined for the surrounding nodes in the grid (c), and averages them with different weights.

New spatio-temporal decorrelation paradigms

Purely predictive schemes in the temporal dimension suffer from serious limitations at low bit-rates because of the avalanche effect derived from strong quantization that we have described in our discussion of controllability. This effect can be softened by the use of B pictures, since a quality decrease of these pictures does not compromise the prediction capability of the system.

Substitution of prediction for transformation in the temporal domain, on the other hand, is also handicapped by delay and lack of efficiency, as has also been explained. In order to overcome this problem, motion compensated 3-D transform has been suggested [29], which incorporates a spatial alignment of the block pictures prior to transformation in order to optimize the temporal correlation between them. Another scheme, referred to as *open loop prediction* [29] consists in the use of a special image as the reference for motion compensation. This image, which is continuously updated, should keep a good quality independently of the distortion showed by the decoded pictures.

5 CONCLUSIONS

We have reviewed in this chapter the paradigms on which pixel-based or first generation image and video coding techniques are based, paying special attention to their crystallization in the form of international standards.

Establishing definitive conclusions on the limits of these paradigms is not, at the moment, a task for a priori analysis because of the difficulty to model, on one hand, the image and video signals, and, on the other, the HVS. Consequently, assessment should be made on an empirical basis. A clear distinction should therefore be made between *standardized* pixel-based techniques and those which have not yet been considered for standardization due to their novelty. For the first, as a result of the deep experience already available on these compression procedures, some limits can be given as those that we provide in Table 6, summarizing partial results given either directly or indirectly in the text.

Quality	Intra	Inter
Very good	0.75–	0.4–
Good	0.25–0.75	0.125–0.4
Moderately good	0.125–0.25	0.05–0.125

Table 6 Compression rates (bpp) of standard intra and interframe pixel-based techniques for different quality levels. The data for the intra case correspond to JPEG. In the inter case, different quality levels correspond to different standards: very good quality above 0.4 bpp is achieved with MPEG-2 for standard TV resolution, while good quality in the range 0.125–0.4 bpp can be obtained with MPEG-1, and the range given for moderately good quality is the one achieved with H.263 for visual telephony video signals.

Regarding the pixel-based approach as a whole, concrete data on achievable quality bounds will have to wait, at least, for the evaluation of the last contributions within the field, such as the new MC schemes and the new transforms, such as those based on wavelets. It has become lately customary to learn from applications of wavelets to image coding which achieve compression ratios of 150:1 and more with hardly noticeable distortion. It will still take some time to know whether these results just apply to picture *Lena* or we are really on the verge of a new generation of VLBR waveform-based video systems with low complexity and delay.

In any case, first generation techniques represent a very narrow world in comparison with the opening world of the second generation techniques, which, consequently, is more likely to provide important breakthroughs.

REFERENCES

[1] N. Ahmed, T. Natarajan & K. R. Rao, "Discrete cosine transform", IEEE Trans. on Computers, vol. 23, no. 1, pp. 90-93, Jan. 1974.

[2] M. Antonini, M. Barlaud, P. Mathieu & I. Daubechies, "Image coding using wavelet transform", IEEE Trans. on Image Processing, vol. 1, no. 2, pp. 205-220, Apr. 1992.

[3] M. Barbero, S. Cucchi & M. Stroppiana, "A bit-rate reduction system for HDTV transmission", IEEE Trans. on Circuits and Systems for Video Technology, vol. 1, no. 1, pp. 4-13, Mar. 1991.

[4] P. J. Burt & E. H. Adelson, "The laplacian pyramid as a compact image code", IEEE Trans. on Communications, vol. 31, no. 4, pp. 532-540, Apr. 1983.

[5] R. E. Crochiere & L. R. Rabiner, "Multirate Digital Processing", Prentice-Hall, Englewood Cliffs, USA, 1983.

[6] E. Dubois, "The sampling and reconstruction of time-varying imagery with application in video systems", Proc. of the IEEE, vol. 73, no. 4, pp. 502-522, Apr. 1985.

[7] F. Dufaux & F. Moscheni, "Motion estimation techniques for digital TV: a review and a new contribution", Proc.of the IEEE, vol. 83, no. 6, pp. 858-876, Jun. 1995.

[8] K. H. Goh, J. J. Soraghan & T. S. Durrani, "Multi-resolution based algorithms for low bit-rate image coding", Proc. of ICIP 94, vol. 3, pp. 285-289, Austin, Nov. 1994.

[9] R. M. Gray, "Vector quantization", IEEE ASSP Magazine, vol. 1, no. 2, pp. 4-29, Apr. 1984.

[10] M. R. Haghiri & F. W. P. Vreeswijk, "HDMAC coding for MAC compatible broadcasting of HDTV signals", IEEE Trans. on Broadcasting, vol. 36, no. 4, pp. 284-288, Dec. 1990.

[11] M. L. Hilton, B. D. Jawerth & A. Sengupta, "Compressing still and moving images with wavelets", Multimedia Systems, vol. 2, no. 5, pp. 218-227, Dec. 1994.

[12] J. Huang & P. Schultheiss, "Block quantization of correlated gaussian random variables", IEEE Trans. on Communications Systems, vol. 11, no. 9, pp. 289-296, Sep. 1963.

[13] ISO/IEC IS 10918 (JPEG), "Information Technology - Digital Compression and Coding of Continuous-Tone Still Images", 1994.

[14] ISO/IEC IS 11172 (MPEG-1), "Information Technology - Coding of Moving Pictures and Associated Audio for Digital Storage Media Up to About 1,5 Mbit/s", 1993.

[15] ISO/IEC DIS 13818 (MPEG-2), "Information Technology - Generic Coding of Moving Pictures and Associated Audio. ITU-T Recommendation H.262", Mar. 1994.

[16] ISO/IEC, "MPEG-2 Press Release", Nov. 1994.

[17] ITU-R Recommendation 601, "Encoding Parameters of Digital Television for Studios", Rev. 2, 1990.

[18] ITU-T Recommendation H.261, "Video Codec for Audiovisual Services at px64 kbit/s", Rev. 2, 1993.

[19] Draft ITU-T Recommendation H.263, "Video Coding for Narrow Telecommunication Channels at $<$ 64 kbit/s", Jul. 1995.

[20] ITU-T Recommendation J.80, "Transmission of Component-Coded Digital Television Signals for Contribution-Quality Applications at Bit Rates Near 140 Mbit/s", Rev. 1, 1993.

[21] ITU-T Recommendation J.811, "Transmission of Component-Coded Television Signals for Contribution-Quality Applications at the Third Hierarchical Level of ITU-T Recommendation G.702", (Former ITU-R CMTT.723), Rev. 1, 1993.

[22] A. K. Jain, "Image data compression: a review", Proc. of the IEEE, vol. 69, no. 3, pp. 349-389, Mar. 1981.

[23] J. R. Jain & A. K. Jain, "Displacement measurement and its application in interframe image coding", IEEE Trans. on Communications, vol. 29, no. 12, pp. 1799-1808, Dec. 1981.

[24] F. Jaureguizar, J. I. Ronda & N. García, "Motion compensated prediction on digital HDTV", Proc. of EUSIPCO 90, pp. 753-756, Barcelona, Sep. 1990.

[25] F. Jaureguizar, "Motion estimation in TV", Ph.D. Thesis (in Spanish), Universidad Politécnica de Madrid, Spain, 1994.

[26] N. S. Jayant & P. Noll, "Digital coding of waveforms", Prentice-Hall, Englewood Cliffs, USA, 1984.

[27] G. Karlsson & M. Vetterli, "Three dimensional subband coding of video", Proc. of ICASSP 88, pp. 1100-1103, New York, Apr. 1988.

[28] A. S. Lewis & G. Knowles, "Video compression using 3D wavelet transform", IEE Electronics Letters, vol. 26, no. 6, pp. 396-398, Mar. 1990.

[29] H. Li, A. Lundmark & R. Forchheimer, "Image sequence coding at very low bitrates: a review", IEEE Trans. on Image Processing, vol. 3, no. 5, pp. 589-609, Sep. 1994.

[30] S. P. Lloyd, "Least squares quantization in PCM", Bell Lab. Memo, Jul. 1957; also in IEEE Trans. on Information Theory, vol. 28, no. 2, pp. 129-137, Mar. 1982.

[31] H. S. Malvar, "Signal Processing with Lapped Block Transform", Artech House, London, UK, 1992.

[32] J. Max, "Quantizing for minimum distortion", IEEE Trans. on Information Theory, vol. 6, no. 1, pp. 7-12, Mar. 1960.

[33] H. G. Musmann, P. Pirsch, & H.-J. Grallert, "Advances in picture coding", Proc. of the IEEE, vol. 73, no. 4, pp. 523-548, Apr. 1985.

[34] N. M. Nasrabadi & R. A. King, "Image coding using vector quantization: a review", IEEE. Trans. on Communications, vol. 36, no. 8, pp. 957-971, Aug. 1988.

[35] A. N. Netravali & J. O. Limb, "Picture coding: a review", Proc. of the IEEE, vol. 68, no. 3, pp. 366-406, Mar. 1980.

[36] J. Oest, F. J. Guirao & N. García, "Digital transmission of component coded HDTV signals using the DCT: design of a visibility threshold matrix", Proc. of EUSIPCO 90, pp. 881-884, Barcelona, Sep. 1990.

[37] M. T. Orchard & G. J. Sullivan, "Overlapped block motion compensation: an estimation-theoretic approach", IEEE Trans. on Image Processing, vol. 3, no. 5, pp. 693-699, Sep. 1994.

[38] A. Ortega, K. Ramchandran & M. Vetterli, "Optimal trellis-based buffered compression and fast approximations", IEEE Trans. on Image Processing, vol. 3, no. 1, pp. 26-40, Jan. 1994.

[39] T. Ozcelik, J. C. Brailean & A. K. Katsaggelos, "Image and video compression algorithms based on recovery techniques using mean field annealing", Proc. of the IEEE, vol. 83, no. 2, pp. 304-316, Feb. 1995.

[40] B. Ramamurthi & A. Gersho, "Nonlinear space-variant postprocessing of block coded images", IEEE Trans. on ASSP, vol. 34, no. 5, pp. 1258-1267, Oct. 1986.

[41] K. Ramchandran, A. Ortega & M. Vetterli, "Bit allocation for dependent quantization with applications to multiresolution and MPEG video coders", IEEE Trans. on Image Processing, vol. 3, no. 5, pp. 533-545, Sep. 1994.

[42] K. R. Rao, "Discrete Cosine Transform, Algorithms, Advantages, Applications", Academic Press, San Diego, USA, 1990.

[43] O. Rioul & M. Vetterli, "Wavelets and signal processing", IEEE Signal Processing Magazine, vol. 8, no. 4, pp. 14-38, Oct. 1991.

[44] J. Roese, W. Pratt & G. Robinson, "Interframe cosine transform image coding", IEEE Trans. on Communications, vol. 25, no. 11, pp. 1329-1339, Nov. 1977.

[45] J. I. Ronda, F. Jaureguizar & N. García, "DCT domain modelization of the TV signal for quantization", Proc. of EUSIPCO 90, pp. 889-892, Barcelona, Sep. 1990.

[46] J. I. Ronda, F. Jaureguizar & N. García, "Optimal bit-rate control of video coders", Proc. of Inter. Workshop on HDTV 93, Ottawa, Oct. 1993.

[47] J. I. Ronda, "Statistical modelling and control of video coders", Ph.D. Thesis (in Spanish), Universidad Politécnica de Madrid, Spain, 1994.

[48] S. Sallent, L. Torres & L. Gils, "Three dimensional adaptive laplacian pyramid image coding", SPIE, Visual Communications and Image Processing 90, pp. 627-638, Lausanne, Oct. 90.

[49] A. Sanz, C. Muñoz & N. García, "Approximation quality improvement techniques in progressive image transmission", IEEE Journal on Selected Areas in Comm., vol. 2, no 2, pp. 359-373, Mar. 1984.

[50] M. J. T. Smith & S. L. Eddins, "Analysis/synthesis techniques for subband image coding", IEEE Trans. on Acoustics, Speech and Signal Processing, vol. 38, no. 8, pp. 1446-1456, Aug. 1990.

[51] G. J. Sullivan & R. L. Baker, "Motion compensation for video compression using video grid interpolation", Proc. of ICASSP 91, pp. 2713-2716, Toronto, May. 1991.

[52] The Grand Alliance, "The US HDTV standard", IEEE Spectrum, vol. 32, no. 4, pp. 36-45, Apr. 1995.

[53] L. Torres & E. Arias, "Stochastic vector quantization of images", Proc. of ICASSP 92, pp. III.385-III.388, San Francisco, Mar. 1992.

[54] K. M. Uz, K. Ramchandran & M. Vetterli, "Multiresolution source and channel coding for digital broadcast of HDTV", Proc. of 4th Inter. Workshop on HDTV and Beyond, Turin, Sep. 1991.

[55] M. Vetterli, "Multidimensional sub-band coding: some theory and algorithms", Signal Processing, vol. 6, no. 2, pp. 97-112, Apr. 1984.

[56] G. K. Wallace, "The JPEG still picture compression standard", IEEE Trans. on Consumer Electronics, vol. 38, no. 1, pp. 18-34, Feb. 1992.

[57] Y. W. Wang & O. Lee, "Active mesh–A feature seeking and tracking image sequence representation scheme", IEEE Trans. on Image Processing, vol. 3, no. 5, pp. 610-626, Sep. 1994.

[58] I. H. Witten, R. M. Neal & J. G. Cleary, "Arithmetic coding for data compression", Comm. of the ACM, vol. 30, no. 6, pp. 520-540, Jun. 1987.

[59] M. Ziegler, "Hierarchical motion estimation using the phase correlation method in 140 Mbit/s HDTV-coding", Proc. of the 3rd Inter. Workshop on HDTV, Turin, Aug. 1989.

3

CODING-ORIENTED SEGMENTATION OF VIDEO SEQUENCES

Ferran Marqués, Montse Pardàs and Philippe Salembier

Department of Signal Theory and Communications,
Universitat Politècnica de Catalunya,
Barcelona, Spain

ABSTRACT

The importance of developing coding-oriented spatial segmentation techniques is stated. The specific problems of image sequence segmentation for coding purposes are analyzed. In order to both overcome such problems and improve the performance of segmentation-based coding schemes, a general segmentation structure is defined. This structure has five main steps: *Partition projection*, *Image modeling*, *Image simplification*, *Marker extraction* and *Decision*. In order to validate it, two different implementations of this structure are presented. The first utilizes a compound random field as image sequence model whereas the second relies on morphological tools.

1 INTRODUCTION

Among the different coding approaches grouped under the name of second generation coding techniques [21], there is an increasing interest in segmentation-based image sequence coding approaches. These coding methods divide the images into a set of connected regions $\{R_i\}$ so that every pixel in the image is related to one, and only one, region. This set of regions is said to form a partition of the image. Each region in the partition receives a different label. Images are described in terms of a partition and of some information related to the interior of each region (e.g.: texture, motion, etc). This information is necessary to reconstruct the image in the receiver.

The reason for the current interest in segmentation-based coding approaches is mainly twofold:

First, they are the basis for new efficient coding methods. Being a second generation technique, segmentation-based coding approaches try to eliminate the redundant information within and between frames, taking into account the special properties of the human visual system. In particular, the segmentation procedure should yield a partition whose regions are homogeneous in some sense (e.g.: gray level, color or motion homogeneity). Due to this homogeneity, the information of each region can be separately coded in a very efficient manner. However, with respect to classical coding techniques, segmentation-based approaches must code an additional information which is the image partition. Nowadays, efficient partition coding techniques are already available. The topic of partition coding is covered in Chapter 4 whereas Chapter 5 describes several techniques for coding the interior of the regions.

Second, they open the door to new functionalities in the coding scheme. In the framework of video coding, new coding schemes allowing functionalities such as content-based multimedia data access or content-based scalability are a very active field of research [19]. Such functionalities demand a description of image sequences in terms of objects which can be grouped into areas of interest by the user. A natural way to describe objects in the scene is by detecting and tracking their boundaries; that is, by segmenting the image sequence. An object can be composed of one or several regions, depending on the main feature that characterizes the object and the homogeneity criteria that have been used in the segmentation.

Image segmentation is a complex task due to the specific nature of the image signal. For instance, images present a degree of uncertainty [50] that may come from different sources: sampling, quantization, etc. This uncertainty yields elements in the image that do not clearly share the characteristics of any of its neighbors; that is, elements corresponding to areas of transition between homogeneous zones. Such areas make difficult the task of partitioning the image into homogeneous regions.

However, most of the difficulty of image segmentation, and of its extension to sequences, comes from the fact that image segmentation is an ill-posed problem [4]. This statement translates into that, given a scene, various partitions can be achieved by changing the homogeneity criteria in the segmentation procedure. Therefore, different descriptions in terms of partitions and of region interior information can be obtained from the same image.

In order to regularize the segmentation problem, constraints related to the specific application should be introduced. That is, the segmentation procedure has to be goal-oriented. In the framework of segmentation-based image sequence coding, the main constraints to be imposed are the quality of the final image representation as well as its coding cost. As a consequence, the segmentation procedure has to take into account the special characteristics of the coding techniques that are used in the segmentation-based coding scheme. These coding techniques have to cope with partitions as well as with region interiors. Thus, the segmentation procedure should yield partitions and region interiors that, not only describe correctly the information in the scene, but also lead to a low cost representation.

Towards this goal, two main segmentation approaches have been proposed in the literature. Both approaches differ in the homogeneity criterion that is used in the segmentation procedure. The main difference between these approaches is the relative importance they assign to the spatial or the motion information. In one case, regions are characterized by their homogeneity in the gray level or color information [37, 49] (called *texture* in the following), whereas in the other case, motion information is used as homogeneity criterion [7, 31, 42]. Some coding schemes have been proposed in the literature that are partially based on a segmentation. These techniques [2, 23, 46] perform a block-based motion compensation of the image and only use segmentation on the compensation error image. As a consequence, images are not completely described in terms of homogeneous regions and such coding schemes do not directly support content-based functionalities. Therefore, these techniques cannot be classified as purely segmentation-based coding approaches.

This chapter addresses the problem of image sequence segmentation techniques for coding purposes that rely basically on spatial information. There are several reasons for devoting this chapter to such a problem. First, as commented above, there are complete coding schemes that only utilize this type of segmentation approach. Second, regardless the chosen approach, any coding technique needs to have an intra-frame coding mode. In order to describe all images in the sequence by means of partitions, an intra-frame coding-oriented segmentation is necessary. Intra-frame segmentations can only utilize spatial information. Third, spatial segmentation helps to solve conflicts that arise when using motion-based segmentation approaches. This is the case, for instance, of new objects appearing in the scene. New objects should not be related to any previous information and, therefore, their boundaries have to be found in the current frame. New region boundaries can be detected in a more reliable way using the spatial information in the current frame. Finally, in order to implement coding schemes with embedded content-based functionalities, objects in the scene have to be somehow selected from the scene and tracked through time. The selection and tracking of objects cannot solely rely on motion information. In order to select an object, it has to be correctly represented in the image.

Therefore, an initial intra-frame representation of the object is necessary, and so is its intra-frame segmentation. In addition, object tracking cannot be based only on motion information since several objects in the scene may share the same motion. Spatial segmentation should be used in order to detect the actual boundaries of the moving objects.

In turn, Chapter 6 deals with the topic of motion estimation for second generation coding techniques and, in particular, it further details the existing methods for motion segmentation. Nevertheless, there are segmentation techniques that directly combine both texture and motion homogeneity in the same partition [3, 15]. These techniques have as central point the spatial information. Furthermore, the technique presented in [14] follows the same philosophy as the spatial segmentation techniques that are discussed in this chapter. Therefore, the discussion will be focussed in such segmentation approaches.

In the context of spatial segmentation of image sequences, three different approaches have been proposed:

1. **Pure 2D:** It consists in segmenting separately each single frame in the sequence. Consequently, the procedure is purely 2D or intra-frame [10, 20, 29]. In a coding-oriented segmentation technique, an additional region matching step is usually necessary in order to relate every region in each frame with a region in following frames [1].
2. **3D block:** This approach deals with the sequence as a 3D (2D plus time) signal and, therefore, performs a 3D segmentation. It implies to split the sequence into 3D blocks of a given number of frames and to segment these 3D blocks. Examples of this approach can be found in [28, 33, 37, 39, 49].
3. **Time recursive:** It is based on a recursive segmentation of the image sequence [25, 28, 35]. This approach still handles the sequence as a 3D signal while relaxing the concept of block of frames. In this case, the segmentation is carried out in piles of a short number of frames. The active pile, where the segmentation is performed, is shifted along the sequence and, at every shift, new images are introduced in the pile and the oldest ones are removed from the pile. In this approach, the partitions of former frames are used to initialize the segmentation procedure of the new frames.

These three different approaches are illustrated in Figure 1.

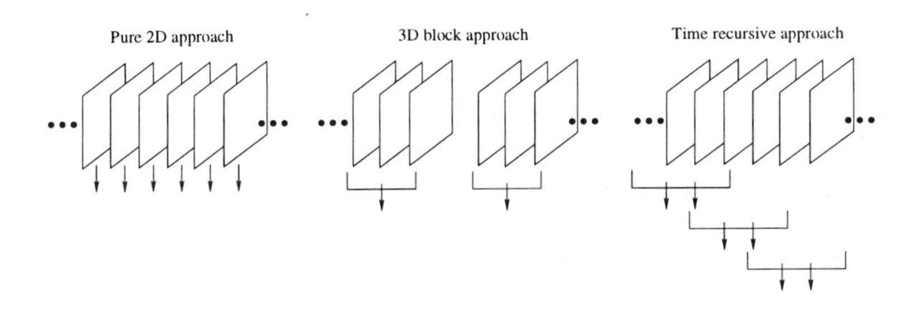

Figure 1 Illustration of the three spatial segmentation approaches

2 PROBLEM STATEMENT

As already pointed out, image segmentation is a complex task that raises many difficulties. The reader is referred to [6, 16, 32, 36, 50] in order to have a deeper view on the general topic. Nevertheless, spatial segmentation of image sequences for segmentation-based coding schemes presents some specific problems. Such problems appear regardless the segmentation approach to be followed although, in some cases, the chosen approach may partially overcome the problem. In this section, these specific problems are discussed.

2.1 Intra-frame image segmentation

Usually, in a coding scheme, an intra-frame coding mode is necessary. It may be used only to handle the first image in a sequence or to periodically introduce a refreshment in the coding scheme. Consequently, an intra-frame segmentation is also necessary in a segmentation-based coding scheme.

Even though the main goal of this segmentation is to yield an intra-frame partition, the fact that this intra-frame segmentation is part of an image sequence coding algorithm has to be taken into account. Partitions describing very accurately the image may not be useful for this application. Such kinds of partitions are very expensive to code and, therefore, very few bits remain for coding the following frames in the sequence. In addition, some details present in the intra-frame coded image may not appear in the following frames and, thus, they can be coarsely described.

The segmentation has to extract the visually relevant objects in the image. Several features such as size, contrast, spectral content or dynamic characterize visually relevant objects. For each feature, a different homogeneity criterion can be applied. In order to control the degree of detail to be obtained by the segmentation procedure, multiscale representations of the image are used. At each scale, a different partition of the image is achieved. This way, the degree of complexity (e.g.: number of regions) of the partition representation can be progressively increased through the multiscale representation. Different multiscale segmentation can be computed using size [11, 38], contrast [26, 38, 41] or dynamic [25] criteria, as well as a combination of some of them [41].

As commented in the Introduction, several partitions can be obtained from a given image. That is, visually relevant objects in a scene can be represented by different partitions. Segmentation should be regularized by introducing some additional constraints related to the final application. In order to improve the coding efficiency, regions in the intra-frame partition should be related to the intra-mode coding techniques available in the coding scheme. Thus, the homogeneity criteria used in the segmentation have to lead to regions whose texture can be coded with a small error using few parameters.

In addition, the final partition should be formed by contours that can be easily coded. Transitions between homogeneous regions may not be clearly defined due to the uncertainty areas. Region boundaries can, therefore, be located inside these uncertainty areas in such a way that contours present good features for coding purposes. This usually translates into contours whose characteristics remain as stable as possible through the whole partition; that is, a partition with homogeneous contours.

The joint fulfillment of these two requirements (texture and contour homogeneity) leads to a trade-off. It can be solved by using a homogeneity criterion combining contour and texture information. Such kind of combination is used in [24, 27, 41].

The previous problems are related to the necessity to restrict the intra-frame coding cost to a given budget. This restriction is not easy to impose directly on the segmentation procedure. It is difficult to know *a priori* the coding cost related to the description of an image in terms of a partition and a set of texture parameters related to each region. In order to overcome this problem, indirect estimations can be computed.

A possibility is to analyze the relationship between some parameters of the partition and the coding cost of the segmented image. The control of these parameters results in the control of the coding cost. To implement this control, multiscale segmentations are very helpful. The relationship between the partition parameters and the coding cost can be adapted to the image characteristics through the multiscale procedure. This way, the partition parameters and the coding cost are more accurately estimated [26, 41]. Such parameters may be the number of regions, the complexity of their texture or the local structure of their contours.

The problem of intra-frame segmentation is of paramount importance when dealing with a *Pure 2D* segmentation approach. In this case, all frames are processed using intra-frame segmentation. On the other hand, a *3D block* approach can totally overcome the problem of the intra-frame segmentation since, in this approach, the first frame in a block is segmented jointly with the other images in the block.

2.2 Partitions with temporal label coherence

A common technique to improve the performance of image sequence coding schemes is to take advantage of the temporal redundancy in the sequence [45, 47]. Most of the information contained in a frame is already present in the previous frame. Therefore, changes can be coded by estimating the motion between the current and the past frames and motion compensating the previous information.

In the framework of segmentation-based coding schemes, such an approach is particularly useful. Indeed, since images have been already segmented into regions with texture homogeneity, motion is very likely to be also homogeneous within these regions. So, motion estimation and compensation can be carried out in a more reliable way. As it will be seen in the sequel, in order to really exploit the advantages of a segmentation-based coding scheme, regions have to be correctly related in the temporal domain. For example, if an object is represented by a region in an initial partition, the region that describes this object in the following partitions should be linked somehow to the initial region. Usually, such a link translates into the fact that regions corresponding to an object in the various partitions share a common label.

In the case of a *Pure 2D* approach, in order to ensure partitions with temporal label coherence, a region matching algorithm has to be applied. Partitions are computed only relying on the current frame information and, therefore, the link between regions has to be found after segmentation. Region matching is a difficult problem and its solutions are usually very time consuming [1]. This problem is directly solved in the *3D block* approach. In this case, a 3D region is created as a volume of any shape within the block and each 3D region has a different label [28, 33, 37, 39, 49]. The label coherence is therefore directly ensured within the block. However, if this coherence has to be preserved when changing from block to block, a region matching procedure is still necessary.

The concept of 3D region is illustrated in Figure 2. In this figure, a 3D region is depicted as being continuous in the time domain. The representation of this region in every frame of the sequence comes from the intersection of the frame plane with the 3D region. In the example, two different blocks of images are used in the segmentation and, therefore, a different label is assigned to the 3D region in each block.

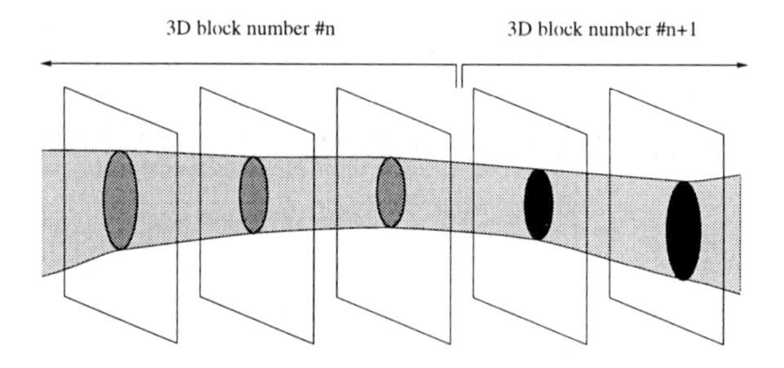

Figure 2 Intersection between the frame planes and a 3D region

A 3D region that preserves the same label through the time domain can also be obtained in the *Time recursive* approach. The only constraint to be imposed is that the pile of frames that are jointly processed should contain some already segmented frames, to initialize the procedure [25, 28, 33]. This way, the label of past frames can be propagated through the current frames.

2.3 Connectivity of regions

There is an additional problem when trying to achieve an image sequence partition with temporal label coherence. In the case of the *3D block* and the *Time recursive* approaches, 3D regions are defined using a specific connectivity. Therefore, the segmentation procedure yields 3D regions that fulfill this connectivity. In the spatial-spatial domain, the connectivity can be directly defined since this concept is related to the idea of compact objects. Assuming a given neighborhood (for instance, 4-connected or 8-connected neighborhoods), two areas that share the same characteristics cannot be merged into a common region if they are not neighbors. That is, if they do not describe a compact object or a compact part of it.

In the time domain, the definition of connectivity is not straightforward. The same object, due to its motion, may be at different locations in consecutive frames. These locations may not be directly connected through the time domain. This effect is more likely to appear if the object is small and moves fast. When the resulting partitions are used for coding purposes, the disconnection of related regions introduces a large overhead of information to send in the coding process. In order to link these two objects, very large neighborhoods should be defined. However, such large neighborhoods extremely complicate the segmentation procedure [28]. To circumvent the connectivity problem, the use of motion compensation in the segmentation procedure has been proposed [34].

Figure 3 shows an example of lack of direct connectivity between the representation of an object in consecutive frames. Note that the problem of linking these regions can be easily overcome in the case of a *Pure 2D* approach. In this case, the region matching algorithm can perform an exhaustive search on the partition, without any connectivity constraints.

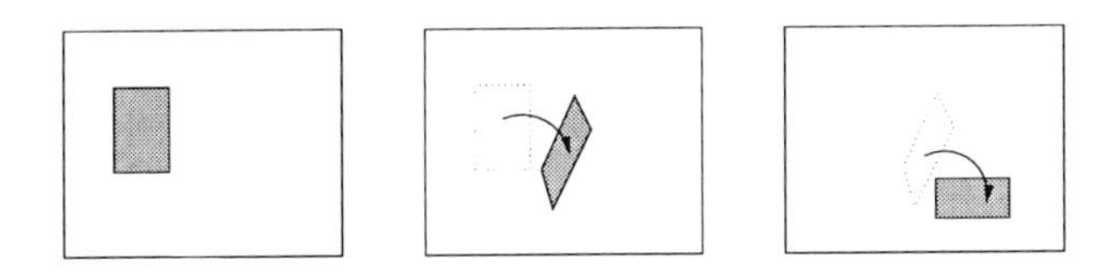

Figure 3 Example of a moving object with no connection in the time domain

The above problem is due to the lack of physical connection in time between related objects. The opposite problem can also appear; that is, a temporal connection of similar but non-related objects. The movement of an object can uncover an area of the scene that may have similar features to those of the moving object. Therefore, there is a time-connection between areas with similar features that, actually, do not represent related objects. A segmentation algorithm using a basic connectivity definition would merge the uncovered areas with the moving object through the time domain. This would be the case of the *3D block* and *Time recursive* approaches. As before, a *Pure 2D* approach shortcuts this problem since it does not directly link regions through the time domain.

In Figure 4, two frames of the sequence *Table-Tennis* are shown. In these frames, the two previous problems appear. The motion of the ball (a small object) is fast enough to disconnect in the temporal domain the object. In addition, the uncovered background presents a noisy texture that can be locally mistaken with a part of the ball; in some areas it has almost the same gray level.

When directly segmenting this sequence with a *3D block* or a *Time recursive* approach, different labels may be assigned to the ball in each frame and, in addition, the ball may be locally connected to the background.

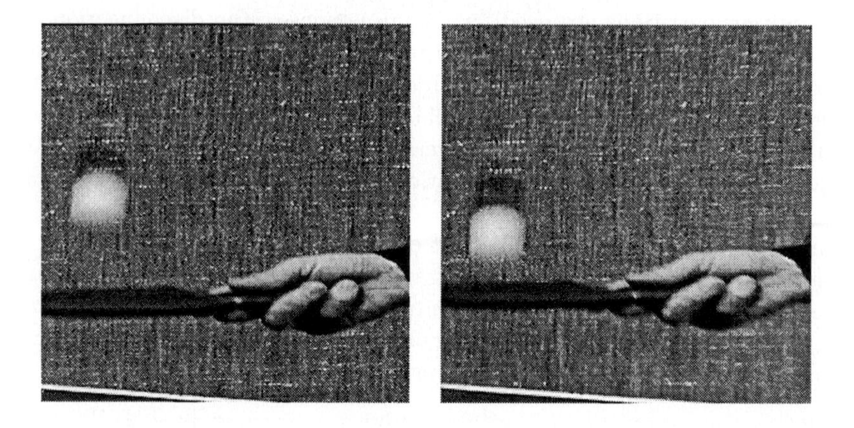

Figure 4 The fast motion of the ball may lead to segmentation errors

Two techniques have been proposed to shortcut the problem of temporal connection of similar but non-related objects. The first technique filters the images before segmentation in order to disconnect such objects [33]. The second technique motion compensates regions from the past partitions in order to connect in the temporal domain the areas corresponding to the same object [34].

An additional problem can also appear due to the spatial-spatial-temporal connectivity of regions. *3D block* and *Time recursive* approaches create 3D connected regions. The representation of 2D regions in a partition comes from the intersection of the frame plane with the 3D connected regions (see Figure 2). However, this intersection may produce from the same 3D connected region more than one unconnected component with a common label in the same frame. Therefore, this intersection does not yield a real partition. This problem is illustrated in Figure 5.

To obtain a partition, unconnected regions sharing the same label in the same frame should be prevented. This is usually done by relabelling the image resulting from the intersection. A second approach is to remove all the components with the same label but one. The areas that are unassigned after the removal are covered by expanding their neighbor regions.

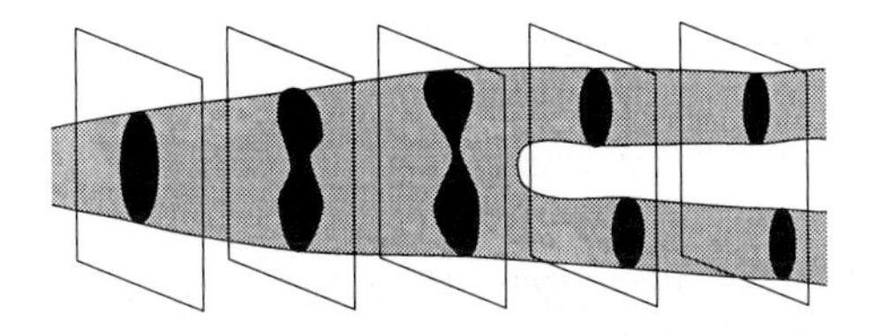

Figure 5 Example of a 3D region that does not lead to a 2D partition

2.4 Coding-oriented inter-segmentation

Even if label coherence is ensured and connectivity through the time domain is correctly handled, partition sequences may not present the best possible features for coding purposes. As commented above, a usual means to improve coding performance is by exploiting the temporal redundancy. This is done by motion compensating the previous frame information, which is already available in the receiver, and coding the compensation error. Consequently, the smaller the compensation error, the higher the coding performance.

In the case of segmentation-based coding schemes, two different kinds of information should be coded: the partition itself and the texture of every region. In this context, motion compensation can be utilized on both types of information. To obtain a small compensation error, partitions and region textures have to preserve as much as possible their features through the temporal domain.

In order to improve the partition coding using motion compensation, current partitions should be controlled by the previous partitions. That is, regions have to remain stable through the partition sequence. In the case of global motion, as many as possible previous regions have to extend to the current partition and their shapes have to vary as little as possible. In the case of objects with different motion, the boundaries of covered and uncovered regions vary from frame to frame. Therefore, the shapes of the regions cannot be forced to remain fixed through the time domain. Nevertheless, some control can be introduced in both cases in order to reduce the cost of the partition coding. In the case of a *Pure 2D* approach, no control on the partition evolution can be introduced. On the other hand, the *3D block* and the *Time recursive* approaches allow a better control of the partition evolution.

Such a control can be viewed as a regularization of the temporal contours of the regions. This regularization can be implemented as an extension of the techniques used in the intra-frame mode. This way, homogeneity criteria have been proposed that take into account the temporal contours as well as the spatial contour and texture information [28, 41]. However, specific techniques have to be applied in order to prevent large random region deformations from frame to frame. In addition, as new regions cannot be motion compensated, new regions should only appear in the partition when they are related to new objects in the scene.

In turn, texture characteristics of regions should be preserved through time. Texture features from the same region in two different frames should be similar so that motion compensation can be exploited. Regardless of the segmentation approach, the texture similarity of related regions is almost ensured if the same homogeneity criteria are used in the different frames.

2.5 Bit rate regulation

The segmentation has to be regulated to achieve a given compression ratio in the coding process. For the intra-frame segmentation, techniques leading to a given cost of the coded image have been previously discussed. In that case, parameters such as the number of regions or the complexity of the contours in the partition are useful to estimate the final coding cost. Moreover, multiscale segmentation approaches help to correctly estimate these parameters.

Fluctuations of the coding cost in the inter-mode can be regulated mainly by two different techniques. First, the amount of regions in the partition can be increased or decreased in order to reach the target bit-rate [41]. That is, new regions are introduced in the partition in order to improve the image quality when the estimated cost is below the target. On the other hand, regions are merged if the coding cost is higher than expected. The merging procedure has to be correctly implemented since merging reduces the number of contour elements in the partition but also yields regions with more complex textures.

A second technique to regulate the coding bit-rate deals with the texture representation [24, 39]. The texture model for the region interiors is adapted to characterize more accurately the original data, if the estimated cost is lower than the target. In an analogous manner, textures are coarsely represented when the estimated cost is above the expected bit-rate.

2.6 Time delay

Finally, in order to obtain a time delay feasible for interactive applications, segmentation approaches should not handle a large amount of future frames at the same time. This leads to the use of small blocks of frames in the *3D block* approach and to reduce the number of future frames in the active pile in the *Time recursive* approach [25, 28, 33]. Moreover, when imposing this constrain, the memory requirements of the coding approach are reduced and so is the computational load.

3 A GENERAL STRUCTURE FOR SEGMENTATION

Following the analysis made in the previous section, image sequence segmentation for coding purposes should extract the visually important regions of the scene, be coherent in time, avoid as much as possible random fluctuations of the partitions, generate partition sequences suitable for coding, allow a regulation of the bit rate and prevent large time delay. With these requirements, it is very difficult to obtain a good segmentation if a *Pure 2D* or a *3D block* approach is used. Indeed, if frames are segmented independently, both the time coherence and the region correspondence problems are difficult to solve. Somehow, the temporal relationship between frames must be exploited. In turn, if the temporal size of the 3D blocks is large, the approach implies huge memory requirements, a high computational load and the introduction of a large processing delay disallowing any interactive application. Besides, if the temporal size of the 3D blocks is small, the temporal correlation between frames will be ignored at each block transition, that is, very often.

Therefore, *Time recursive* approaches are the most suitable for coding purposes. In order to reduce the complexity of the segmentation as well as the time delay, the pile of frames to be jointly processed can be reduced. Most of the information contained in a frame is already present in the previous frame. This point of view leads to a segmentation process only involving those previously segmented and the current images. This segmentation approach involves two modes of operation: intra-frame and inter-frame segmentation. During the intra-frame mode, a single frame is segmented. It is a purely 2D process. Then, during the inter-frame mode, each frame is recursively segmented using the partition of the previous frame.

In the general segmentation scheme that is proposed, both modes make use of the same basic structure. This structure, which is an extension of the structure presented in [38], involves five basic steps: *Partition projection*, *Image modeling*, *Image simplification*, *Marker extraction* and *Decision*. This general structure is presented in Figure 6.

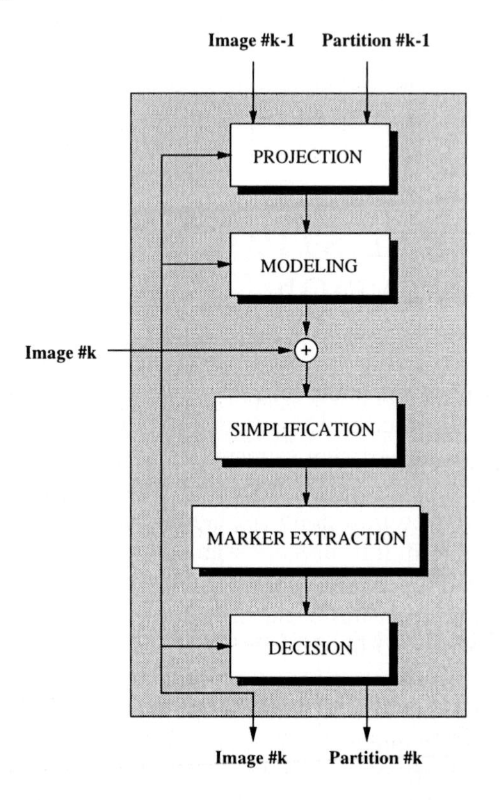

Figure 6 Block diagram of the basic segmentation structure

In the case of an intra-frame mode, this basic structure is applied several times in a hierarchical manner. This hierarchy allows the control of the detail representation in the partition. On the inter-frame mode, only one level of the hierarchy is necessary. Actually, the basic structure is used to transfer the partition information from the previous to the current image. Consequently, it is an iterative procedure. In the case of an intra-frame mode, partition indeces in Figure 6 denote the hierarchical level whereas, in an inter-frame mode, they denote the frame number in the sequence. The relationship between the general segmentation structure and the two modes of operation is shown in Figure 7.

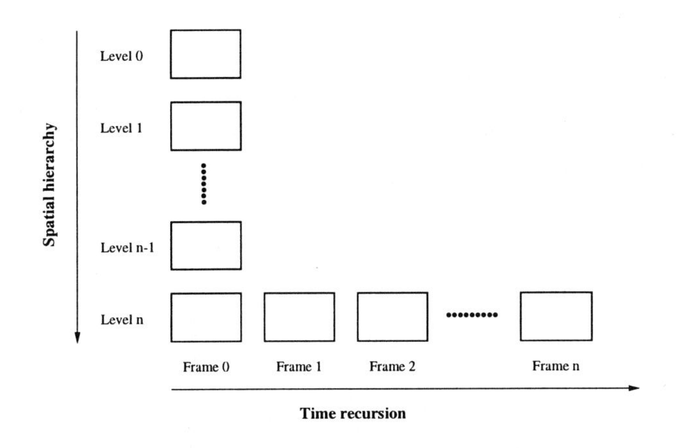

Figure 7 Relationship between the intra-mode (hierarchical) and the inter-mode (recursive)

In the sequel, each one of the five basic steps is described. In this description, the relationship of these steps with the different problems presented in the previous section is pointed out. Note that in the following, the segmentation is assumed to be performed on the luminance signal because most of the visually important transitions are defined by the luminance signal.

3.1 Partition projection

Assume that the partition at level $k-1$ (in the intra-frame mode) or of the frame $k-1$ (in the inter-frame mode) is already available. Most of the information already present in image $k-1$ should be also in image k. Therefore, the segmentation of image k should rely on the partition of image $k-1$. The *Partition projection* step detects the actual distribution of the previous regions in the current frame. It accomodates the partition of image $k-1$ to the data of image k. That is, it gives a first approximation of the final partition of image k. Since it is based on the previous partition, new regions cannot be introduced in the partition at this step.

Intra-frame mode

In this case, the procedure is initialized by imposing a fixed partition in the coarsest level of the hierarchy. For instance, the whole image is considered as a single region. The *Partition projection* should take into account the possible deformations or displacement of a region due to the change of scale in the multiscale hierarchy. For example, if the hierarchical multiscale representation is a pyramid, the partition of the previous image $(k-1)$ should be interpolated in order to have a first approximation of the current image partition (k). On the other hand, if there are not such changes, the partition projection can be the identity operator.

Inter-frame mode

In the inter-frame mode, the procedure is always initialized by an intra-frame segmentation. The *Partition projection* step has to account, in this case, for the motion of the objects in the scene. Regions in the previous partition $(k-1)$ have to be correctly located in the current partition (k). In addition to the change of position, possible region deformations have to also be detected in this step.

In order to correctly project moving regions, the motion in the scene is estimated. After motion estimation, each region is motion compensated. A set of motion compensated regions is not likely to result in a partition. Depending on the motion estimation and compensation techniques various problems may appear. After motion compensation, some areas of the current image may be covered by more than one region whereas a region from the previous partition may yield more than one connected component in the current image. Furthermore, some areas of the current image may not be covered by any compensated region and, therefore, some pixels may have no label assigned.

The first two problems are solved by a cleaning procedure that renders an image only composed of unassigned pixels and, at most, one connected component for each region from the initial partition. These connected components ensure that temporal label coherence is preserved in the sequence segmentation. In addition, the problem of region connectivity through the time domain is shortcut by means of the motion compensation. The effect of motion compensating moving regions is that of temporal connectivity being locally adapted to the motion of the different objects in the scene. The final result is an image containing, for each region coming from the previous image, a connected component with the same label as in the previous image. Such a component roughly marks the position of the region core in the current frame.

The areas of unassigned pixels are covered by extending these connected components. In these areas, the actual boundaries of the projected regions should be found. Note that such areas arise due to the lack of accuracy of the estimated motion. The region motion cannot be completely modelled and, therefore, cannot be completely estimated. In such cases, region boundaries have to be defined relying on texture information. Indeed, region boundaries are obtained by extending the above connected components and their growth is controlled by texture information.

A block diagram of the *Partition projection* step in the inter-frame mode is shown in Figure 8. Note that the cleaning procedure does not make use of the original image since it only utilizes the compensated label information.

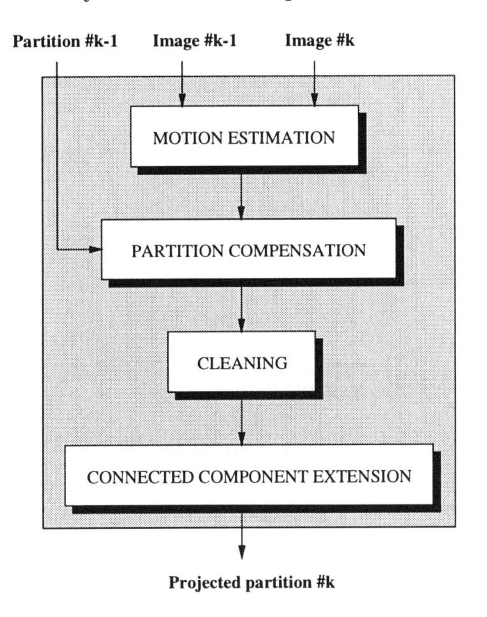

Figure 8 Block diagram of the *Partition projection* step in inter-frame mode

Note that the *Partition projection* step does not introduce additional time delay since no future frames are used. Furthermore, only information about the previous image is necessary and, therefore, it does not imply huge memory requirements.

3.2 Image modeling

Once the projected partition has been computed, the following steps have to improve this first estimation of the current image partition. Since this segmentation procedure is coding-oriented, such improvements should deal with those regions that cannot be correctly represented by the coding process using the projected partition. Such regions are related to new objects in the scene, to areas that have not been correctly compensated or to regions that were not already correctly represented in the previous coded image.

To have information about these regions, in the *Image modeling* step each projected region is actually coded. In this step, any kind of partition and texture coding technique can be used. This way, the segmentation algorithm detects which regions cannot be correctly represented with the available coding techniques. Moreover, if a new coding technique with higher performance is developed, it can be directly included in the segmentation procedure. Note that the image modeling step is the same for the intra and the inter modes.

The difference between the coded and the original images is then computed. This image is refered to as the modeling residue. It concentrates all the information about the areas in the image which are poorly coded using the projected partition. Therefore, this is the signal that is used to drive the segmentation in the following steps.

3.3 Image simplification

The modeling residue contains a large amount of information. It contains information about new regions that are not present in the projected partition as well as information about the textures of projected regions that have not been correctly represented by the modeling. In the *Image simplification* step, the modeling residue is simplified in order to make it easier to segment. The simplification controls the amount and nature of the information that is kept. Different simplification tools can be used depending on the segmentation criterion. For example, the simplification can aim at reducing the complexity of textured areas, at removing regions smaller than a given limit or at keeping regions with a contrast higher than a given value. Regardless of the simplification criterion, the *Image simplification* should yield areas that can be easily segmented.

Intra-frame mode

In the intra-frame mode, the simplification defines the notion of multiscale hierarchy. The use of a multiscale image segmentation in the intra-frame mode allows the control of the amount of detail to be represented in the intra-frame image partition. Coarse partitions composed of a few regions are derived in the coarsest scales. These partitions result in coded images of moderate visual quality but very high compression ratios. Coarse level partitions govern the segmentation procedure of the finest level images. New regions are added in the fine levels of the hierarchy, increasing the visual quality of the coded images at expenses of the compression ratio. It is therefore a top-down procedure.

Since the simplification parameters can be adapted at each level, the influence of these parameters in the coding cost can be analyzed through the multiscale hierarchy. This way, a partition leading to a target coding cost can be easily reached in the finest level of the hierarchy.

For coding applications, various visually important criteria can be used. In practice, a combination of different criteria is utilized to build the multiscale hierarchy given that most of the image structure is represented by large regions but an important part of the meaningful information is defined by small but contrasted details.

Inter-frame mode

In this case, the main task of the *Image simplification* step is to make easier the detection of new regions in the scene. Assuming that the *Partition projection* has correctly tracked past regions, the modeling residue should mostly contain information about new regions. As in the intra-mode case, several criteria can be used to detect such regions. Nevertheless, contrast information is the most suitable in order to detect visually relevant new regions.

The parameter of simplification should be adapted frame by frame in order to control the bit stream. The estimation of the coding cost of the projected partition indicates the amount of bits that can be used to code new regions and, therefore, the number of new regions that can be introduced. The simplification parameter is set in order to preserve after simplification this estimated number of regions.

3.4 Marker extraction

The goal of this step is to detect, in the simplified modeling residue, the presence of relevant regions. For each relevant region, a marker is obtained. Markers are connected components with a specific label identifying the presence of homogeneous regions. The technique to extract markers depends on the simplification criterion that has been used. Since different simplification criteria may be used in intra or inter-frame mode, the *Marker extraction* step may also vary.

A marker defines a set of pixels that surely belong to the region. Although a marker may contain the major part of the region, the actual boundary of the region has yet to be found. The fact that the marker does not completely define the region is related to the presence of uncertainty in the boundary position. As commented in the introduction, this uncertainty leads to areas of transitions between homogeneous areas that cannot be directly assigned to any marker.

3.5 Decision

In order to finally obtain a partition, pixels laying on the uncertainty areas have to be assigned to some region. After *Marker extraction*, the number and the interior of the regions to be segmented are known. Therefore, markers have to be extended in order to correctly cover the uncertainty areas. The precise shape of every region is obtained in the *Decision* step.

Assigning uncertainty areas to a set of given regions can be viewed as a decision process. For coding purposes, this decision cannot be made only accounting for texture information. Indeed, for coding applications, if the gray level transition between two regions is not very strong, it is useful to have simpler contours, even if some precision in the boundary position is lost. Simple contours lead to partitions with lower coding cost. This decrease in the partition coding cost compensates for the increase in the texture coding cost due to the lack of boundary precision.

3.6 Comparison between the intra and inter-frame modes

Note that both modes of segmentation are very similar. The main difference is in the projection step which has to take account of possible variations of images through the multiscale representation in the intra-mode and of motion in the inter-frame mode. In intra-frame mode, it can be considered that the algorithm segments a sequence of still images and that the segmentation parameters are modified to increase the number of regions. In inter-frame mode, the hierarchy is not on the spatial space but on the temporal space which leads to a time iteration. The sequence is moving and the segmentation parameters are mainly updated to maintain the partition characteristics. In Table 1 the main characteristics of both modes are summarized and compared.

Mode	Partition projection	Image modeling	Image simplification	Marker extraction	Decision
Intra	Multiscale variations	Texture and partition coding	Combination of criteria	Criterion dependent	Contour/Texture complexity
Inter	Movement of objects	Texture and partition coding	Contrast criterion	Contrast criterion	Contour/Texture complexity

Table 1 Main characteristics of the intra-frame and inter-frame modes

4 AN IMPLEMENTATION BASED ON COMPOUND RANDOM FIELDS

The general structure presented in the previous section can have different implementations. In this section, a first implementation is presented. It exploits the fact that image sequences can be modeled by means of compound random fields [13, 18]. This model assumes that image sequences are composed of a set of 3D regions, each one characterized by an independent random field. The union of these random fields, named the upper level of the model, forms the observed sequences. Each random field characterizes the texture of a region in the spatial-spatial-time domain. In turn, the location of these random fields within the sequence (position and shape of the 3D regions) is governed by an underlying random field, named the lower level of the model. The sequence partition is defined by the lower level of the model. Therefore, compound random fields directly describe images in terms of a partition sequence and texture information, as it is assumed in segmentation-based coding techniques.

Compound random fields enable, first, the separated characterization of partition and texture information and, second, their combination in a single model. This way, the trade-off between contour and texture homogeneity in the segmentation procedure can be easily addressed. In this implementation, images are modelled by means of a compound random field whose upper level X is a set of independent white Gaussian random fields whereas the lower level Q is a 3D second order Strauss process [12].

Strauss processes are a type of Markov random field and, therefore, are governed by a Gibbs Distribution [12, 13]. For a complete review of the topic of

Markov random fields, the reader is referred to [9]. In Strauss processes, clique potentials are defined depending on the clique homogeneity. For partition characterization, only those cliques related to region bondaries are useful; that is, cliques with elements belonging to different regions in the partition. Therefore, zero potential is assigned to cliques with equally valued elements (homogeneous cliques). Non-homogeneous cliques have associated potentials whose values vary depending on the type of clique c:

$$V_c(x) = \begin{cases} 0 & \text{if all elements in c are equal} \\ \beta_c & \text{otherwise} \end{cases} \tag{3.1}$$

Partitions can be characterized only using two-pixel cliques [29, 27]. Since 3D second-order neighborhoods are assumed, there are twenty different two-pixel clique configurations [13]. Moreover, the potential definition should be invariant by rotation if isotropy is assumed. However, the different nature of the temporal and spatial domains has to be taken into account. Thus, only four different types of cliques are considered: two involving pixels in the same image (spatial cliques) and two involving pixels in different images (temporal cliques). Spatial cliques of type 1 contain horizontal and vertical cliques, whereas spatial cliques of type 2 contain diagonal cliques. The same classification is followed for the temporal cliques. A 2D version of this model has been proposed for segmentation of still images in [29], applied to coding applications in [27] and extended to image sequence segmentation for coding purposes in [28].

Relying on this model, images are segmented using a deterministic maximization algorithm based on the Iterated Conditional Modes approach presented in [5]. It obtains the realization of the two-level model (partition and texture information) that better fits the given sequence. Model parameters can be easily adapted in order to penalize or to favour partition information with respect to texture information. The segmentation can, therefore, be tuned so that partitions following not only texture features but also a given contour behavior are achieved. In practice, such partitions should yield low coding cost representations of the sequence, in terms of partition and texture information.

In the sequel, the implementation of the previous general structure based on compound random fields is described. The description follows the coding procedure: first, the hierarchical intra-frame segmentation that initializes the sequence segmentation is presented and, afterwards, the time iteration of the inter-frame mode is discussed.

4.1 Intra-frame mode

The intra-frame segmentation is a top-down procedure. Images to be segmented at each level of the hierarchical procedure correspond to the various levels of the Gaussian pyramid [8] of the original image. Coarse levels in the hierarchy contain little texture areas that yield a reduced number of regions, due to the filtering and decimation procedures involved in the computation of the Gaussian pyramid. Coarse level partitions lead the segmentation of fine level images in which, progressively, new regions are introduced. At each hierarchical level of the segmentation, the basic structure presented in the previous section is applied.

A 2D image model is used in the intra-frame segmentation. The upper level of the model is formed by a set of 2D independent white Gaussian random fields. In turn, the Strauss process that characterizes the partition only defines spatial cliques [27].

Image projection

The procedure is initialized assuming that, at the very top of the pyramid, each pixel belongs to a different region. This initial partition is projected into the following level in order to accomodate it to the data of the following image. The projection is divided into two steps.

First, the partition of image $k-1$ (that is, the $k-1$ level in the Gaussian pyramid) is interpolated in order to correspond to the size of the image at level k. This interpolation is carried out by repeating at level k the label value of the pixels in level $k-1$. More sophisticated interpolation techniques are worthless since, after interpolation, the resulting partition has to be refined.

Second, the interpolated partition q is refined so that it comforms better to the data at level k. This is done relying on the previous stochastic model. The input image x is assumed to be a realization of the upper random field $(X = x)$. The partition refinement is carried out by seeking the lower level realization $(Q = q)$ which, more likely, has given rise to x. That is, the position of the contours in the interpolated partition are refined so that they correspond to transitions in the current image. This can be expressed as a *maximum a posteriori* estimation $P(Q = q/X = x)$. By Bayes' rule, the previous maximization leads to the same result as maximizing $P(Q = q, X = x)$, which is used since it is simpler to maximize. Given that Q is a Markov random field and that X is formed by a set of independent white Gaussian random fields, the expression to maximize is:

$$P(Q = q/X = x) = \frac{1}{Z}\ exp(-U(x)) \prod_{R_i} P(I_i/\mu_i, \sigma_i), \tag{3.2}$$

where Z is a normalizing constant, R_i is the $i-th$ region in the partition q, $\{I_i\}$ represents the set of pixels belonging to R_i, and μ_i and σ_i are the mean and the variance of the set $\{I_i\}$, respectively. The energy function $U(x)$ is given by

$$U(x) = \frac{1}{T} \sum_{c \in C} V_c(x) = \frac{1}{T}(n_1 V_1 + n_2 V_2) \tag{3.3}$$

where T is a constant, n_1 and n_2 represent the amount of cliques of type 1 and type 2 that are in the partition and V_1 and V_2 their associated potentials, respectively. The three parameters that define the lower level model (T, V_1, V_2) can be reduced to two:

$$\frac{1}{T}(n_1 V_1 + n_2 V_2) = \frac{V_1}{T}(n_1 + n_2 \frac{V_2}{V_1}) = \frac{1}{T^*}(n_1 + n_2 V^*) \tag{3.4}$$

T^* controls the relative importance between the upper and lower levels of the model; that is, between the texture and partition information. As T^* decreases, final segmentations are governed by the contour information and region shapes are very alike through the whole partition. T^* represents the trade-off between the homogeneity of contours through the image and the homogeneity of textures inside each region. In turn, V^* controls the relationship between potentials of cliques of type 1 and type 2. Modifications of this relation result in giving priority to some contour configurations with respect to others. For instance, the situation $V^* \to 0$ leads to overlooking the local compactness of regions. As a consequence, regions having zones of small width are more likely to appear. These contour configurations are very expensive to code and, therefore, they should be avoided.

The partition refinement translates into the maximization of expression (3.2); that is, a function that might be multimodal. To avoid local maxima, stochastic maximization techniques should be used [13]. Since the projected partition is utilized as initial state of the maximization procedure a deterministic approach is used [27]. This approach is a special case of the so-called Iterated Conditional Modes algorithm [5].

The partition refinement is performed by testing the label of those pixels lying on the boundary of a region. Given an initial partition ($Q = q'$), its joint likelihood is computed ($P' = P(Q = q', X = x)$). For every pixel lying on the boundary of a region (i, j), all its neighbors (k, l) are checked. The label of pixel (k, l) is given to pixel (i, j) and the joint likelihood due to this new partition is computed (P''). If $P'' > P'$, the label of pixel (i, j) is changed ($q_{ij} \Leftarrow q_{kl}$).

Image modeling

In the *Image modeling* step, the image at level k of the Gaussian pyramid is coded based on the projected partition. Here, a simplified version of the texture and partition coding techniques can be used. Actual coding techniques slow down the segmentation procedure and do not drastically improve the segmentation results. This way, the partition is assumed to be coded with a lossless technique. In turn, the texture of each region is coded utilizing a simplification of the actual gray-level function to be used in the coding procedure. The modeling residue is computed using this coded image and the image at level k of the Gaussian pyramid.

Image simplification

The modeling residue has to be simplified in order to help the segmentation procedure. In this implementation, a first frequency-oriented simplification has been aplied to the image. Note that, the image to be segmented at level k in the hierarchical procedure is the level k of the Gaussian pyramid of the original image. Therefore, image k is a low-pass version of the original image.

A contrast simplification is then applied to the modeling residue. The modeling residue mainly contains information about regions not represented in the projected partition and areas that have not been correctly coded. In order to remove the information not related to new regions, contrast extractor operators are used that simplify those areas that do not present a high contrast [26]:

$$Positive\ contrast\ extractor = Max(\varphi\gamma, I), \tag{3.5}$$

$$Negative\ contrast\ extractor = Min(\gamma\varphi, I), \tag{3.6}$$

where $\varphi\gamma$ is the open_close filter, $\gamma\varphi$ is the close_open filter and I stands for the identity operator. The combination of both contrast extractors enables the detection of both positive and negative contrasted regions.

Marker extraction

From the simplified image, a set of markers is obtained. The simplified image is formed basically by almost flat regions and contrasted areas showing the presence of new regions. Therefore, markers can be selected by detecting such contrasted areas. In order to remove very small markers, which are usually related to texture coding errors rather than to real new regions, a morphological opening is applied to the thresholded image.

Decision

Markers are labeled and directly introduced as new regions in the projected partition. Then, the partition refinement that is used in the *Image projection* step is applied to this new partition. This way, the actual shape of the regions that have been marked in the *Marker extraction* step are obtained. The partition refinement is allowed to decide which is the correct shape and location of these new regions and whether they should be expanded, shrunk or removed.

In addition, the shape of those regions that were already present in the projected partition is refined. This is necessary since new regions have been detected in the interior of the projected regions, whose statistics may have changed.

Results

Figure 9 illustrates the procedure of applying the previous intra-frame segmentation to the frame number 100 of the sequence *Foreman.* In the first row, the Gaussian pyramid of the original image is shown, as well as the result of the segmentation of the first levels of the hierarchy. The partition at each level is presented by means of an image in which the contours between the different regions have been marked. In addition, the modelling of the final partition at each level is also presented. In the second row, the original image is shown, jointly with the results of the *Partition projection* and *Image modelling* steps in the finest level of the hierarchy. Finally, the third row presents an image containing the set of markers obtained after the *Marker extraction* step, the final partition and its modeling. The final partition contains 65 regions and, for the modeling, orthogonal functions have been used.

Note that the hierarchical approach permits to achieve, after the *Partition projection* step a good first estimate of the final partition. In addition, the *Marker extraction* step selects markers for those areas that are not well represented in the projected partition. This is the case of some parts in the face (e.g.: the left eyebrow) or in the building (e.g.: the separation between the floors). Such markers result in new regions in the final partition, so that these areas are finally represented in the partition.

4.2 Inter-frame mode

In this mode, the 3D image sequence model previously presented is used. With respect to the intra-frame mode, the formulation of the upper level of the model

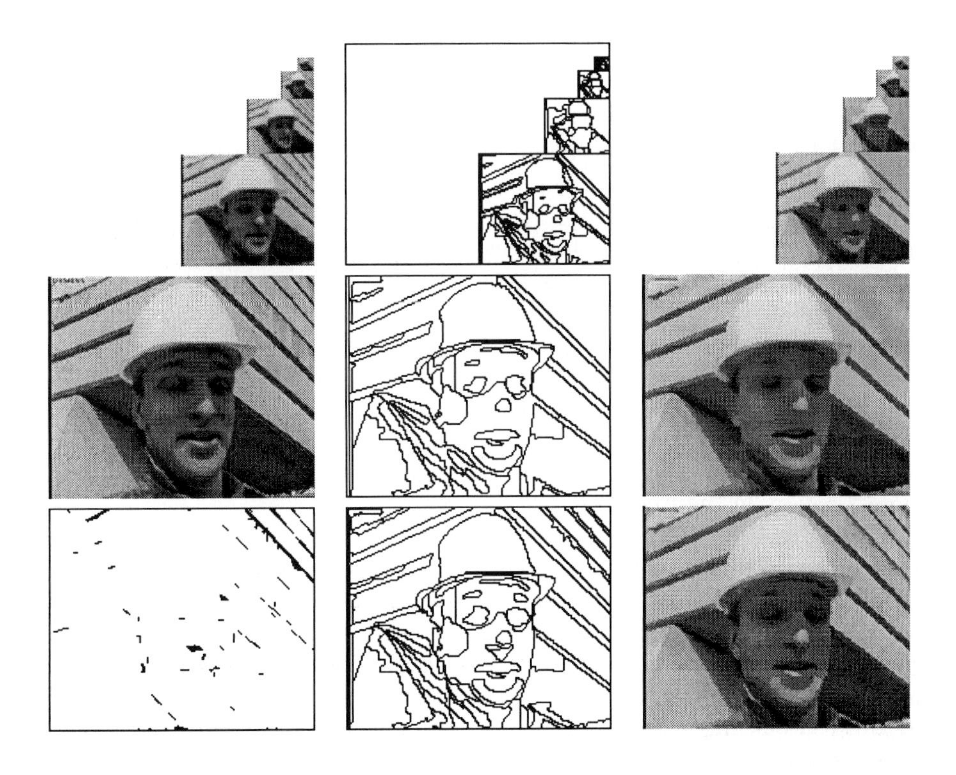

Figure 9 Example of intra-frame segmentation

does not vary. In the lower level model, new types of cliques are necessary to characterize the time evolution of the partition. This way, the energy function $U(x)$ is defined by means of four different types of clique:

$$U(x) = \frac{1}{T} \sum_{c \in C} V_c(x) = \frac{1}{T}(n_{s1}V_{s1} + n_{s2}V_{s2} + n_{t1}V_{t1} + n_{t2}V_{t2}) \tag{3.7}$$

where n_{is} and n_{it} represent the amount of cliques of type i in the spatial and temporal domains, respectively. In turn, V_{is} and V_{it} represent their associated potentials. This set of parameters can, as in the intra-case, be reduced:

$$U(x) = \frac{1}{T^*}(n_{s1} + n_{s2}V_s^* + R_{st}^*(n_{t1} + n_{t2}V_t^*)) \tag{3.8}$$

In this 3D model, V_s^* sets the behaviour of the spatial contours whereas V_t^* sets that of the temporal ones. R_{st}^* penalizes or favours the importance of the spatial contours with respect to the temporal contours.

Image projection

In order to project the partition of image $k-1$, the motion between images $k-1$ and k is estimated. Since motion information is used to provide a rough estimation of the position of the previous regions in the current frame, motion estimation does not need to be extremely accurate. A block-matching algorithm is therefore utilized. It is applied backwards so that no overlapping problems appear in the compensated image [17].

After motion compensation, some regions may yield more than one connected component in the current image. Only the largest connected component for each region is kept. The areas that become unassigned after this cleaning step are covered by extension of the remaining connected components. This extension is purely geometrical, since unassigned pixels receive the label of their closest region, in the Euclidean distance sense.

This first estimation of the projected partition is further improved applying the partition refinement procedure presented in the intra-frame segmentation. In this case, the parameters related to the texture model in expression (3.2) (μ_n and σ_n) are estimated using the gray-level values of a subset of the pixels belonging to region R_n. This subset contains all the pixels inside the region in image $k-1$ but only those pixels laying inside the largest connected component of region R_n after motion compensating the previous partition.

The refinement procedure is only allowed to change the position of the contours in the current image. The partition of image $k-1$ leads the projection but, since in the coding procedure the previous partition has been already sent, there is no use in refining it. Therefore, only pixels belonging to image k may vary their label. The result of the refinement procedure is the projected partition. The *Partition projection* step for the inter-frame mode is ilustrated in Figure 10. In it, the 3D signals formed by the union of two original frames or two partitions are denoted by $\mathcal{F}$ and $\mathcal{P}$, respectively.

Extraction of new regions

The following steps in the segmentation structure do not change substantially from the intra to the inter-mode. Indeed, the *Image modeling*, *Image simplification* and *Marker extraction* steps are purely 2D. These three steps only rely on the current image and the projected partition, as in the intra-mode segmentation. The unique difference is that, in the inter-mode segmentation, there

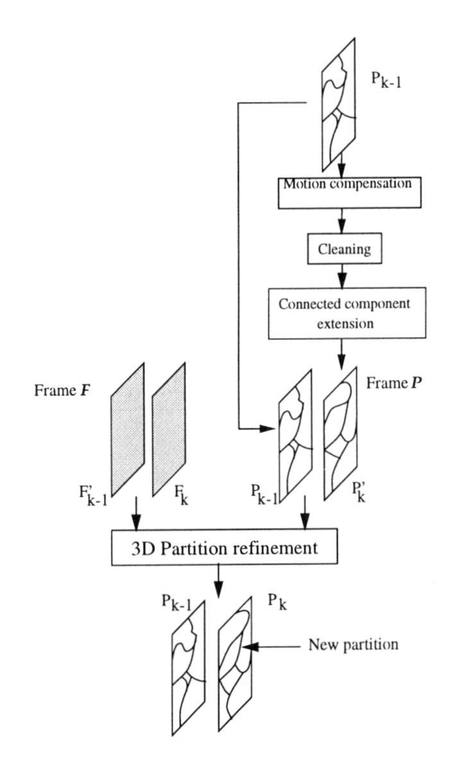

Figure 10 Projection of markers of frame $k-1$ into frame k

is no frequency simplification since the Gaussian pyramid of the image is not computed.

However, the *Decision* step makes use of 3D information. The final shape and position of all regions are defined based on the information in the current as well as in the previous image. As pointed out in the intra-frame mode, the projected partition and the new regions, that have been introduced in the projected partition as a set of markers, have to be jointly refined. If such a refinement is purely 2D, projected regions would vary their shape without taking into account the previous partition. That is, the final partition would be computed overlooking the motion information. Therefore, it is necessary to perform a final partition refinement based on the previous and current images and partitions.

Results

Figure 11 illustrates the procedure of applying the previous inter-frame segmentation to the frame number 105 of the sequence *Foreman*. The previous segmented image is the frame number 100. In the first row, the two original images are presented. In the second row, the partition obtained for the previous image is shown as well as the result of the *Partition projection* step. In this case, partitions are represented giving to each region a different value. This is done in order to shown the coherence of the labels through the time domain. The final image in this row is the modeling of the projected partition. In the third row the modeling residue is presented, jointly with the final partition and its modeling. The final partition contains 65 regions and, for the modeling, orthogonal functions have been used.

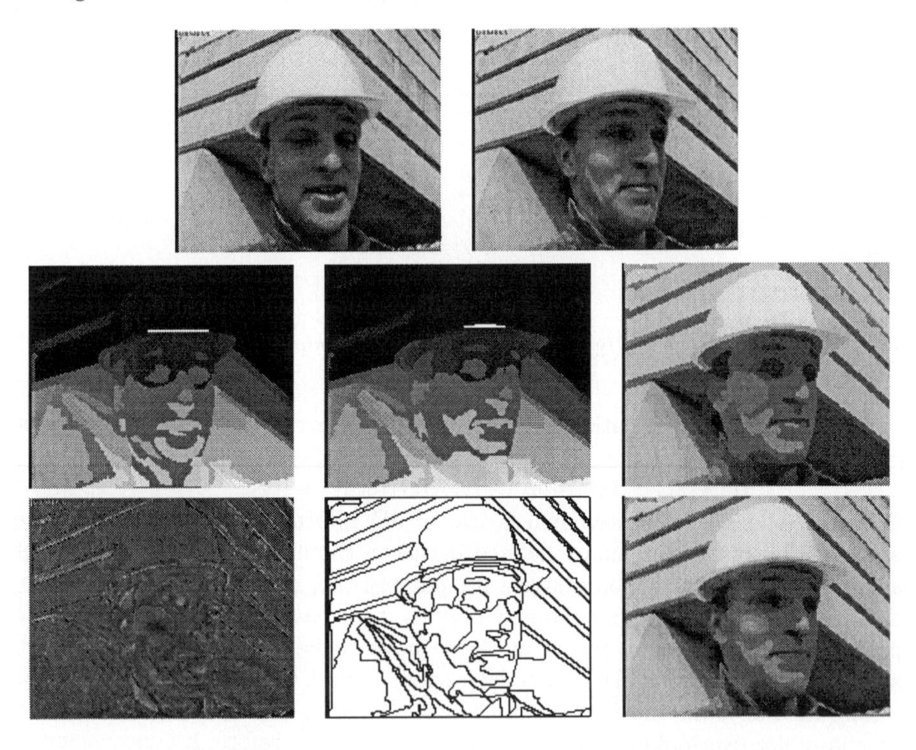

Figure 11 Example of inter-frame segmentation

The major point to be highlighted in this example is the ability of the segmentation algorithm to track the evolution of the regions in the scene. Note that the projected partition correctly represents the image to be segmented while preserving the temporal coherence of the labels. Finally, only a few details (e.g.: in the ear and in the neck) are introduced as new regions given that the evolution of the image has been correctly represented.

5 AN IMPLEMENTATION BASED ON MORPHOLOGICAL TOOLS

The scheme presented in this section relies entirely on tools from Mathematical Morphology. It is clear that geometrical information is of prime importance in a segmentation process. Classical linear signal processing tools are not well suited for a geometrical approach. On the contrary, Mathematical Morphology has been developed as a geometrical approach to signal processing [43, 44]. This is the reason to present a segmentation scheme developed on this basis. In the sequel, an implementation using morphological tools of the general structure described in Section 3 is analyzed. The usefulness of Mathematical Morphology in this context will be demonstrated. First, the segmentation performed on the intra-frame mode to initialize the segmentation of the sequence is seen. Afterwards the time iteration of the scheme to lead to the time recursive segmentation is described.

5.1 Intra-frame mode

The intra-frame segmentation is a top-down procedure which first produces a coarse segmentation in the sense that it involves a reduced number of regions. Due to the properties of the morphological operators used, these regions have their contours correctly located. Then, the segmentation is progressively improved only by introducing more regions. Typically four segmentation levels are used. Each segmentation level involves the four basic steps described in the previous section.

Partition projection

The projection step adapts the partition of the previous level to the current level. In the intra-mode the previous level corresponds to a partition of the same image. That is, in the scheme presented in Figure 6, the image k is always the original image. In the scheme proposed the spatial hierarchy is based on a simplification of the image by morphological filters. Due to the properties of these filters, which will be seen in the simplification step, it is not necessary to modify the contours obtained in previous levels, and there is no need to make a change of scale through the different levels. For these reasons, the *Partition projection* can be the identity operator. To initialize the procedure, the whole image is considered as a single region, and the successive levels introduce the visually important regions in a top-down procedure.

Image Modeling

In this step the projected partition is coded in order to obtain information of the regions which cannot be correctly represented by the coding process. Ideally, the real coding algorithm should be used to code the partition and the texture. However, most of the time, a simplified version of the coding can be used. Usually only a texture coding technique is used (the contour coding process is assumed to be lossless). In practice, each region is filled with a gray-level function which approximates the coded version of the region the receiver will have. The modeling residue is computed as the difference between this coded image and the original one.

Image simplification

This *Image simplification* step controls the nature and amount of information that is kept for segmentation at this level of the hierarchy. Morphological tools are very useful in this context, because they allow us to simplify images with visually important criteria such as size, contrast or dynamic. In the implementation presented here, size and contrast criteria are used, because most of the image structure is represented by large regions but an important part of the meaningful information is defined by small but contrasted details. The operators used for size simplification are morphological filters by reconstruction or area filters. For contrast we use h-maxima and h-minima operators. The interesting property of these operators is that they simplify the signal by removing small (size simplification) or poorly contrasted (contrast simplification) regions without corrupting the contour information. Moreover, they produce flat zones on the interior of the regions, which is very useful in the *Marker extraction* step. Let us briefly describe these operators.

Let us denote f an input signal and $\delta_n(f)$ its dilation with a structuring element of size n. The geodesic dilation of size one of f with respect to a reference function r is defined as [22]:

$$\text{Geodesic dilation of size 1:} \quad \delta^1(f,r) = Min\{\delta_1(f), r\} \tag{3.9}$$

By iteration of the geodesic dilation, the reconstruction by dilation $\gamma^{rec}(f,r)$ is defined:

$$\gamma^{rec}(f,r) = \delta^{\infty}(f,r) = ...\delta^{1}(...\delta^{1}(f,r)...,r) \qquad (3.10)$$

By duality, the geodesic erosion and the reconstruction by erosion can be defined. The interesting filters for size simplification are the opening by reconstruction of erosion $\gamma^{rec}(\epsilon_n(f),f)$ and its dual, the closing $\varphi^{rec}(\delta_n(f),f)$[41]. These filters have a size-oriented simplification effect on the signal but preserve the contour information. However, the size of the simplified components in this kind of filters depends on the shape of the structuring element which has been used. An improvement consists of using the area filters [48], which have the same properties of the opening and closing by reconstruction, but do not depend on the structuring element shape. The area opening of size n can be defined as the maximum of the morphological openings with any connected structuring element of size larger or equal to n. By duality, area closings are defined. It has to be mentioned that reconstruction processes and area filters can be implemented very efficiently by using queues which avoid any iterating process and lead to extremely fast algorithms.

For contrast-oriented simplification, the h-maxima and h-minima operators are used [41]. They can also be defined in terms of reconstruction. If h is a constant,

$$h - max(f) = \gamma^{rec}(f-h,f) \qquad (3.11)$$

$$h - min(f) = \varphi^{rec}(f+h,f) \qquad (3.12)$$

All these operators belong to the class *connected operators* [40], which leads to the interesting properties previously mentioned: fundamentally, they interact with the signal producing and merging flat zones.

Marker extraction

The goal of this step is to produce markers identifying the interior of the regions that will be segmented. For size-oriented segmentation, the marker extraction consists in labeling the interior of large flat zones after the simplification [41]. This can be easily done thanks to the simplification process which has produced flat zones. For contrast-oriented segmentation, the marker extraction relies on the labeling of the extremal flat zones produced by the h-maxima/minima operators [41].

Decision

After *Marker extraction*, a decision about the uncertainty areas which have not been assigned to any region must be taken. The classical morphological decision tool is the watershed [30]. It is generally used on the morphological gradient of the image to segment. However, the use of the morphological gradient results in a loss of information on the contour position of ± 1 pixel which is generally too high for coding applications. In the case of still image coding and depending on the application, one may consider that it is not a serious problem. However, for image sequences, the loss of information about the contour position depends on the region motion and can become very large. Therefore, the use of the gradient should be avoided.

To solve this problem, a different version of the watershed algorithm working on the original signal has been used. This algorithm results in a region growing process: the set of markers is extended until they occupy all the available space. During the extension, pixels of the uncertainty areas are assigned to a given marker. A point is assigned to a specific region because it is in the neighborhood of the marker of this region and it is more similar (in the sense given by a specific criterion) to the area defined by this marker than to any other area corresponding to another marker of its neighborhood.

A possible similarity criterion is the gray tone difference between the pixel under consideration and the mean of the pixels that have already been assigned to the region. However, in order to take into account the complexity of the contours, this basic similarity measure has been modified. For this aim, the similarity criterion is defined as the weighted sum of the gray-level difference between the pixel and the mean of the region plus a penalty term corresponding to the contour complexity [33]:

$$\text{Similarity } = \alpha \text{ Difference in gray-tone } + (1-\alpha) \text{ Contour complexity} \quad (3.13)$$

The measure of contour complexity is made by counting the number of contour points that are added if the pixel is assigned to that region. The weighting factor α allows for giving more importance to the gray level measure or to the contour complexity. The decision produces the new partition k which can be used as input to the following segmentation level.

Typically four hierarchical levels of segmentation are used for the intra-frame processing. The first three deal with a size criterion and the last with a contrast criterion. At each level, the same procedure is repeated and the only difference is the simplification parameter which is decreased to allow the progressive introduction of small or low-contrasted regions. In this hierarchical procedure the main advantages of this structure can be seen: it produces a good segmentation of the image taking into account the possibilities of the texture coding (it segments the modeling residue). It allows the progressive estimation of the segmentation parameters, size and contrast, to get a segmentation result compatible with the coding objective (appropriate number of regions or of contour points, etc.).

Results

The intra-frame segmentation is illustrated in Figure 12. For each level the residue, the simplified residue, the partition and the modeled partition are shown. The first, third and forth hierarchical levels are shown respectively in columns first, second and third. The first two columns correspond to a size simplification criterion and the last one to a contrast simplification. The residue is modeled using forth order orthogonal functions.

This example illustrates how the information is progressively extracted from the original frame and introduced in the segmentation. Note in particular the evolution of the modeling error which becomes simpler at each level. Ideally, this error should tend towards zero. Comparison between the modeling error and its simplification illustrates the usefulness of using connected operators such as filters by reconstruction. For example, in the first level, the simplified image is really a coarse approximation of the original. Only large objects are present. However, the contours of the remaining objects are well defined. This simplified image is an 'easy' image to segment. The simplified image involves a large number of flat zones which are very useful for marker extraction. The marker extraction step assigns a label to these flat zones. The number of segmented regions is respectively 10, 44, and 62 for each level.

5.2 Inter-frame mode

In the inter-frame mode a recursion is performed to segment the whole sequence in a time-recursive way. That is, the partition of every image is computed taking as the starting point the partition of the previous image. For this aim, this previous partition is projected and then the new regions corresponding to the current frame are extracted. In this implementation, the clue of this procedure is on the projection step. The extraction of new regions is performed as in the intra-frame mode.

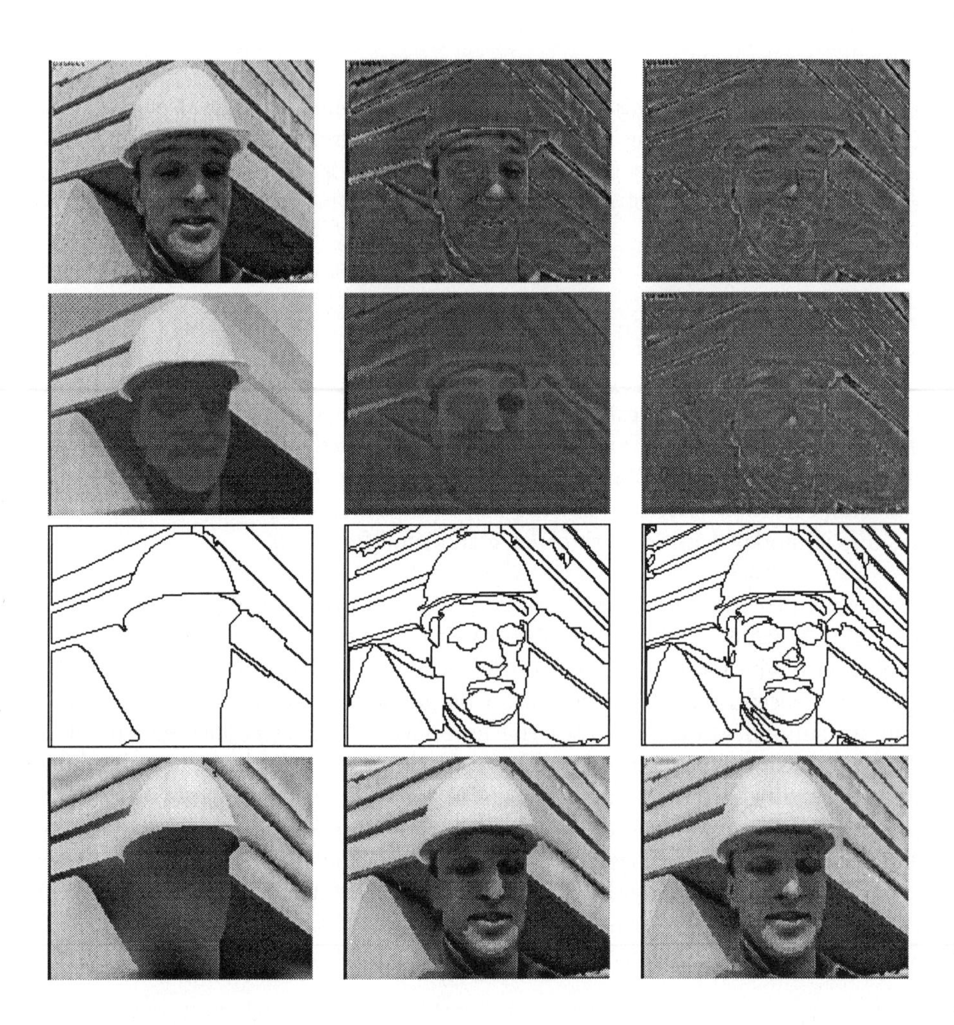

Figure 12 Example of intra-frame segmentation

Partition projection in a time-recursion

As it is shown in the sequel, this projection can be performed with the watershed algorithm applied on the appropriate 3D images. However, to avoid disconnection of objects in successive frames due to large motion, and also to produce more continuous contours, motion estimation is estimated before segmentation. The whole process is illustrated in Figure 13

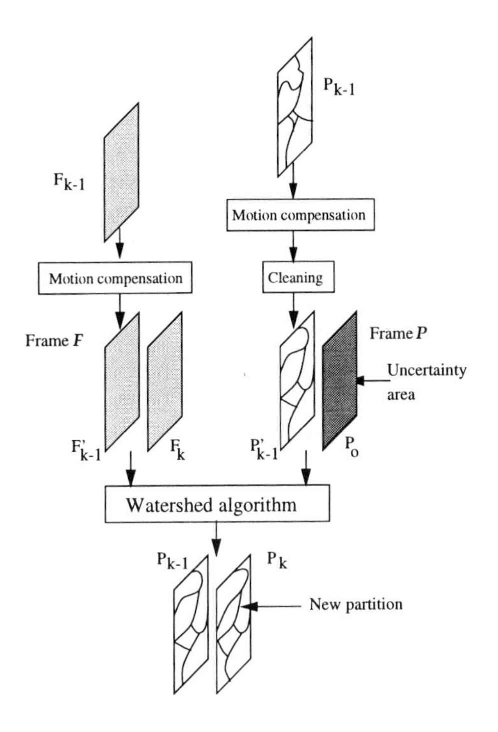

Figure 13 Projection of markers of frame $k-1$ into frame k

The motion is estimated between the frame which has just been segmented and the next frame which is going to be segmented. Let us call these frames F_{k-1} and F_k, respectively. The procedure which has been used is a block-matching algorithm applied backwards. This algorithm involves the division of F_k into small square blocks. For each block of pixels, a search is conducted within a confined window in F_{k-1}, to locate a best matching block [17]. Even though this algorithm assumes uniform motion in every block of pixels, which is not strictly true specially on the blocks involving more than one object, the approximation is good enough for our purposes. Moreover, small blocks are used in order to have a more accurate estimation (8x8 pixels for QCIF format). Once the motion vectors have been estimated for all the blocks of the image, the first image F_{k-1} and its partition P_{k-1} are motion compensated. That is, the motion vectors found for the original frame are also applied to the partition. Let us call the compensated images F'_{f-1} and P'_{k-1}. It is important to remark that, using a backward motion estimation, every pixel of the current image will be covered by one, and only one, pixel from the previous image. However, as the motion estimation has been produced in a block basis, the compensation of the regions can produce disconnected regions in the compensated image P'_{k-1}. This effect can be seen particularly as small disconnected components close to the contours of the regions. As the compensated segmentation is going to be used for the extension of the labels into the new image, small disconnected parts could produce wrong extensions of these labels. Thus, small disconnected objects have to be eliminated. This is simply implemented by keeping from every region in P_{k-1} only its largest connected component in P'_{k-1}. Of course, this produces non-labeled pixels in the compensated partition P'_{t-1}, but, as this image is only used as initialization for the extension of the regions in the new image, it does not involve any problem. Once the previous partition has been motion compensated, its projection on the current image can be performed.

Assuming that the partition of the image $k-1$, P_{k-1}, is known, our goal is to find the segmentation, P_k, at image k without introducing new regions. For this purpose, two 3D signals are constructed using the compensated images generated before. Frames F'_{k-1} and F_k are grouped together to form a temporal block $\mathcal{F}$ of size 2 in the time dimension. Similarly, the frame P'_{k-1} is grouped with an empty frame P_o representing an entire frame of uncertainty. The resulting 3D signal denoted $\mathcal{P}$ is considered as the set of markers that should be used to segment the signal $\mathcal{F}$. The marker extension itself is achieved by the same decision algorithm used in the intra-frame segmentation, that is the watershed. The only difference is the nature of the signals that are now 3D signals. The watershed extends the markers defined by P'_{k-1} into the empty frame P_o. Each pixel of the uncertainty area (that is of image P_o) is assigned to a region of image P'_{k-1} based on a similarity criterion.

As in the intra-frame mode, the similarity criterion combines a gray tone distance with a contour complexity measure. The contour complexity is assessed by counting the number of contour points that are added if the pixel is assigned to a particular region. A 6 connected-neighborhood is assumed for this purpose, 4 pixels in the spatial dimension and 2 for the temporal. The similarity in the time dimension plays an important role with respect to the temporal stability of the contours.

Extraction of new regions

Once the labels of the previously segmented regions have been propagated into the current frame, new regions have to be extracted. From now on, the process is purely intra-frame. First of all, the residue image is computed. As for the intra-frame segmentation discussed previously, the residue image is obtained by texture coding of the partition obtained from the extension of the past regions. In practice, a simplified version of the texture coding is used. Then, the difference with the original image gives the residue, where new regions can be detected. The segmentation of new regions is performed using the same steps as in the intra-frame segmentation: *Simplification*, *Marker extraction* and *Decision*. The segmentation can be performed according to a size or a contrast criterion. However, only contrast segmentation is used since, most of the time, large and non-contrasted regions of the residue are not visually important. In the inter-frame mode the segmentation parameters are only updated to maintain the partition characteristics; that is, the number of regions and contour points.

Results

The inter-frame mode of segmentation is illustrated in Figure 14. The first row presents the original images where the inter-frame segmentation is applied. In the second row the partition of the first frame and its projection on the current frame are shown. The last image in this row is the modeling of the projected partition. Finally, the last row gives the modeling error, the final partition and its modeling. It can be observed that the projection step correctly follows the time evolution of the regions, and only a few new regions are introduced, improving some details (e.g.: the nose).

6 CONCLUSIONS

Among the so-called second generation coding techniques for video signals, those relying on a previous segmentation of the video sequences open very attractive possibilities. Segmentation helps to code in a more efficiently manner video signals since it divides images into a set of homogeneous regions. Different homogeneity criteria can be used to drive the segmentation procedure. Segmentation techniques using spatial homogeneity criteria are of paramount importance since, regardless of the final segmentation-based coding approach, there is always a need for such kind of approaches (e.g.: the case of intra-frame coding). In addition, they give support to new contain-based functionalities.

Several problems arise when computing a coding-oriented segmentation of a video sequence using spatial information. These problems lead to a set of requirements that a coding-oriented segmentation should fulfill: it should provide a coherent intra-frame segmentation mode, it should achieve partition sequences with temporal label coherence, it should solve the lack of connectivity between

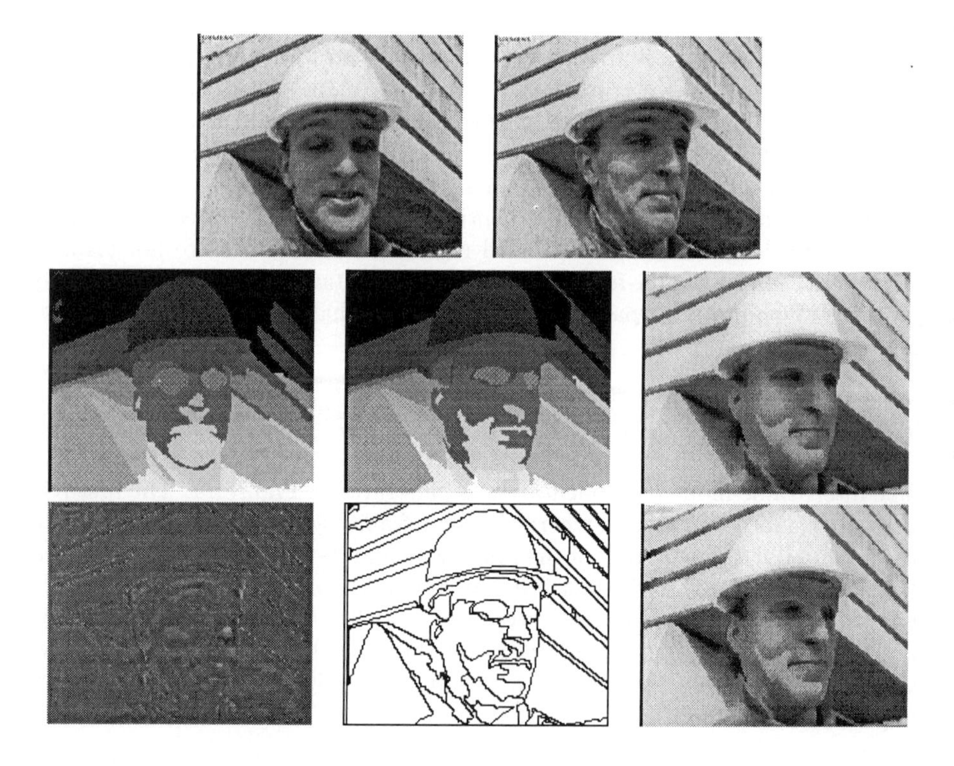

Figure 14 Example of inter-frame segmentation

areas related to the same object in different frames, it should yield partitions as suitable as possible for coding applications, it should allow a bit rate regulation and it should introduce a time delay as short as possible.

A general scheme of video-segmentation for coding applications is proposed that tries to fulfill the above requirements. It relies on a basic structure that is divided into five different steps: *Image projection*, *Image modeling*, *Image simplification*, *Marker extraction* and *Decision*. This basic structure is used for both the intra and the inter-frame mode. In the intra-mode, it leads to a hierarchical procedure whereas in the inter-mode it translates into a time recursion.

To evaluate this general scheme, two different implementations are presented. The first one utilizes a compound random field as an image sequence model, while the second makes use of morphological tools. In both implementations, the above requirements are fulfilled and the segmentation results are very suitable for coding-oriented applications.

Acknowledgements

Part of this work has been supported by the Morpheco project of the European Race Program and by the project of the Spanish goverment CICYT TIC 92-1319-C03-01-PB.

The authors would like to thank Montse Cortit for her comments and assistance in obtaining the results.

REFERENCES

[1] J. K. Aggarwal and W. N. Martin. Analizing dynamic scenes containing multiple moving objects. In T. S. Huang, editor, *Image sequence analysis*, pages 355–380. New York, 1981.

[2] F. Bartolini, V. Capellini, A. Mecocci, and R. Vagheggi. A segmentation-based motion-compensated scheme for low-rate video coding. In *Proceedings of the First IEEE International Conference on Image Processing*, volume II, pages 457–461, Texas, U.S.A., November 1994.

[3] J. Benois, L. Wu, and D. Barba. Joint contour-based and motion-based image sequences segmentation for TV image coding at low bit rate. In *Visual Communication and Image Processing*, pages 1074–1085, Chicago, USA, September 1994.

[4] M. Bertero, T. A. Poggio, and V. Torre. Ill-posed problems in early vision. *Proceedings of the IEEE*, 76:869–887, 1988.

[5] J. Besag. Spatial interaction and the statistical analysis of lattice systems (with discussion). *J. Royal Statisc. Soc., series B*, 34:75–83, 1972.

[6] S. Beucher and F. Meyer. The morphological approach to segmentation: the watershed transformation. In E. Dougherty, editor, *Mathematical morphology in image processing*, chapter 12, pages 433–481. Marcel Dekker, 1993.

[7] P. Bouthemy and E. François. Motion segmentation and qualitative dynamic scene analysis from an image sequence. *International Journal of Computer Vision*, 10(2):157–182, 1993.

[8] P. J. Burt and E. Adelson. The Laplacian pyramid as a compact image code. *IEEE Transactions on Communications*, 31:532–540, 1983.

[9] R. Chellappa and A. Jain. *Markov random fields. Theory and application.* Academic Press, Inc., 1993.

[10] D. Cortez, P. Nunes, M. M. de Sequeira, and F. Pereira. Image segmentation towards new image representation methods. *EURASIP Image Communication*, 6(6):485–498, February 1995.

[11] J. Crespo. *Morphological connected filters and intra-region smoothing for image segmentation.* PhD thesis, School of Electrical Engineering, Georgia Institute of Technology, 1993.

[12] H. Derin and P. Kelly. Discrete-index Markov-type random processes. *Proceedings IEEE*, 77:1485–1510, 1989.

[13] S. Geman and D. Geman. Stochastic relaxation, Gibbs distributions, and Bayesian restoration of images. *IEEE Transactions on Pattern Analysis and Machine Intelligence*, 6(6):721–741, November 1984.

[14] C. Gu and M. Kunt. 3D contour image sequence coding based on morphological filters and motion compensation. In *International Workshop on Coding Techniques for Very Low Bit-rate Video*, Colchester, U.K., April 1994.

[15] C. Gu and M. Kunt. Very low bit-rate video coding using multi-criterion segmentation. In *First IEEE International Conference on Image Processing*, volume II, pages 418–422, Texas, U.S.A., November 1994.

[16] R. M. Haralick and L.G. Shapiro. Image segmentation techniques. *Computer Vision, Graphics and Image Processing*, 29:100–132, January 1985.

[17] R. Jain and A.K. Jain. Displacement measurement and its application in interframe coding. *IEEE Transactions on Communications*, 29(12):1799–1808, 1981.

[18] F. C. Jeng and J. W. Woods. Compound GaussMarkov random fields for image estimation. *IEEE Trans. on Signal Processing*, 39:683–697, 1991.

[19] ISO/IEC JTC1/SC29/WG11. MPEG-4 Proposal Package Description (PPD). July 1995.

[20] M. Kocher and M. Kunt. A contour-texture approach to picture coding. In *Proc. of the 1982 IEEE Int. Conf. on Acoustics, Speech and Signal Processing*, pages 436–440, Paris, France, May 1982.

[21] M. Kunt, A. Ikonomopoulos, and M. Kocher. Second generation image coding techniques. *Proceedings of the IEEE*, 73(4):549–575, April 1985.

[22] C. Lantuejoul and F. Maisonneuve. Geodesic methods in image analysis. *Pattern Recognition*, 17(2):117–187, 1984.

[23] W. Li and M. Kunt. Morphological segmentation applied to displaced frame difference coding. *EURASIP Signal Processing*, 38(1):45–56, September 1994.

[24] B. Marcotegui, J. Crespo, and F. Meyer. Morphological segmentation using texture and coding cost. In *IEEE Workshop on Nonlinera Signal and Image Processing*, pages 246–249, Halkidiki, Greece, June 1995.

[25] B. Marcotegui and F. Meyer. Morphological segmentation of image sequences. In J. Serra and P. Soille, editors, *Mathematical morphology and its applications to image processing*, pages 101–108. Kluwer Academic Publishers, 1994.

[26] F. Marqués, J. Cunillera, and A. Gasull. Unsupervised segmentation controlled by morphological contrast extraction. In *Proc. IEEE Int. Conf. Acoust., Speech and Signal Processing*, pages 5.17–5.20, May 1993.

[27] F. Marqués, A. Gasull, T. Reed, and M. Kunt. Coding-oriented segmentation based on Gibbs-Markov random fields and human visual system knowledge. In *Proc. IEEE Int. Conf. Acoust., Speech and Signal Processing*, pages 2749–2752, May 1991.

[28] F. Marqués, V. Vera, and A. Gasull. A hierarchical image sequence model for segmentation: Application to object-based sequence coding. In *Proc. SPIE Visual Communication and Signal Processing-94 Conference*, pages 554–563, Oct 1994.

[29] R. Mester and U. Franke. Statistical model based image segmentation using region growing, edge relaxation and classification. In T. R. Hsing, editor, *Visual Communications and Image Processing '88*, pages 616–624, Cambridge, Massachuset, November 9-11 1988.

[30] F. Meyer and S. Beucher. Morphological segmentation. *Journal of Visual Communication and Image Representation*, 1(1):21–46, September 1990.

[31] H.G. Musmann, M. Hotter, and J. Ostermann. Object-oriented analysis-synthesis coding of moving images. *Signal Processing, Image Communications*, 1(2):117–138, October 1989.

[32] R. Nevatia. Image segmentation. In T. Y. Young and K. S. Fu, editors, *Handbook of Pattern Recognition and Image Processing*, chapter 9, pages 215–231. Academic Press, New York, 1986.

[33] M. Pardàs and P. Salembier. 3D morphological segmentation and motion estimation for image sequences. *EURASIP Signal Processing*, 38(1):31–43, September 1994.

[34] M. Pardàs and P. Salembier. Joint region and motion estimation with morphological tools. In J. Serra and P. Soille, editors, *Second Workshop on Mathematical Morphology and its Applications to Signal Processing*, Fontainebleau, France, September 1994. Kluwer Academic Press.

[35] M. Pardàs and P. Salembier. Time-recursive segmentation of image sequences. In EURASIP, editor, *EUSIPCO 94, VII European Signal Processing Conference*, pages 18–21, Edinburgh, U.K., September 13-16 1994.

[36] T. Pavlidis. *Structural Pattern Recognition*, chapter 4. Springer-Verlag, New York, 1977.

[37] S. Rajala, M. Civanlar, and W. Lee. Video data compression using three-dimensional segmentation based on HVS properties. In *International Conference on Acoustics, Speech and Signal Processing*, pages 1092–1095, New York (NY), USA, 1988.

[38] P. Salembier. Morphological multiscale segmentation for image coding. *EURASIP Signal Processing*, 38(3):359–386, September 1994.

[39] P. Salembier and M. Pardàs. Hierarchical morphological segmentation for image sequence coding. *IEEE Transactions on Image Processing*, 3(5):639–651, Sept. 1994.

[40] P. Salembier and J. Serra. Flat zones filtering, connected operators and filters by reconstruction. *IEEE Transactions on Image Processing*, 3(8):1153–1159, August 1995.

[41] P. Salembier, L. Torres, F. Meyer, and C. Gu. Region-based video coding using mathematical morphology. *Proceedings of IEEE (invited paper)*, 83(6):843–857, June 1995.

[42] H. Sanson. Joint estimation and segmentation of motion video coding at very low bitrates. In *Proc. COST 211ter European Workshop on New Techniques for Coding of Video Signals at Very Low Bitrates*, pages 2.2.1–2.2.8, Dec 1993.

[43] J. Serra. *Image Analysis and Mathematical Morphology.* Academic Press, 1982.

[44] J. Serra. *Image Analysis and Mathematical Morphology, Vol II: Theoretical advances.* Academic Press, 1988.

[45] I. Sezan and R. L. Lagendijk. *Motion Analysis and Image Sequence Processing.* Kluwer Academic Publishers, Boston, 1993.

[46] M. Soryani and R. J. Clarke. Coding colour image sequences by segmentation of difference frames and motion-adaptive frame interpolation. In *Picture Coding Symposium*, page 2.7, Cambridge, MA, March 1990.

[47] G. Tziritas and C. Labit. *Motion Analysis for Image Sequence Coding.* Elsevier Science B. V., The Netherlands, 1994.

[48] L. Vincent. Grayscale area openings and closings, their efficient implementation and applications. In J. Serra and P. Salembier, editors, *First Workshop on Mathematical Morphology and its Applications to Signal Processing*, pages 22–27, Barcelona, Spain, May 1993. UPC.

[49] P. Willemin, T. Reed, and M. Kunt. Image sequence coding by split and merge. *IEEE Transactions on Communications*, 39(12):1845–1855, 1991.

[50] R. Wilson and M. Spann. *Image segmentation and uncertainty.* Research Studies Press Ltd., Letchmore, Hertfordshire, England, 1988.

4

CODING OF PARTITION SEQUENCES

Philippe Salembier, Ferran Marqués and Antoni Gasull

Department of Signal Theory and Communications,
Universitat Politècnica de Catalunya,
Barcelona, Spain

ABSTRACT

This chapter deals the coding of the partition information resulting from a segmentation of video sequences. Both intra-frame and inter-frame coding modes are discussed. For intra-frame mode, lossless and lossy coding techniques are presented. Motion compensation of partition sequences is described as an efficient inter-frame mode of coding. It involves the prediction of the partition, the computation of the partition compensation error, the simplification of the error and its transmission. The major issues and processing steps of a general motion compensation loop for partitions are presented and discussed.

1 INTRODUCTION

Segmentation-based coding of video sequences is rapidly becoming a very active field of research [36, 25, 31, 22]. Its basic idea is to define a partition of the image sequence that is more appropriate for coding than the block partition used in more classical coding schemes. The coding algorithms involve at least three steps: 1) the definition of the partition, 2) the coding and transmission of the partition and 3) the coding and transmission of the pixel values inside each region defined by the partition. Several segmentation criteria can be used to define the partition. Let us mention in particular the homogeneity in gray level value [36, 31], the classification in static or moving region [25], the homogeneity in motion [2], etc. Chapters 3 and 6 discuss the issues of video segmentation and describe possible coding-oriented techniques. In all cases, the resulting partition depends on the input video sequence and has to be transmitted to the receiver. This transmission is the objective of the so-called *partition coding* sometimes referred as *contour coding.*

As in any video coding scheme, two modes of transmission can be distinguished: *intra-frame* and *inter-frame.* The intra-frame mode consists of sending the information about the partition of each frame independently. This mode has to be used for example for still images, for the first frame of a sequence or even periodically in order to totally refresh the information in the receiver. By contrast, the inter-frame mode relies on the characterization and coding of the time evolution of the partition from one frame to the next one. For video sequences, this second mode of transmission is generally much more efficient than the intra-frame one, especially if the segmentation has been able, not only to define the partition of each frame, but also to solve the region correspondence problem (see Chapter 3).

The organization of this chapter is as follows: the following Section will precisely define the partition coding problem. Section 3 focuses on techniques to code the partition in intra-frame mode. Section 4 is devoted to the inter-frame mode. Finally, Section 5 gives the conclusions.

2 PROBLEM STATEMENT

In this chapter, we consider that the video sequence has been segmented and that the resulting partition sequence has to be transmitted to the receiver. The partition sequence is a sequence defining the shape of the regions and their time evolution. The most simple way to define a partition sequence is to use the same representation as the original sequence, that is to assign to each pixel of the sequence a specific number, called a *label,* defining the region it belongs to. The labels should be coherent in time if the temporal redundancy has to be taken into account, that is if an efficient inter-frame mode has to be defined. Therefore, following the terminology used in Chapter 3, we will assume that the segmentation is temporally coherent.

The information contained in a partition sequence is composed of the shape of each region, their position within each frame and their label (for the inter-frame mode). Note that the problem is therefore more complex than the pure *contour coding* problem which deals only with coding of the shape of the regions. In the sequel, the term *partition coding* will refer to the coding of the shape, position and labels. Note also that partition coding is a different problem from the coding of binary images (or sequences) because, in the case of partitions, the notion of foreground objects and background areas have no meaning. In particular, some object-based schemes use binary masks to define the coding strategy (for example to separate the moving area with respect to a static background [25]). The coding of this kind of binary mask is not the purpose of this chapter, although some shape and position coding techniques presented in the following sections may be used.

3 INTRA-FRAME CODING

Intra-frame coding of partitions can be viewed as a simplification of the general problem presented in the previous Section. Indeed, only the shape and the position of the regions have to be coded and the process cannot use any information previously sent to the receiver. It is not necessary to code the label information because the same labeling techniques can be used on the transmitter and receiver sides and because, in intra-frame mode, we do not want to create a temporal coherence with respect to any previously sent partition.

As any coding scheme, the partition coding can be lossless or lossy. However, in practice, it turns out that the human visual system is quite sensitive to errors resulting from the coding process (shape and position in particular). So, in the case of lossy techniques, the amount of error should be strictly controlled and bounded.

Figure 1 describes a classification of the most popular intra-frame partition coding techniques. The organization of this Section follows this classification. In the case of lossless techniques, two main approaches can be used: *contour-oriented* leading to *Chain Code* techniques and *shape-oriented* leading to *Skeleton* and *Quadtree* decompositions. Lossy techniques can be derived from simplification or extension of a lossless coding approach. In this class, the so-called *Multi-Grid Chain Code* can be found. Finally, the last set of lossy techniques relies on curve approximation or modeling. In practice, the approximation can be done either globally to approximate the whole shape of a region (*Fourier Descriptors*) or locally (*Polygon, Spline* approximations). Let us start by describing lossless techniques.

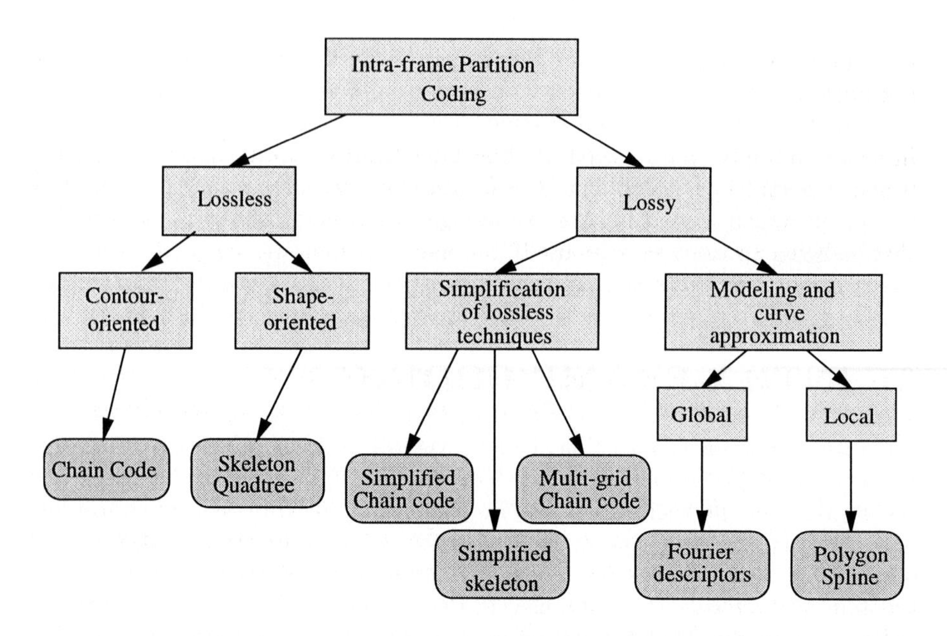

Figure 1 Classification of intra-frame coding techniques

3.1 Lossless partition coding

As mentioned previously, partition coding techniques can be classified regarding the approach they follow to code the regions shape. Two classical approaches can be used: *contour-oriented* or *shape-oriented.*

Contour-oriented approach: Chain code

Contour-oriented approaches represent regions by their boundaries and the coding procedure consists on tracking and encoding the boundaries. Among them, *Chain Code* (CC) techniques [11] allow lossless coding of boundaries. This technique exploits the fact that two consecutive boundary points in a discrete grid are neighbors. Therefore, in order to track a boundary, only a few movements are possible. A boundary is coded by chaining the movements needed to completely describe it. In the case of a 4-connected boundary, only 4 different movements are possible. The amount of possible movements decreases if the so-called *Derivative Chain Code* (DCC) is used [11]. Derivative chain code describes the possible movements with respect to the last one. In a 4-connected boundary, only 3 different movements are allowed: turn right, turn left and straight ahead. The fourth movement, going back to the previous position, is not possible.

In order to apply chain code techniques, partitions have to be represented in terms of boundary information. A common approach is to define the boundary of a given region as those pixels of the region that have, at least, one neighbor that belongs to another region. If the neighborhood system is 8-connected, the resulting boundary is 4-connected, and vice-versa. However, this method does not ensure a one-to-one relationship between partitions and their boundary representations. In particular, different partitions containing regions with one-pixel width elongations may lead to the same boundary representation. Therefore, they cannot be considered as being lossless coding techniques.

In order to solve this problem, the work presented in [10] does not allow the regions to be coded to have one-pixel width elongations. Contour segments fulfilling this constraint are coded using a derivative chain code. Following this approach, a lower bound of 1.27 bits per movement is derived. It is also shown that, by grouping the movements in triplets and applying Huffman coding, better results can be obtained. However, these figures do not represent the total cost of coding a partition. Indeed, the work in [10] only deals with the coding of contour segments (that is, shape information) and it does not cope with the position information.

A complete partition coding scheme is presented in [18]. In this work, the position information is coded by addressing the initial pixel of each contour segment. The first contour segment is addressed by the coordinates of its initial pixel. The remaining contour segments are addressed in a relative way with respect to the first pixel of the previous coded contour segment. Similar techniques are also used in other contexts [6]. The mean cost reported for each initial point is of 9 bits (images of 256x256 are assumed). However, this figure strongly depends on the density of initial points in the partition. In [18], the shape of the regions is represented using an 8-connected boundary. Such boundaries are coded by a DCC and pairs of consecutive movements are grouped. This technique cannot strictly be said to be lossless since, previous to the encoding process, partitions are filtered to remove all one-pixel width elongations.

In order to allow the coding of arbitrary partitions, a correct boundary representation should be used. An hexagonal contour grid [21] allows a one-to-one relationship between partitions and their boundary representation. This contour grid is related to the partition grid in the following way: a contour grid site separates each pair of neighbor sites in the partition grid, assuming a

4-connected neighborhood system for the partition grid. This concept is illustrated in the first example of Figure 2, where square and line segment elements represent sites from the partition and contour grids, respectively. A contour grid site is active if its two closest partition grid sites have different labels. An example of the relationship between both grids is shown in the second example of Figure 2, where both partitions are represented together in a single matrix.

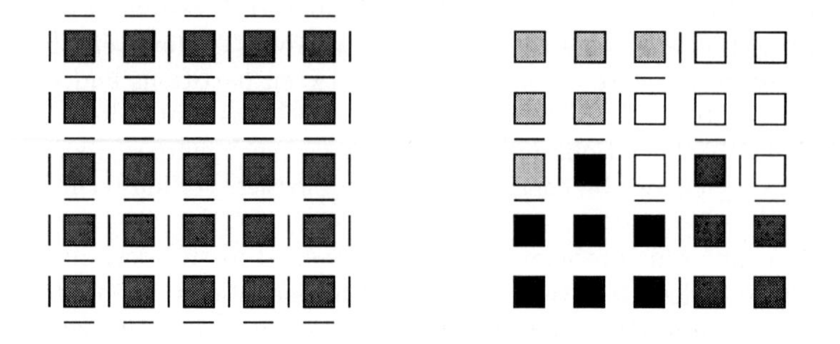

Figure 2 Relationship between partition and contour grid sites

In this contour grid, the neighborhood system is 6-connected and, therefore, there are 6 basic movements. Nevertheless, when using a derivative chain code, the amount of possible movements reduces to 3 as in the 4-connected case: turn right, turn left and straight ahead (r, l, s). Figure 3 compares the possible movements that may appear in both grids.

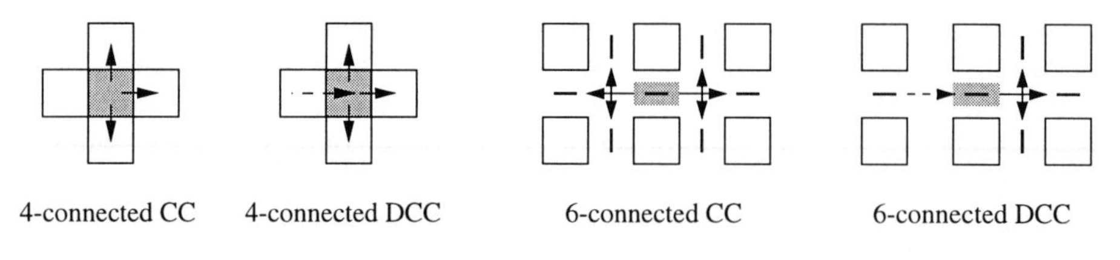

Figure 3 Possible movements in 4- and 6-connected grids using CC and DCC

Based on this hexagonal contour grid, a partition coding technique is presented in [21] which makes no assumption about the shape of the regions. Shape and position information are jointly coded by introducing the location information in the chain code itself. This method relies on the concept of triple point. A triple point in the contour grid is a site that has at least three active neighbor sites. Therefore, triple points are found at the intersection of two different contours and can be used to locate the initial point of a new contour segment. Triple points can be efficiently coded by introducing a new symbol, say t, in the chain itself. Moreover, not all triple points are useful to describe the whole set of contours. Indeed, using classical contour following techniques, one can see that only a subset of the triple points are actually necessary. These triple points are called *active* triple points and only active triple points are marked in the chain by the symbol t.

However, not all the initial points of contour segments can be addressed using triple points. Indeed, regions (or a cluster of regions) that have no contact point with other regions do not produce any triple point. Therefore, interior regions (or clusters of interior regions) have to be directly addressed by the coordinates of their initial points. Nevertheless, such configurations appear quite seldom (in average, 5% of the total number of regions).

Once the whole partition has been coded, symbols in the chain should be entropy coded. One possibility is to group them by triplets and use a Huffman coder. This technique yields, as average results, 1.36 bits per boundary element, including both shape and position information. The cost of the position information is of 7 bits per initial point in average.

Following the same philosophy, a partition coding technique is presented in [23]. It initially transforms the image partition into a hexagonal grid. With this technique, there are only two possible movements (turn right and turn left) and, therefore, because of triple points, there are three basic symbols in the chain (r, l, t). However, several additional symbols are necessary to mark situations such as ending a contour segment or reaching the image border. For lossless coding, the coding performances are comparable to that achieved in [21].

Shape-oriented approach: Morphological skeleton and Quadtree

Shape-oriented approaches decompose regions into simple forms and the coding procedure consists in finding and encoding this set of simple forms. The coding of each simple form implies encoding its type, its center and its size. Techniques based on the Morphological Skeleton ($\mathcal{MS}$) [33] allow easy implementations of the decomposition and reconstruction procedures.

Morphological skeletons are *a priori* defined for binary images. They rely on the notions of erosion ε and dilation δ by a structuring element B [33]:

$$\delta_B(X) = \bigcup_{b\in B, x\in X} \{x+b\} \quad \text{and} \quad \varepsilon_B(X) = \bigcap_{b\in B} \{X-b\} \qquad (4.1)$$

In practice, to compute a skeleton, one has to chose an elementary structuring element of size one, for example a square of 3*3 pixels or a cross of 5 pixels. This structuring element allows the definition of the erosion ε and dilation δ of size one. Erosions and dilations of size i, noted by ε^i and δ^i are obtained by iteration. The morphological skeleton of a binary set of points X defining a region is given by:

$$\mathcal{MS}(X) = \bigcup_i \varepsilon^i / \delta\varepsilon^{i+1}(X) = \bigcup_i \mathcal{MS}^i(X) \tag{4.2}$$

The operator / represents the residue between the two input sets:

$$\psi(X)/\theta(X) = \psi(X) \cap [\theta(X)]^c. \tag{4.3}$$

Points resulting from the residue computation compose the morphological skeleton and represent the centers of all *maximal balls* inscribed in the region. A ball is maximal if there is no other larger ball that could be inscribed in the region and totally contains the first ball. The shape of the balls is given by the structuring element. On turn, the function assigning to each skeleton points the value of its skeleton level $\mathcal{MS}^i$ is called the *quench function*. The set X can be restored from the skeleton and the quench function by dilating each skeleton point by a structuring element whose size should be the value of the quench function at this skeleton point [17]:

$$X = \bigcup_{i \geq 0} \delta^i(\mathcal{MS}^i(X)) \tag{4.4}$$

As can be seen, it is a shape-oriented coding approach because the type of form used in the decomposition is characterized by the structuring element, the information about the centers of maximal balls is stored in the skeleton, and the size of each maximal ball is contained in the quench function.

Morphological skeleton was first used for binary coding purposes in [19] where it is shown that the morphological skeleton is a redundant representation. To solve this problem, an algorithm obtaining a non-redundant skeleton is proposed. A binary image containing the points from the non-redundant skeleton is coded as a spare matrix using Elias coding and the values of the quench function are coded using a Huffman code. However, in [19], only binary image coding is addressed. The extension to partition coding is proposed in [29], where a binary image is created for each region belonging to a subset of regions allowing the reconstruction of the partition (in practice the subset corresponds to 80% of the total number of regions). The binary images are coded using the previous technique. With this approach, a large number of contour segments are coded twice since they belong to the boundary of two different regions. Therefore, this skeleton based approach leads to coding schemes with lower performance than contour-oriented approaches.

A way to overcome the drawback of coding twice common boundaries between two regions is by forcing the structuring elements used in the skeleton computation of a region not to take into account those contour segments that have been already coded. This concept leads to two different implementations: the *geodesic* and the *overlapping* skeleton.

The Geodesic Skeleton ($\mathcal{GS}$) [5] allows structuring elements to be deformable with respect to already coded regions. This is implemented using geodesic dilations δ_R and erosions ε_R which are defined, for a structuring element of size 1 (square or cross) and a reference set R as:

$$\delta_R(X) = \delta(x) \cap R \qquad \varepsilon_R(X) = \varepsilon(X \cup R^c) \cap R \tag{4.5}$$

Geodesic dilations and erosions of sizes larger than 1 are computed by iterating the respective operations of unitary size. Thus, the definition of the geodesic skeleton is:

$$\mathcal{GS}_R(X) = \bigcup_i \varepsilon_R^i / \delta_R \varepsilon_R^{i+1}(X) = \bigcup_i \mathcal{GS}_R^i(X) \tag{4.6}$$

The reference set R is made of the union of already coded regions.

The overlapping skeleton ($\mathcal{OS}$) [8, 20] allows structuring elements to cover already coded regions as well as areas outside the partition. Therefore, larger structuring elements can be used to represent a region and the number of skeleton points reduces. The overlapping skeleton of a set X_n is defined, with respect to the previous coded sets $\{X_{i=1\ldots n-1}\}$, as

$$\mathcal{OS}(X_n) = \mathcal{MS}(\bigcup_{i=0}^{n} X_i) \tag{4.7}$$

Main differences between the geodesic and overlapping skeletons are illustrated in Figure 4. It can be seen how, in the geodesic case, the structuring element adapts its shape to the previously coded regions. However, for the same size and position, the structuring element in the overlapping case covers more area.

In order to code the information obtained by the skeleton decomposition, both methods use extensions of the technique presented in [19] to remove redundancy. Although the overlapping skeleton improves the coding performance obtained by using the geodesic skeleton [20], none of them outperform contour-based approaches.

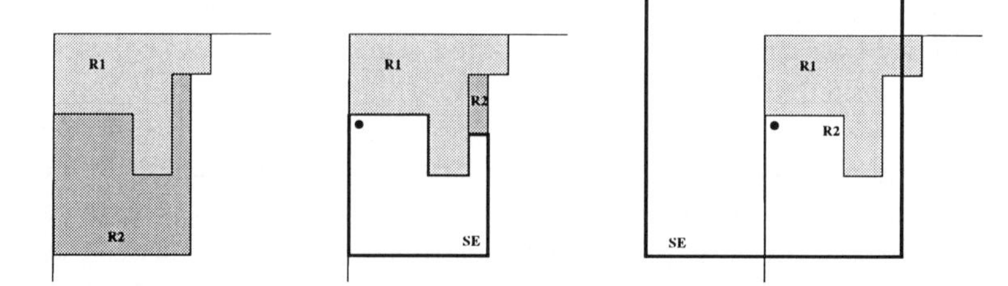

Figure 4 Comparison of structuring elements using $\mathcal{GS}$ and $\mathcal{OS}$

A way to improve skeleton-based coding techniques is by combining, in the same skeleton definition, different families of structuring elements [14]. This idea has been applied to both techniques: geodesic [4] and overlapping [8] skeleton, further improving previous results. New theoretical results concerning the morphological skeleton have recently been presented and applied to the case of binary image coding [15]. These results show that the main part of the quench function is not necessary for a perfect reconstruction of the region. In addition, the area of the image that may contain skeleton points can be tightly bounded and, therefore, the skeleton points may be more efficiently coded. These very interesting results have not been yet applied to the case of partition coding, but combined with the geodesic or overlapping approach, they may lead to comparable or even better performances than the contour-based approaches.

Finally, let us mention that a classical way to describe shapes is to use a *quad-tree* decomposition. Recently, it has been shown that this approach can be considered as a skeleton approach. The only difference with respect to the previous skeletons is the definition of the basic dilation. Assume that the basic dilation of a point (i, j) is composed of four points: $\{(2i, 2j), (2i+1, 2j), (2i, 2j+1), (2i+1, 2j+1)\}$. In [16], it is shown that the application of the skeleton formula with this basic dilation leads to the classical quadtree decomposition. Therefore, the problems and issues related to the use of the quadtree for coding are the same as the ones discussed for the skeleton.

3.2 Lossy partition coding

There are at least two different ways of developing a lossy partition coding technique: first, simplify a lossless technique so that a certain amount of loss is allowed or, second, derive a specific lossy partition coding technique. In the second case, the algorithm can be developed so that losses are introduced either locally or globally.

In any case, the amount of losses introduced by the lossy coding technique has to be strictly controlled. The human visual system being very sensitive to partition information, strong approximations or simplifications in the partition degrade rapidly the visual quality of the decoded image.

Simplification of lossless techniques

Some techniques presented in the previous section can be used within lossy partition coding methods. This is the case of the contour-oriented approach presented in [18] where partitions are filtered before coding. Another simplification method consists in subsampling the partition and, then, applying the contour-oriented coding approach on the subsampled partition. However, with this method, there is little control on the type of losses that are introduced in the partition coding.

A popular method to simplify contour-oriented approaches is to extend the chain code technique in such a way that larger movements might be coded with a single symbol [11]. In addition, this extended chain code sets the basis of the so-called *Multi-Grid Chain code* that is described in the following section.

Shape-oriented partition coding techniques can also be simplified by removing from the region description a given set of forms. This can be easily carried out when using a skeleton approach since the skeleton decomposition allows to reconstruct the original region as well as all its possible openings. The simplification is obtained removing some skeleton levels from the reconstruction procedure (4.4) [19]. If the first k levels are removed, the opening by a structuring element of size k is obtained:

$$\gamma^k(X) = \bigcup_{i \geq k+1} \delta^i(\mathcal{MS}^i(X)) \tag{4.8}$$

This way, less skeleton points have to be coded as the skeleton representation does not contain all the maximal balls. In the decoding step, some points of the decoded partition are not initially assigned to any region. In order to obtain a complete partition, a hole-filling technique has to be applied.

Multi-Grid Chain code

Lossy partition coding techniques can also be derived using the hexagonal contour grid that allows lossless coding for contour-oriented approaches (see Figure 2). The so-called *Multi-Grid Chain Code* (MCC) [24, 12] utilizes a contour grid whose basic cell is shown in Figure 5. A single movement is used to go through the cell. In the example of Figure 5, from the input contour element, marked with 0, seven possible output contour elements can be reached to go through the cell $\{1 \ldots 7\}$. Each one of these output contour elements represents a different movement. If a derivative chain code is used, movements can be represented independently of the input contour element. The input contour element is denoted by the symbol 1 and the symbol 1 corresponds to the element which is on the same side as the symbol 0. This symbol 1 defines the rule of symbol assignation which can be either clockwise or counter-clockwise as can be seen in Figure 5.

However, a symbol does not uniquely represent a contour configuration. As illustrated in Figure 5, a movement (from 0 to 3 in the example) can correspond to two different contour configurations. Therefore, the central pixel may belong to two different regions. The decoding algorithm has to decide between two contour configurations. Errors in the decision introduce losses in the coding procedure. Note that all possible movements, except 7, may introduce errors and, for each movement, the only possible error concerns the central pixel of the cell.

Some movements in this technique represent larger steps than others. This is the case of symbols 3 or 4 with respect to 1 or 7. This is an interesting point because if we can use symbol 3 or 4 with a high probability, the average step size will be large. As a result, the number of steps necessary to describe the contour, that is the number of symbols to code will be low. In order to increase the probability of large steps occurrence, the multi-grid chain code technique changes the position of its basic cells through the grid [24]. More precisely this technique combines four grids and, at each position, a cell from one of these grids is selected. The selection relies on a prediction of the contour trajectory at each site. The prediction assumes boundary smoothness.

An example of this technique is shown in Figure 6. In this example, a contour segment has been coded by means of the symbol 4. Following the boundary smoothness assumption, the trajectory of this contour segment should not change. If cells of the same grid are used to code it, symbols 7 and 2 will be used.

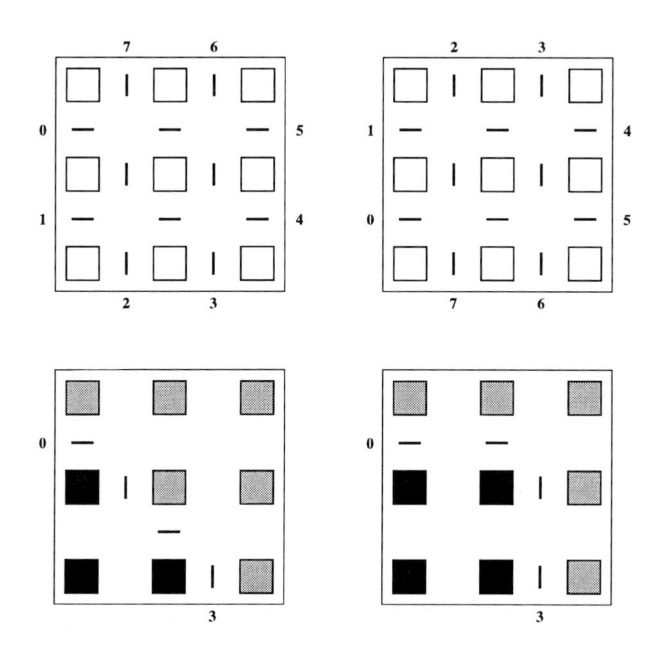

Figure 5 Basic cell of the MCC and two possible contour configurations for symbol 3

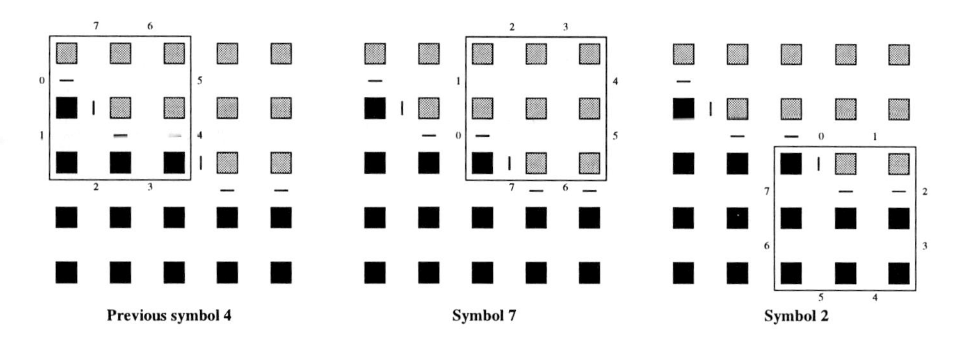

Figure 6 Coding with a fixed grid

Nevertheless, if the grid (that is, if the position of the center of the basic cell) is changed, the contour segment can be coded only using one additional symbol 4. This second possibility leading to a more compact representation is shown in Figure 7. In order to cover all possible contour configurations, four different grids are necessary. A simple grid change algorithm was proposed in [24].

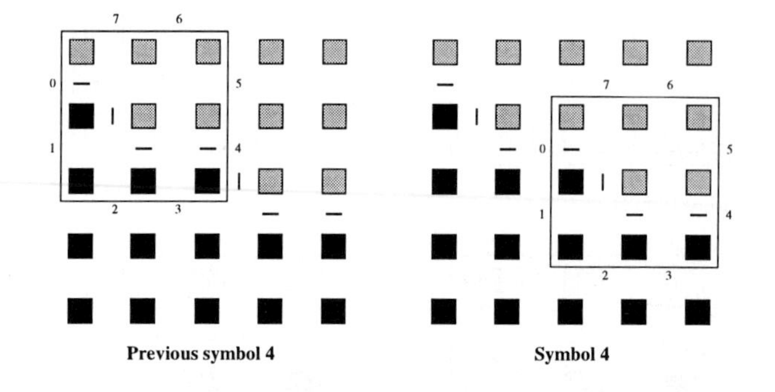

Figure 7 Variation of the grid with respect to the previous movement

This lossy shape coding technique can be combined with the position coding approach based on the concept of triple points in order to have a complete partition coding technique [12]. In this case, cells having two outputs are marked as being triple points. If a cell presents k different outputs, $k-1$ marks of triple points are introduced in the chain.

In the decoding step, a special policy has to be followed so that new regions are not created. Although only the label of one pixel may change at each cell, this change may split a region into two parts. Nevertheless, such problems can be solved in the decoder side.

In practice, the use of multi-grid chain code results in an average improvement of 25% on the coding figures with respect to the chain code approach. The size of the alphabet is greater than the one used for the chain code, however, the average step size is much lower and the probability of some symbols is much higher than the probability of other symbols and efficient entropy coding can be achieved. This results in an average bit cost of 1 bit per boundary element, including shape and location information. Furthermore, losses introduced by this technique are almost unnoticeable as it is shown in Figure 8. Although the number of erroneous points seems to be high, they are hardly visible in the resulting partition. This is because errors are isolated points that mainly smooth the contours.

Figure 8 A frame of the sequence *Miss America*, a boundary representation of its partition, the points of this boundary representation that are wrongly decoded when using MCC and the final decoded boundary representation.

Global modeling of the contours: Fourier descriptors

For each region, a function can be derived containing the boundary information. Although this function describes the boundary in a local manner, its Fourier descriptors convey global information of the region shape. There are several ways to define the Fourier descriptors of a region [35]. For partition coding purposes, the definition that gives a direct relationship between the position function $z[n]$ and its Fourier transform $Z[k]$ is very useful:

$$Z[k] = \frac{1}{N} \sum_{n=0}^{N-1} z[n] \; e^{-j\frac{2\pi}{N}kn} \tag{4.9}$$

The position function $z[n]$ is a complex function that contains the position of each contour point ordered and represented by its coordinates as a complex number.

All the information contained in the Fourier descriptors is not equally relevant for coding. Usually, only a few descriptors are sufficient to represent a region. Therefore, the coding procedure only takes into account this set and the remaining descriptors are removed. However, the type of losses introduced by the removal of some Fourier descriptors is difficult to control and may lead to visually annoying results. This is in particular the case of regions containing straight line segments or corners. Furthermore, an additional distortion appears in the coding process due to the quantization of the Fourier descriptors. This distortion worses the performance of this technique since, in order to avoid it, a large number of bits is required in the quantization process. Figure 9 compares an original region with the result of coding it using a subset of its Fourier descriptors with and without quantization. In this example, the effect of the quantization leads to a complex curve with crossing points that cannot represent a meaningful region.

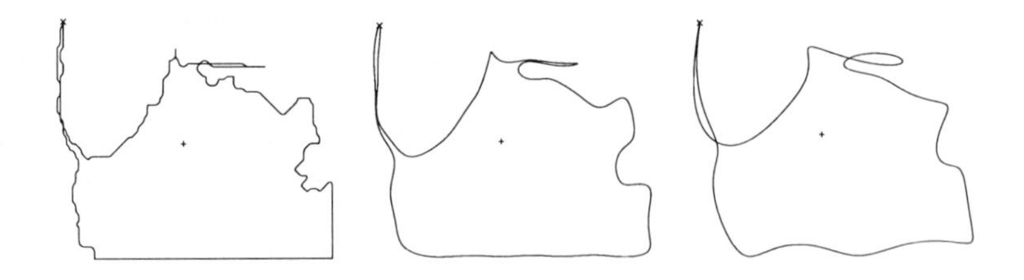

Figure 9 An original region and its decoded version using 10% of its Fourier descriptors before and after quantization

Finally, since the partition is coded region by region, some problems arise in the decoding process. First, due to the simplification procedure, a pixel may be covered by more than one region after decoding. Second, as in the skeleton simplification case, the decoded partition may have pixels that are not assigned to any region. Therefore, a hole-filling technique is necessary to obtain the final partition.

Local modeling of the contours

The last set of intra-frame coding techniques relies on local approximation of the shape information. In contrast to the Fourier descriptors, they segment the position function and approximate locally the shape information with simple functions.

The segmentation step can be viewed as a 'dominant or control point extraction'. Most of the time, dominant points are defined by the geometrical characteristics of the shape to code. They mainly correspond to points of high curvature. The detection of high curvature points on digital curves is not a simple task [28, 34]. Two sets of approaches have been reported in the literature. The first one directly extracts dominant points by the analysis of the position function. They rely mainly on the notion of angle and corner. The reader is referred to [34] for a review of the most popular algorithms of dominant points extraction. The second approach is iterative and successively adds control points to improve the curve approximation until a certain error criterion is met. A review of these technique can be found in [28, 9]. A shape coding technique following this second approach is used in the 'Analysis-Synthesis' codec proposed in [25]. Finally, let us mention that some control points extraction algorithms take into account not only the geometrical characteristics of the region but also the gray level distribution of the pixels [26]. With these techniques, it is therefore possible to accept larger position error if they do not result in visible artefacts.

Once the control or dominant points have been extracted, the position function should be interpolated. The most popular interpolation curves rely on classical mathematical functions: lines [7], Splines [32] or Bernstein polynomials [1], etc. Finally, let us mention that some schemes [25] combine several interpolation functions.

These techniques can achieve a high coding efficiency at the expense of a rather high error in the shape information. Typical figures are a cost of 0.5 bit per contour point for an approximation of $\pm$ 2 pixels. Therefore, the use of these techniques depends on the application and also on the segmentation criterion that has been used to create the partition. Regarding the application, one has to take into account in particular the format of the image to process. Indeed, a shape error of $\pm$ 2 pixels may be quite acceptable for an HDTV image of 1440*1152 pixels, however, it may not be acceptable for a QCIF image of 176*144 pixels. The segmentation criterion is also to be considered: in practice, if the partition has been defined by the gray level homogeneity, errors in the shape representation will be rapidly visible. By contrast, if the segmentation is only motion oriented, and depending on the coding strategy, the shape approximation may tolerate higher error.

4 INTER-FRAME CODING

This section describes the inter-frame coding of a partition sequence. As in the case of intra-frame coding, the shape and the position of the regions in each frame should be coded. However, as explained at the beginning of this chapter, the coding of partition sequences also requires the transmission of the label of each region.

The objective of this section is to describe a general motion compensation strategy for partitions. We will present and discuss the main issues of this coding strategy, in particular, the type of motion information to be used, the problem of overlapping regions and the structure of the compensation loop. Our objective is to describe a general scheme that does not rely on a specific contour representation and does not imply a specific segmentation scheme.

4.1 Motion compensation of partitions

As illustrated in Figure 10, the inter-frame mode of partition coding relies on the estimation of the motion of the partition between two successive frames, the prediction by motion compensation of the previous partition, the computation of the prediction error, the simplification of this prediction error (in the case of lossy coding) and the coding of the simplified error. This scheme is similar to the one classically used for motion compensation of texture, but it is applied here on the partition information, that is on a frame of label.

The problem of partition compensation is illustrated in Figure 11. Before the compensation, the motion of the various regions of the current partition *seg*

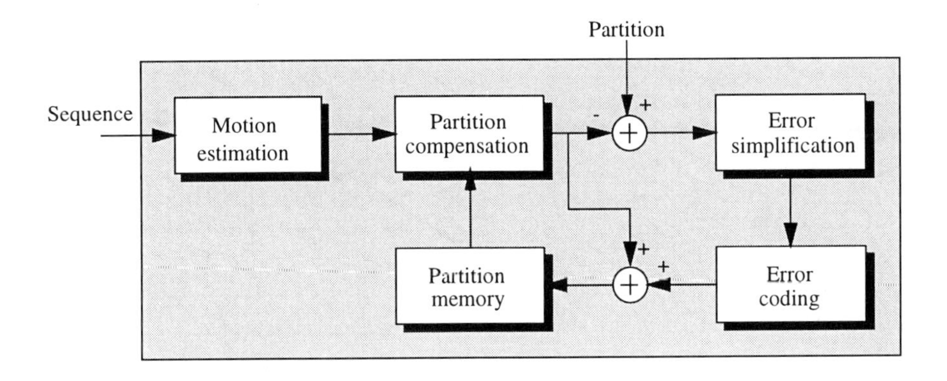

Figure 10 Motion compensated partition coding

have been estimated (Figure 11.A). The estimation gives a set of parameters describing the time evolution of each region in a backward mode. In Figure 11.A only one vector has been represented for each region, but the estimation generally gives more complex models defining one motion vector for each pixel of each region. The first question to answer is to define which kind of information can be used to motion compensate the partition. Note that, the motion of the pixels inside a region (texture motion) and the motion of the shape of a region may not be equivalent. Both motions coincide in the case of a foreground region. Indeed, for such regions, the pixels of interior and contours follow the same motion. This is, however, not the case for a background region because the modifications of its shape or of its contours are defined by the motion of the regions in its foreground. Following the discussion of [27], in the sequel, we will assume that the texture motion is used to compensate both the partition and the texture. The partition compensation problem is shown in Figure 11.B: based on the previously coded partition $rec(T-1)$, that has been stored in the receiver, and on the transmitted motion parameters, the compensation should define a prediction of the current partition $rec(T)$.

The compensation itself can work either in a *forward* mode or in a *backward* mode. In the forward mode (see in Figure 12), the pixels of $rec(T-1)$ are projected towards $rec(T)$. This projection can be done if the motion vectors, as defined by the estimation done in a backward mode, are inverted. As can be seen, two problems may result from the transformation of the regions. Some pixels of $rec(T)$ may have no label, they constitute the so-called *empty areas*. By contrast, some pixels may have several label candidates, these conflicting areas are called *overlapping areas*. To solve the conflicts, an extra information called the *order* can be used [27]. The order information is used to decide

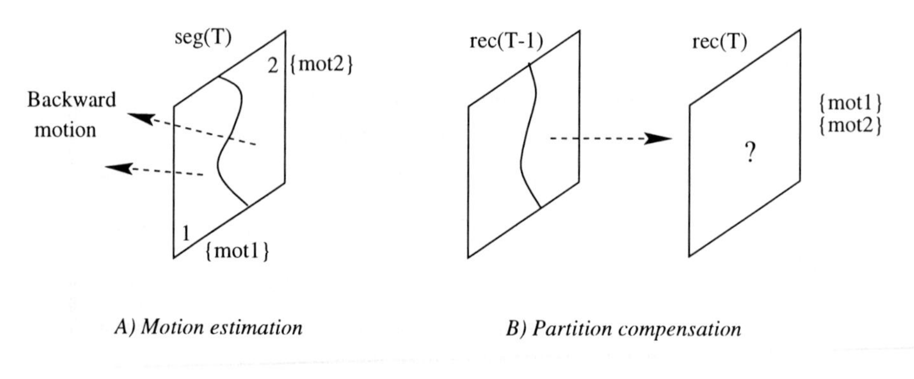

Figure 11 Partition compensation

which region is considered to be in the foreground of which region. In case of conflicts between labels, the foreground region gives the correct label. The problem of overlapping areas is specially important if the texture motion is used to compensate the partition because of the issue of foreground / background relationship between regions commented above. However, the use of the texture motion and of the order is a quite efficient solution because the texture motion information leads to a good compensation of the texture and the order only represents a small amount of information. Finally, the empty areas are left without label and are processed as compensation errors.

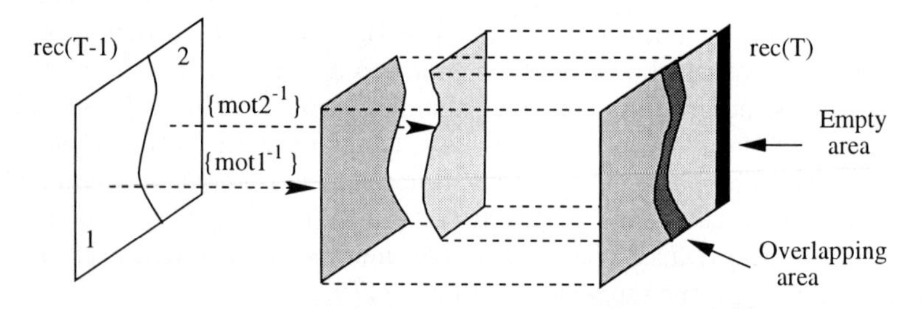

Figure 12 Forward motion compensation of partitions

The dual mode of compensation is illustrated in Figure 13. It is a backward mode in the sense that, for each pixel of $rec(T)$, one tries to look on $rec(T-1)$, which label has to be selected. In this case, the main problem is to define which motion vector has to be used when the pixel (i, j) of $rec(T)$ is considered. Indeed, since the current partition is not yet known on the receiver side, one does not know the region the pixel belongs to and therefore its corresponding motion model. The solution consists in considering all possible vectors defined by all possible regions. In the case of Figure 13, there are two regions, therefore, two vectors are considered for each point: one given by region 1 ($\{mot1\}$) and one given by region 2 ($\{mot2\}$). Each time a vector as defined by region n does not point to a pixel belonging to region n in $rec(T-1)$, the compensation is considered as being invalid and is discarded. In Figure 13, this is the case for two vectors. However, as in the case of forward motion compensation, some pixels have no valid compensation (*empty areas*) and some others have more than one candidate (*overlapping areas*). As previously, the order information is used to solve the conflicting areas.

The main difference between the forward and backward modes of compensation deals with the quantization of the pixel locations. Indeed, generally, the motion vectors start from an integer pixel location but point to a non-integer location. In the forward case, it is the locations of pixels of $rec(T)$ that have to be quantized whereas in the backward case, the locations of pixels of $rec(T-1)$ have to be quantized. There are some more difficulties related to the forward mode in the case of motion models involving modifications of the scale (zoom in particular). Indeed, in the case of region expansion, the modification of the distance between two pixels may create more empty areas in the compensated frame. These problems can be solved [27] but, in general, the backward mode is more simple.

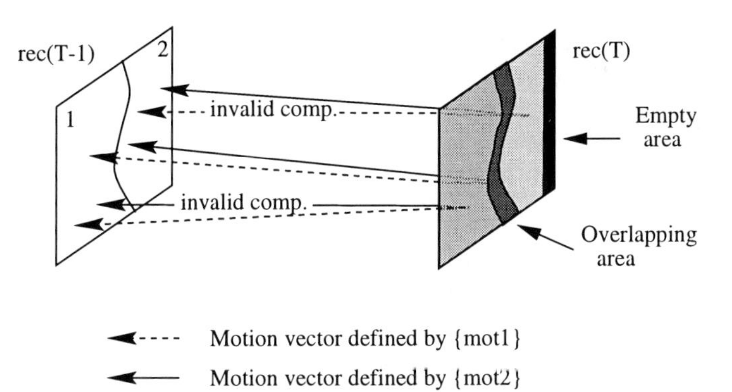

Figure 13 Backward motion compensation of partitions

Two examples of partition coding using forward motion compensation can be found in [3, 13]. The main difference between these two works consists in the coding of the error which is done by the coding of its contours (chain code) in [13] or of its shape (Geodesic skeleton) in [3]. Both techniques rely on the estimation of the motion of each region and the processing of regions individually. That is, the motion of the shape of each region is estimated (not the texture motion), then each region is compensated and finally, the error resulting from the region compensation is coded and transmitted individually for each region.

In [25], the texture motion is used to compensate the mask and no order information is necessary because the model assumes *a priori* that there are only one foreground moving region (the mask to compensate) and one static background region. Note that the motion compensation only deals with a binary mask. This situation can be viewed as a simplified version of the general problem that is discussed here: we want to deal with complex partitions, each region of which can have its motion, and various foreground/background relations with respect to other regions.

The objective of this section is to describe precisely the various steps of the motion compensation loop of Figure 10, in particular the estimation of the order between regions and the various steps of the algorithm for partition coding: partition compensation, partition error computation, error simplification and error coding [30].

In the sequel, we will assume that the goal of the inter-frame partition coding is to transmit the information about a time varying partition with the help of the motion of the pixel values. The partitions are represented by label images (images where the gray level values of the pixels define the region number). Let us assume the following notations:

- *seg* represents the current (at time T) partition.
- *rec* denotes the previous (at time $T-1$) partition reconstructed on the receiver side. Note that, in general, the partition coding process implies some losses and the reconstructed partitions are not exactly the same as the ones defined by the segmentation.
- *motion* and *order* are the motion and order information that characterize the evolution of the partition between time $T-1$ and T.
- *original* is the original frame at time T.

The objective of this chapter is not to address the problem of motion estimation of the pixels (see Chapter 6 for more details). Therefore, we will assume that the motion of the pixels value has been estimated (that is *motion* is known). In general this motion information is represented by a set of parameters that are assigned to each region. Moreover, the objective of this chapter is not to describe a segmentation algorithm (see Chapters 3 and 6), therefore *seg* is assumed to be known.

The objective of the inter-frame mode of partition coding is twofold:

1. **Order estimation**: Based on *seg*, *rec* and *motion*, how one can estimate the order information *order*?
2. **Partition coding**: Based on *seg*, *rec*, *motion*, *order* and *original*, how one can produce the new reconstructed partition?

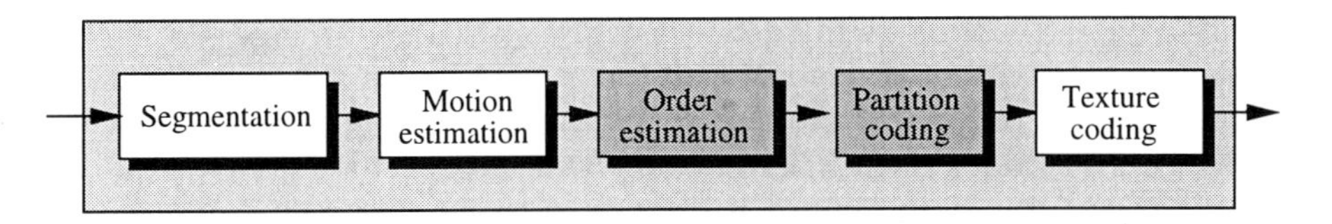

Figure 14 Structure of the codec

The structure of the codec is presented in Figure 14. The two dark blocks represent the parts of codec that are concerned with the present section. In the following we will describe separately the two blocks since they are highly independent.

4.2 Order and transmission mode estimation

The goal of the order estimation is of course to estimate the order for the partition compensation. In practice, this order will be used on the receiver side to solve the conflicts between compensated regions. Conflicts appear on locations where several region labels overlap. This notion is implicitly used in [13, 31] where a transmission order is defined. A transmitted region is supposed to be in the background of all previously transmitted regions. The transmission order is defined by estimating the prediction error for each region individually and the region creating the lowest error is transmitted first. This solution is not optimum because it assigns one order to each region and does not try to solve the possible conflicts between pairs of regions. For example, it cannot deal with situations such as an object A being in the foreground of an object B which is in the foreground of C which, itself, has some parts in the foreground of A. Depending on the segmentation criteria, such situations may happen for scenes with 3D objects and several depth levels. So, the real objective of the order estimation is to solve all the possible conflicts between pairs of regions.

One may think that the problem of overlapping regions may be solved by choosing a specific strategy in the receiver side and by processing the resulting errors as compensation errors. From our experience, this is not true because, first, the order information only represents a small overhead, and second, the motion information that is used does not characterize the shape motion. Finally, let us mention that the order information seems to be of vital importance for motion models that are more complex than the simple translation.

Moreover, in practice, it may not be always efficient to send all regions in inter-frame mode (that is by compensation). For example, if the motion estimation has not produced reliable results or if the motion of the region cannot be modeled by the motion estimation, then the prediction error of the region shape may be so large that it may be less expensive to directly code the entire region. In this case one has to switch from inter-frame mode to intra-frame mode. This idea leads to the definition of a transmission mode (intra or inter) for each region [30]. This solution is in contrast to the work reported in [13, 3] where all regions of the partition are motion compensated. As a result, the gray level and color values of one region can be sent in inter-frame mode and its shape and location can be sent either in intra-frame or in inter-frame.

The 'Order computation' and the 'Mode estimation' are interdependent. Indeed, the 'Order computation' needs to know the transmission mode of each region and the 'Mode estimation' needs to know how to compensate each region. That is, it needs to know the output of the 'Order estimation'. This leads to the decision-directed loop shown in Figure 15. The input data are the current partition *seg*, the previous reconstructed partition *rec* as well as the motion information *motion*. To start the process, all new regions (that is, regions present in *seg* but not present in *rec*) are assumed to be transmitted in intra. Based on this assumption, the order is estimated. Then, assuming that the estimated order is correct, the transmission mode is estimated and some transmission modes may be modified. If this is the case, the order has to be estimated again, and so on. In practice, a few iterations (typically two or three) are performed. The results are the order and a list of transmission modes. In the following we describe more precisely the two blocks of Figure 15.

Define intra/inter-mode

The objective of this block is to define the transmission mode (intra or inter) of each region. This decision is taken on the basis of the cost of the contour. Note that, one region can be sent in intra-frame mode for partition coding and in inter-frame mode for texture coding. To deal with this situation, one has

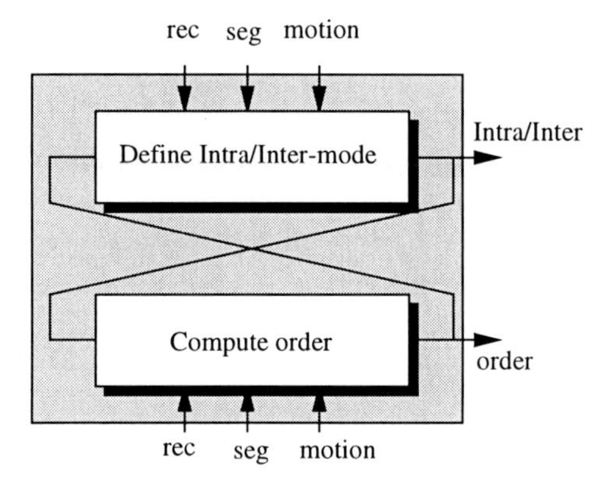

Figure 15 Decision-directed loop for the order estimation: order computation and intra/inter-mode estimation

to add a binary information to say to the receiver whether or not the region is transmitted in intra-frame mode for partition coding. This information can be stored with the motion information (*motion* will therefore involve a set of motion parameters for each region plus a binary information indicating the transmission mode of the contours).

The decision on the transmission mode is simply achieved by comparing the cost of the region shape if it is sent in intra-frame mode or in inter-frame mode (that is by motion compensation). The procedure is illustrated in Figure 16. Based on the previous partition *rec*, *order* and *motion*, the whole partition is compensated and the compensation error *error* is computed. Finally, the 'Select mode' block has to compare for each region the cost of transmission of the region as defined in *seg* (intra-frame mode) or as defined by *error* (inter-frame mode).

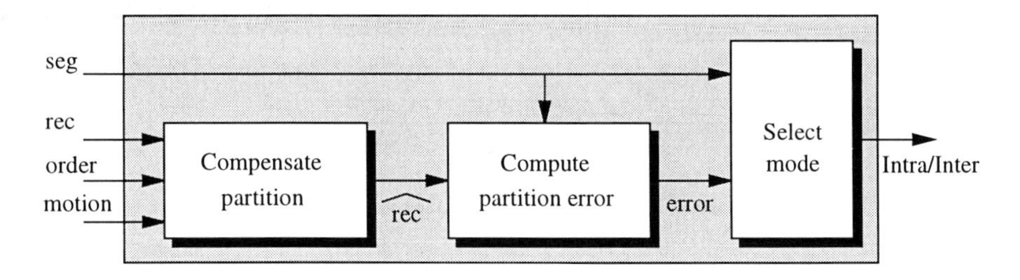

Figure 16 Description of the 'Define intra/inter-mode' block of Fig. 15

The 'Select mode' deals with each region individually. For each region, the coding costs as defined by *seg* (intra-frame mode) and by *error* (inter-frame mode) are estimated. This estimation can be done either by actually coding the information and measuring the resulting amount of information or by using any other technique providing an approximation of the cost. Once the two estimations have been performed, the comparison can be done and the resulting decision is stored as a motion information.

Compute order

The order computation can be decomposed in two steps as shown in Figure 17. The first step ('Estimate overlapping') estimates the conflicts between regions during the compensation and the second one ('Define order') achieves a quantization of the order information for the transmission. In the following we will assume a backward mode of compensation, but the principles remain the same for a forward mode.

The 'Estimate overlapping' relies on a double loop: a first one scanning all possible regions (defined by a number *region_num*) and a second one scanning the positions (i, j) of the image space. The first loop is more precisely described in Figure 18. It involves two steps respectively called 'Expand motion' and 'Update overlap'. The objective of the 'Expand motion' block is to define a motion vector for each position of the space (i, j) if the motion parameters of region *region_num* are used (backward motion compensation). In general, this can be easily done because the motion information gives for each region a motion model which is function of the position (i, j). For example, if the motion model for a given region is a translation (δ_x, δ_y) then this motion vector is assigned to all points (i, j) of the image. If the motion model is an affine model $(\alpha_x, \beta_x, \delta_x, \alpha_y, \beta_y, \delta_y)$, then the motion field at the point (i, j) is defined as $V_x(i, j) = \alpha_x i + \beta_x j + \delta_x$ and $V_y(i, j) = \alpha_y i + \beta_y j + \delta_y$. Of course, if the region has to be transmitted in intra-frame mode no motion field expansion is performed.

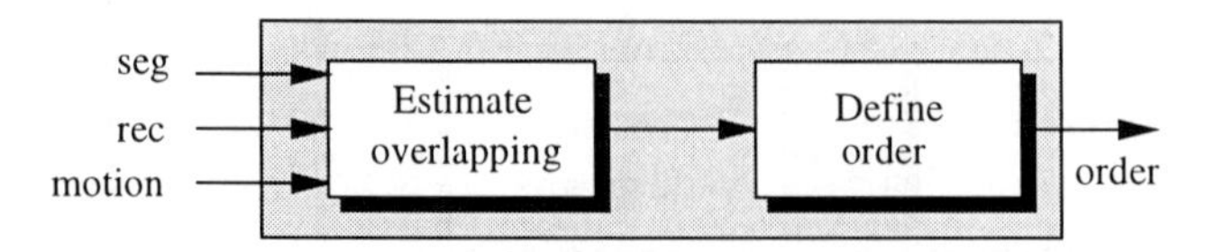

Figure 17 Description of the 'Compute order' block of Fig. 15

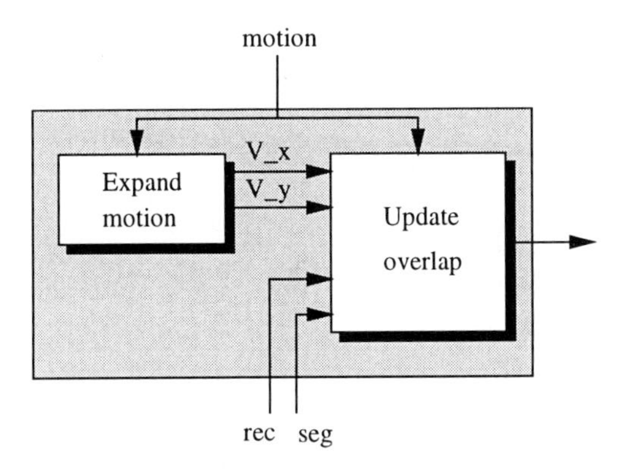

Figure 18 Description of the 'Estimate overlapping' block of Fig. 17

Once the motion field has been defined for the entire space, the overlap corresponding to region *region_num* is updated by the 'Update overlap'. The update can be achieved in the following way: each pixel of the image is scanned; denote by (i, j) the position of the current pixel and assume that its motion defined by the 'Expand motion' block is equal to $(V_x(i, j), V_y(i, j))$. Moreover, denote by *Current_label* and *Previous_label* the region numbers of the pixel (i, j) in the current partition *seg* and of the pixel $(i - V_x(i, j), j - V_y(i, j))$ in the previous partition *rec* (see Figure 19).

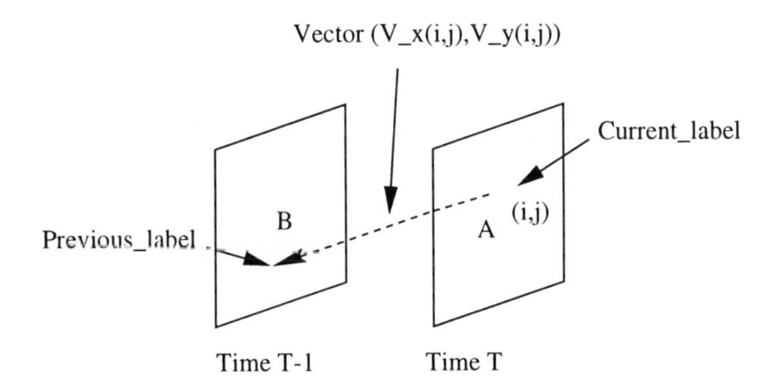

Figure 19 Definition of the parameters for the 'Update overlap' block

The overlap information consists of a list of possible regions in conflict, that is, of regions of *rec* that may overlap with other regions during the compensation. The list is set to zero at the beginning of the process and the list entry corresponding to the conflict between region *region_num* and *Current_label* is updated if the following conditions are fulfilled:

1. The *Current_label* does not correspond to a region transmitted in intra-frame mode. Indeed, if the point (i, j) in *seg* corresponds to a region which is transmitted in intra-frame mode, any compensated label that falls at that position is discarded (that is why the order estimation needs to know the transmission mode).

2. *Previous_label* is equal to *region_num*. The point B of Figure 19 corresponds to *region_num*, this means that the compensation is valid and that the point (i, j) is a candidate to receive the label *region_num*.

3. *Current_label* is not equal to *region_num*. The correct label of (i, j) is not *region_num*. Therefore, *Current_label* and *region_num* (= *Previous_label*) are conflicting regions and for this location (i, j), *Current_label* is in the foreground of *region_num*. Therefore, the list entry corresponding to the conflict between *Current_label* and *region_num* is incremented by one unity.

This procedure is iterated for all points (i, j) and for all regions *region_num*. At the end, the overlap list gives the number of occurrences where it has been possible to declare that a given region is in the foreground of another region. Finally, the order has to be quantized because the receiver only needs a binary decision in order to be able to resolve the conflicting labels during the compensation.

The quantization is achieved by examining each pair of labels and by comparing the number of occurrences where the first one has been declared to be in the foreground of the second one with the number of occurrences where the second one has been declared to be in the foreground of the first one. The final order between the two regions is defined as the one corresponding to the largest number of occurrences.

Finally, *order* is entropy coded and sent through the transmission channel. The order, being a binary information, can be very efficiently coded (see Section 4.5)

4.3 Inter-frame partition coding

Once the order has been estimated, the actual coding of the partition can be done. The structure of the encoding process is illustrated by Figure 20. First, all regions which have to be sent in intra-frame mode are processed. As explained previously, these regions are either new regions (region that are not present in *rec*) or regions that cannot be well compensated and are more efficiently sent in intra-frame mode. The second step consists in computing a partition involving the regions sent in intra-frame mode and the compensated regions. This is the objective of the 'Compensate partition' block. Finally, the partition errors are extracted, simplified and coded by the 'Code inter-regions' block. In the following, we describe more precisely these three blocks.

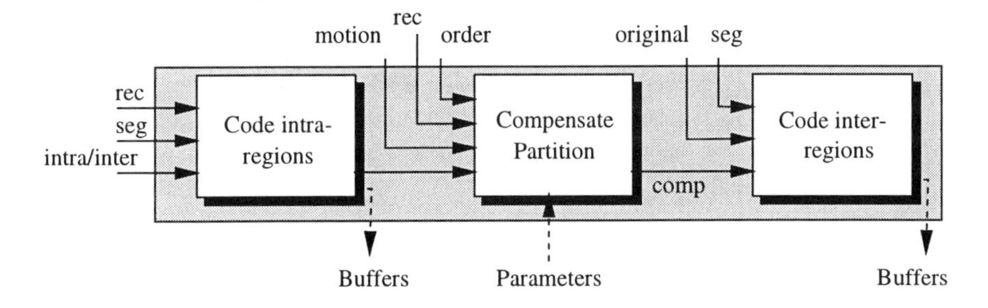

Figure 20 Structure of the inter-frame partition coding

Code intra-regions

The objective of this block is twofold: first, to send the contour information of the regions transmitted in intra-frame mode and second to create a binary mask defining where the partition is considered as being already defined. This mask will be useful during the compensation because each time a label is compensated at a location corresponding to a region transmitted in intra-frame mode then the compensated label will not be taken into account.

The coding in intra is performed in three steps as described in Figure 21. The first step consists in extracting the set of regions which have to be sent in intra. The set is composed of regions that are present in *seg* but not in *rec* (new regions) plus regions which were defined during the order estimation as to be sent (for contour coding) in intra. The 'Select regions' block defines a partition of the space which has to be sent to the receiver. Almost all partition coding techniques (lossy or lossless) described in the first Section of this Chapter may be used. As can be seen in Figure 21, the information which is transmitted by

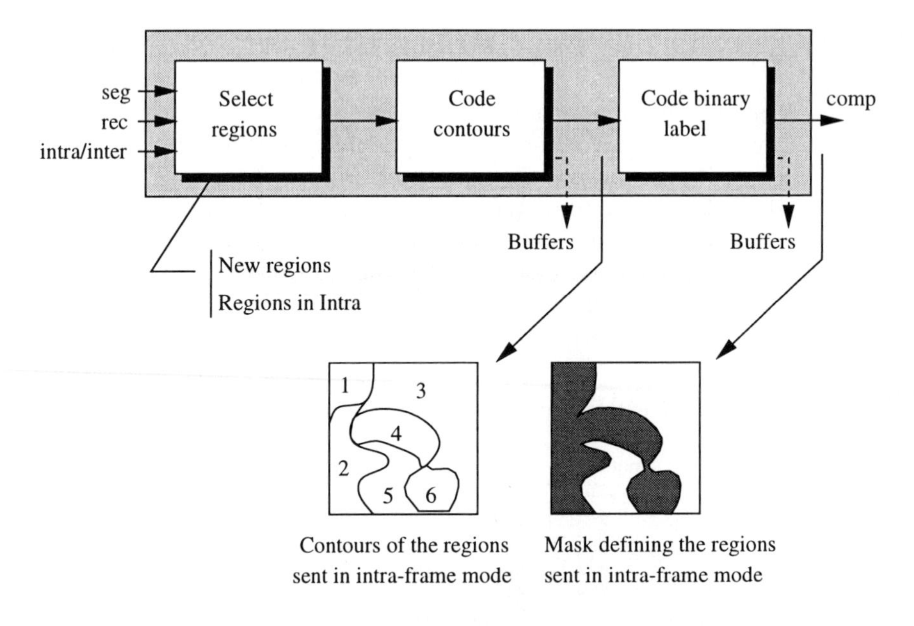

Figure 21 Description of the 'Code intra-regions' block of Fig. 20

the 'Code intra-regions' block allows the restoration on the receiver side of the contours of the intra-regions. However, it does not say which region is transmitted in intra. For example, in Figure 21, one can see that the transmission of the contour information creates a partition of 6 regions. Assume, for example, that from these 6 regions, only 4 are sent in intra-frame mode: region number 1, 2, 4 and 6. In this case, we want to create a mask indicating the locations where the label has been sent in intra-frame mode. This is the purpose of the 'Code binary label'. It sends different codes for each region depending on its transmission mode. In the following, the mask information is assumed to be stored in an image called *comp*, that is the first version of the compensated partition.

Compensate partition

The compensation of the previous partition *rec* is illustrated in Figure 22. Figure 22.A describes the compensation of a region *region_num*. The procedure is similar to the one used for the 'Estimate overlap' block described in Figure 18. Here also, a backward mode of compensation is assumed, but the same approach can be used for a forward mode. The first step is to define a motion field for the entire image using the motion parameters assigned to region *region_num*. This is done by the 'Expand motion' block (see Section 4.2 for more details).

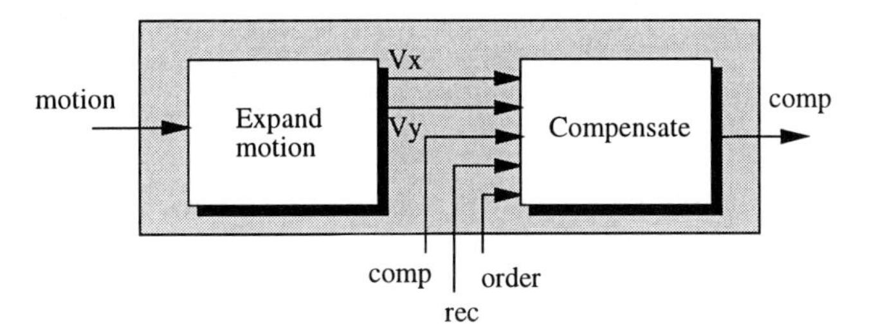

A) Compensation of region "region_num"

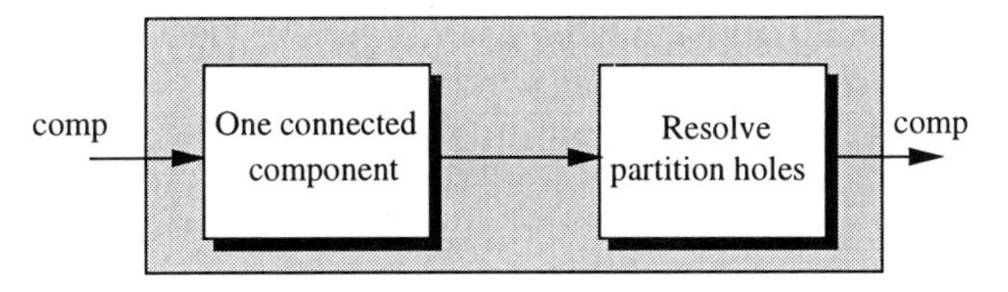

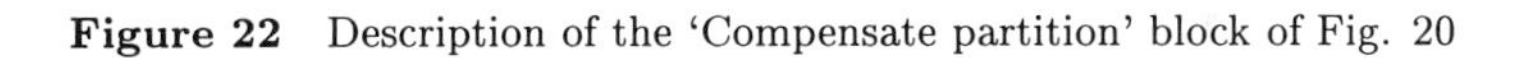

B) Postprocessing of the compensated partition

Figure 22 Description of the 'Compensate partition' block of Fig. 20

The second step is the compensation itself which has strong similarities to the 'Update overlapping' block of Figure 18. As in Section 4.2, denote by (i, j) the position of the current pixel and assume that its motion is equal to $(V_x(i, j), V_y(i, j))$. Moreover, denote by *Compensate_mask* the value of the mask (*comp*) defined by the 'Code intra-regions' block at location (i, j) and *Previous_label* the region number of the pixel $(i - V_x(i, j), j - V_y(i, j))$ in the previous partition *rec* (see Figure 19). For each pixel (i, j) of the compensated partition *comp*, the compensated label is defined as follows:

- If *Current_mask* corresponds to a region transmitted in intra-frame mode, the label is not compensated. Indeed, if the point (i, j) corresponds to a region which has been transmitted in intra-frame mode, any compensated label that falls in that region is discarded.
- If *Previous_label* is equal to *region_num*, the compensation is valid and the point (i, j) is a candidate to receive the label *region_num*.

- If the location (i, j) of the compensated partition is empty, then assign the label *Previous_label* to that position.

- If the location (i, j) has already been assigned a label *Compensated_label*, then *Previous_label* and *Compensated_label* are in conflict. In this case, the conflict can be solved by the order information stored in *order*.

This procedure is iterated for all points (i, j) and for all regions *region_num*. At the end, the compensated information can be postprocessed in order to create a partition. Indeed, the previous compensation technique does not guarantee the definition of connected regions and in practice several connected components may have the same label. In some applications, it is desirable to have only one connected component per label. In this case, the two step procedure described in Figure 22.B may be used. The first step selects one connected component for each label. Several criteria may be used. A simple one consists of selecting the largest component (note that this selection can be done on the receiver side without the need of transmission of any overhead information). The elimination of some connected component will create some 'holes' in the partition, that is regions which correspond neither to a region sent in intra-frame mode nor to a compensated label. These holes can be left and processed as individual regions by the following step or they can be removed by the 'Resolve partition holes' block (see Figure 22.B). Here also several techniques may be used, in particular geometrical techniques are attractive because they can be implemented in the receiver with no transmission cost. For example, the hole can be eliminated by a propagation of the neighboring labels or by assigning it to the largest neighboring region, etc.

At the end of this procedure, one has the final compensated image (stored in *comp*). The final step of the encoding process deals with the compensation errors.

Code inter-regions

Figure 23.A describes the organization of the partition coding in inter-frame mode. It consists of computing the error of the compensated partition, in simplifying part of the error, and in sending the information about the contours and labels of the error.

- *Compute error partition:* The error computation consists in extracting the locations of the space where the compensated label *comp* is different from

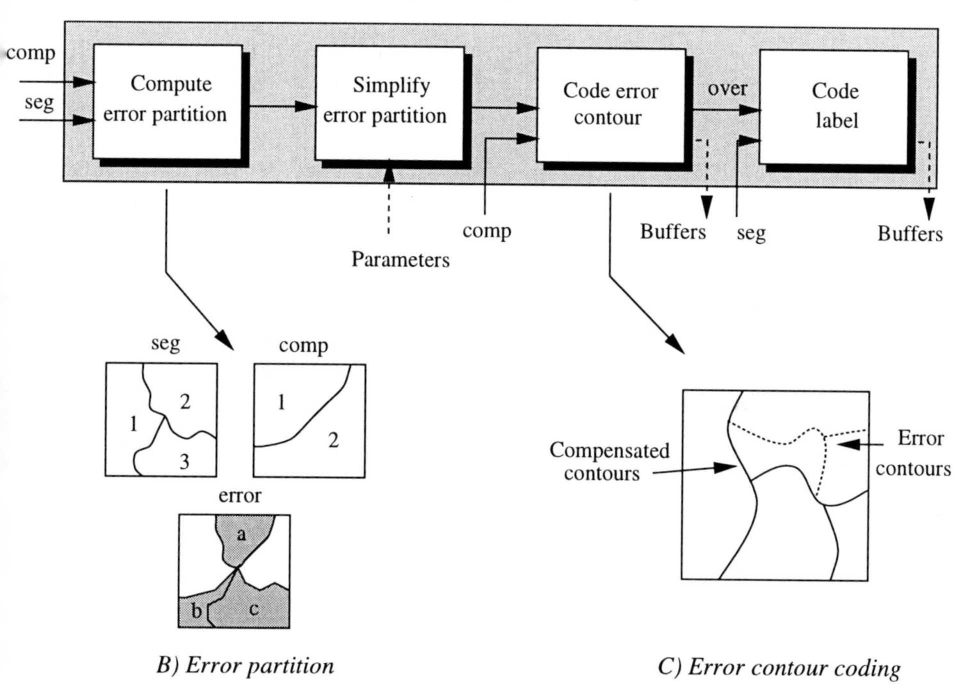

Figure 23 Description of the 'Code inter-regions' block of Fig. 20

the current label defined by *seg*. Note that the locations corresponding to regions sent in intra-frame mode are not considered. This results in an error mask which has to be labeled. The error mask is labeled in such a way that in each resulting region of the error, both the labels of the compensated partition *comp* and of the current segmentation *seg* be constant values. Figure 23.B illustrates a simple example where the segmentation is made of three regions and the compensated partition only involves two regions. The error partition is also shown and three regions of error have been created with label a, b and c.

- *Simplify error partition:* This step introduces the major part of the losses in the the coding process. Each region of the error is examined to know whether it has to be preserved and coded or discarded. In practice, the most useful criteria are:

  1. Geometrical criterion: All error regions that are smaller than a given size may be removed [13, 3, 30, 25]. Note that the size of the region can be considered as its area or its maximum elongation or its maximum width, etc.
  2. Gray level criterion: An error region can be discarded if it will not introduce a strong modification of the gray level (or color) value [30]. The decision can be taken by considering the values assigned to the error region if it is assumed to belong to the region defined by *comp*, on the one hand side, and by *seg*, on the other hand side. If the difference is small, the error region can be discarded.

- *Code error contour:* The actual coding of the error is done in two steps. The first step consists in coding the contours of the error regions. This coding step is presented in Figure 23.C. It relies on a differential coding of the contours. Indeed, the receiver already knows the contours that correspond to the compensated partition *comp*. Therefore, only, the new contours (dotted lines) have to be sent. Techniques similar to the one used for the intra-frame mode ('Code partition intra' of Figure 20) can be used. Classical examples can be found in [13, 3]. The only difference is that one has to take into account the presence of already known contours. In particular, the starting points of the new contours are always located on already known contours. This information can be used to efficiently code these starting points. Moreover, new contours end when they reach a contour of the compensated partition. The resulting partition *over* can be seen as an 'over-partition' because it involves contours of the compensated partition and of the segmentation.

- *Code label:* The final coding step is to send the label of each region of the over-partition *over*. As commented above, the label information has to be restored on the receiver side in order to be able to define and to track the time evolution of the various regions. The procedure is illustrated in Figure 24. First, the labels assigned on the receiver side are extracted. These labels, called *compensated_label*, are the labels which were defined by the compensation process. Second, the most probable labels of the segmentation *seg* are estimated. Note that, because of the simplification step, each region of the over-partition *over* may involve several labels of the segmentation *seg* (this is not the case for the compensated partition *comp*). The most probable label is defined as the dominant label (label with the highest number of pixels) of *seg* that falls within the over-partition region, say *dominant_label*.

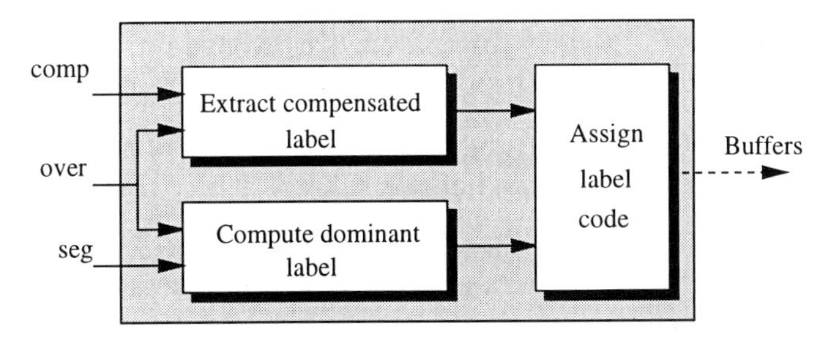

Figure 24 Description of the 'Code label' block of Fig. 23

The labels can be sent directly in the transmission channel but, in general, this will result in an excessive amount of information. To reduce this amount of information one can use the following coding strategy: For a given region, consider the set of compensated labels of its neighboring regions. Denote by $\{compensated\_neigh\_label\}$ this set of labels and create an ordered list of these labels. The order can be defined simply by the numerical values of the labels or by the geometrical position of the neighboring regions. Finally, one can assign the following code to define the label:

- If *dominant_label* is equal to *compensated_label*, then, assign the label code 0.
- If *dominant_label* is equal to one of the labels in the ordered list $\{compensated\_neigh\_label\}_n$, assign a label code indicating the position of the label in the order list.
- In all other cases, send an escape symbol followed by the actual label *dominant_label*.

This strategy is in general quite efficient because, most of the time, the *dominant_label* is either the *compensated_label* or a label of the neighboring regions. Moreover, the number of the neighboring regions is rather small (typically 4 or 5) which means that the entropy of the symbol (position of the label in the ordered list) is small. These symbols are finally entropy coded and sent.

4.4 Structure of the decoder

The decoding process is presented in Figure 25. As can be seen, it follows the same steps as the encoding process.

1. The contours of the regions sent in intra are restored by the 'Decode intra-regions' block and the mask indicating these regions is created.
2. The partition *rec* is motion compensated using the information of motion and order extracted from the buffers.
3. Finally, new contours are decoded to create the over-partition and the labels of the resulting regions are decoded. This produces the reconstructed partition at time T.

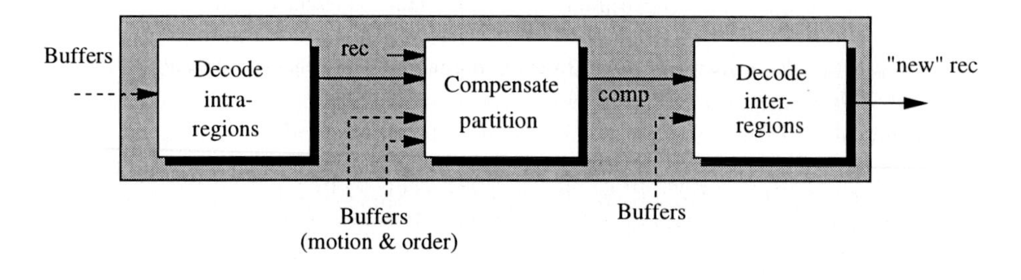

Figure 25 Structure of the decoder

4.5 Results

The purpose of this section is to describe the behavior of the motion compensated partition coding technique that has been described previously. Note that the performance of the method depends on two important points: the quality of the motion estimation and the kind of contour representation. It is not our purpose here to describe a particular motion estimation technique. However, in order to code the partition, we have used a motion estimation technique that assumes an affine model and estimates a set of six parameters for each region. Moreover, the general strategy of motion compensation of partition can be used with almost any kind of contour representation. For the practical coding of the partition, we have used a chain code representation. The results given in this section will therefore be related to this representation, however, most of the behavior of the approach (not the actual figures) can be translated to other representations.

As a reference, let us consider the case of intra-frame transmission mode. Tables 1 and 2 describe the performances of a modified chain code technique [21] to code two partition sequences: *Foreman* and *News*. The figures are given in *bits per contour points (bpcp)*. As can be seen, in both cases, most of the information is devoted to the coding of the individual movements of the chain. In the case of the *News* sequence, the results are better than for the *Foreman* sequence. This is due to the complexity of the contours that are simpler in the first case. In fact, the segmentation of *News* involves a large number of straight lines, this increases the probability of one movement and the entropy coding takes advantage of this higher probability.

Sequence	Chain moves	Starting points	Labels	Total
Foreman	1.25	0.01	0.05	1.31
News	1.08	0.02	0.03	1.13

Table 1 Coding result in the intra-frame mode (figures are in bits per contour points)

Figure 26 illustrates the whole process of partition coding by motion compensation. First, the intra-frame regions are transmitted (Figure 26.A). Second, the previously coded partition is motion compensated (Figure 26.B). In case of conflict between two labels, the order information is used to take a decision. Figure 26.C shows the compensation error after simplification. The simplification has removed the components of size lower than five pixels as well as the components that will not produce a gray level error higher than five. The contours of the error are sent to the receiver (Figure 27) to create the over-partition. Finally, the coded partition can be seen in Figure 26.D. Coding figures about this sequence can be found in table 2. As can be seen, the overwhole cost has been severely reduced.

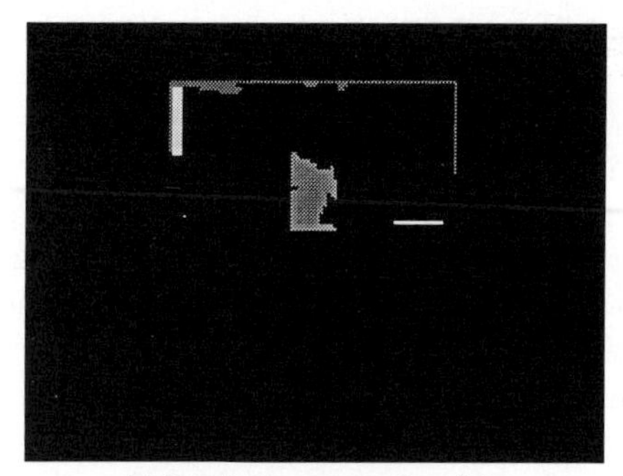

A) Regions transmitted in intra-frame mode

B) Compensated partition (pixels with no label are in black)

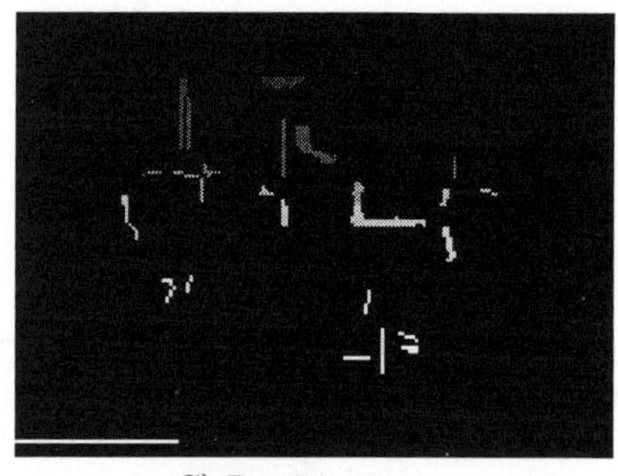

C) Partition error

D) Final partition

Figure 26 Partition coding steps

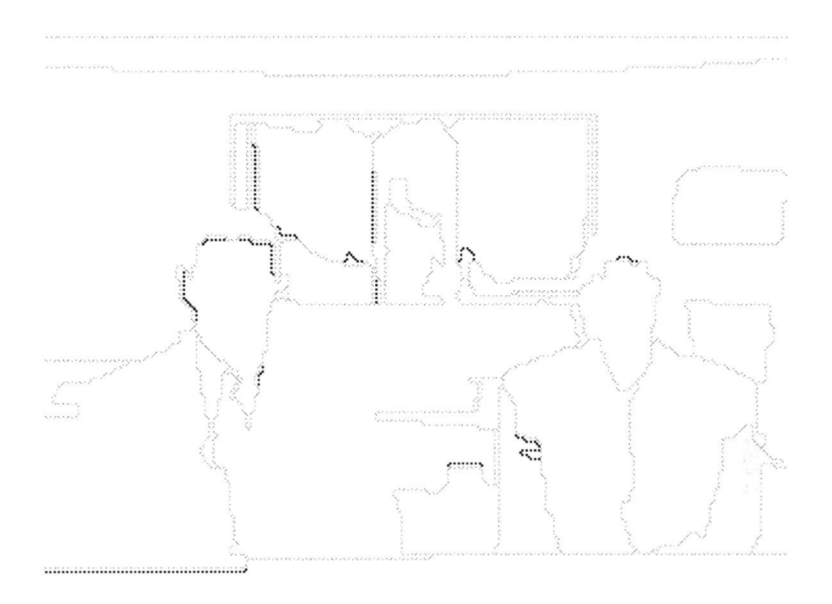

Figure 27 Contours of the transmitted errors, black: transmitted contours, gray: compensated contours

Sequence	Intra-frame	Order	Chain moves in inter	Starting points	Label	Total
News	0.26	0.12	0.14	0.10	0.10	0.72

Table 2 Coding result in the inter-frame mode (figures are in bits per contour points)

Figure 28 illustrates the importance of the order and transmission mode estimation. The sequence is the so-called *Mother and Daughter*. Comparison between Figures 28.A and 28.B shows that, without estimation, the number of contour points to send is quite high.

Finally, Tables 3 and 4 illustrate the influence of the error simplification criterion (sequence *Mother and Daughter*). Important reduction on the number of bits per contour point can be obtained. Of course, the definition of a proper simplification value should depend on the final application.

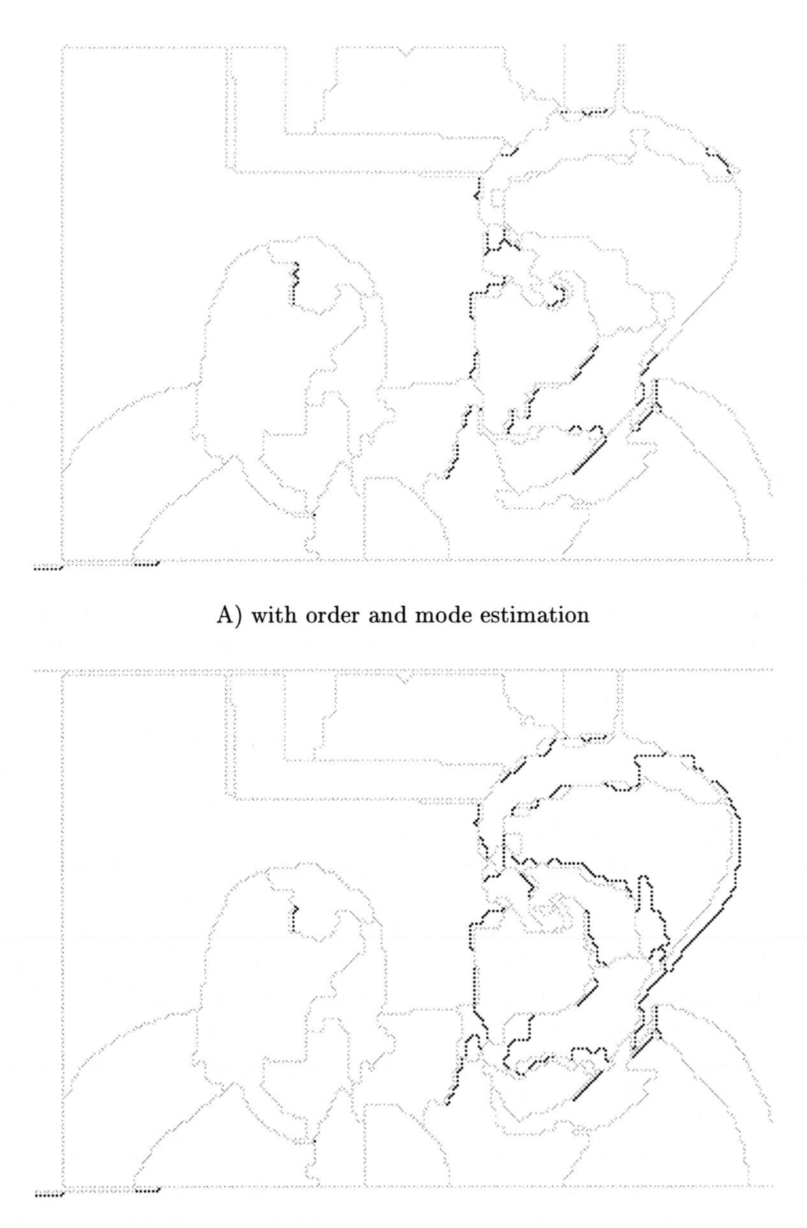

A) with order and mode estimation

B) without order and mode estimation
black: Transmitted contour points, Gray: Compensated contour points

Figure 28 Influence of the order and transmission mode estimation

Size	0	5 pixels	10 pixels
bpcp	0.88	0.46	0.32

Table 3 Size-oriented simplification of the partition error

Gray level	0	10	20
bpcp	0.88	0.82	0.64

Table 4 'Gray level'-oriented simplification of the partition error

5 CONCLUSIONS

Coding of partition sequences is one of the major issues of second generation video coding. Creating a signal-dependent partition suitable for efficient coding or for the so-called functionalities is a viable solution only if its resulting transmission cost is not prohibitive. This chapter has focused on various partition coding techniques.

The first set of techniques is devoted to the intra-frame mode where the shape and position information is coded. Lossless coding is an active field of research, however, it still seems difficult to derive techniques that are much more efficient than the classical *Chain Code*. However, techniques relying on skeleton decompositions are progressing very rapidly and may provide real alternatives in the near future. For a large set of applications, one has still to rely on lossy techniques. In this case, the major problem is to reach a given coding efficiency while controlling very precisely the resulting error. In this framework, techniques such as the *Multi-Grid chain Code* and its possible extensions seem to offer a very interesting trade-off.

It is very likely that the major issue of partition sequence coding concerns the inter-frame mode. This is a rather new field of research and few contributions have been made. In this chapter, we have tried to deal with the major issues of motion compensation of partition sequences. It is believed that motion compensation is a good solution to take benefit of the temporal correlation that exists between frames in a partition sequence. As a consequence, we have described the various steps of a general motion compensation loop. Each step has been discussed in a formal way without relying on a specific contour representation or a specific partition. Current results have shown that the use of compensation can divide at least by a factor of two the partition coding cost with respect to the intra-frame mode.

Acknowledgements

Part of this work has been supported by the Morpheco and Mavt projects of the European Race Program.

REFERENCES

[1] Sambhunath Biswas and Sankar K. Pal. Approximate coding of digital contours. *IEEE Transactions on Systems, Man, and Cybernetics*, SMC-18(6):1056–1066, November/December 1988.

[2] P. Bouthemy and E. François. Motion segmentation and qualitative dynamic scene analysis from an image sequence. *International Journal of Computer Vision*, 10(2):157–182, 1993.

[3] P. Brigger and M. Kunt. Contour image sequence coding using the geodesic morphological skeleton. In *International Workshop on Coding Techniques for Very Low Bit-rate Video*, pages 3.1–3.2, Essex, Colchester, April 1994.

[4] P. Brigger and M. Kunt. Geodesic skeleton decomposition using several distance measures: Application for shape representation. In *International Workshop on HDTV'94*, Torino, Italy, October 1994.

[5] P. Brigger, F. Meyer, and M. Kunt. The geodesic morphological skeleton and its fast reconstruction. In J. Serra and P. Soille, editors, *Second Workshop on Mathematical Morphology and its Applications to Signal Processing*, pages 133–140, Fontainebleau, France, September 1994. Kluwer Academic Publishers.

[6] S. Carlsson. Sketch based coding of grey level images. *EURASIP, Signal Processing*, 15(1):57–83, July 1988.

[7] B.B. Chaudhuri and M.K. Kundu. Digital line segment coding: a new efficient contour coding scheme. *Inst. Elec. Eng., Proc. pt. E*, 131(4):143–147, 1984.

[8] M. Van Droogenbroeck. *Traitement d'images numérique au moyen d'algorithmes utilisant la morphologie mathématique et la notion d'objet: Application au codage.* PhD thesis, Université Catholique de Louvain, Belgium, and Ecole Nationale Supérieure des Mines de Paris, France, 1994.

[9] J. G. Dunham. Optimum uniform piecewise linear approximation of planar curves. *IEEE Transactions Pattern Analysis and Machine Intelligence*, 8:67–75, January 1986.

[10] M. Eden and M. Kocher. On the performance of contour coding algorithm in the context of image coding. Part 1: Contour segment coding. *EURASIP, Signal Processing*, 8:381–386, 1985.

[11] H. Freeman. On the coding of arbitrary geometric configurations. *IRE Trans. Electronic Comp.*, EC(10):260–268, June 1961.

[12] A. Gasull, F. Marqués, and J. A. García. Lossy image contour coding with multiple grid chain code. In *Workshop on Image Analysis and Synthesis in Image Coding94, WIASIC'94*, pages B4–1–B4–4, Berlin, Germany, October 1994.

[13] C. Gu and M. Kunt. 3D contour image sequence coding based on morphological filters and motion compensation. In *International Workshop on Coding Techniques for Very Low Bit-rate Video*, Colchester, U.K., April 1994.

[14] R. Kresch and D. Malah. Multi–parameter skeleton decomposition. In J. Serra and P. Soille, editors, *Second Workshop on Mathematical Morphology and its Applications to Image Processing*, pages 141–148. Kluwer Academic Publishers, 1994.

[15] R. Kresch and D. Malah. New morphological skeleton properties applicable to its efficient coding. In IEEE, editor, *1995 IEEE Workshop on Nonlinear Signal and Image Processing*, pages 262–265, Halkidiki, Greece, June 20-22 1995.

[16] R. Kresch and D. Malah. Quadtree and bitplane decompositions as particular cases of the generalized morphological skeleton. In IEEE, editor, *1995 IEEE Workshop on Nonlinear Signal and Image Processing*, pages 995–998, Halkidiki, Greece, June 20-22 1995.

[17] C. Lantuejoul and F. Maisonneuve. Geodesic methods in image analysis. *Pattern Recognition*, 17(2):117–187, 1984.

[18] R. Leonardi. *Segmentation adaptative pour le codage d'images.* PhD thesis, École Polytechnique Fédérale de Lausanne, Lausanne, Switzerland, 1987.

[19] P. A. Maragos and R. W. Schafer. Morphological skeleton representation and coding of binary images. *IEEE Transactions on Acoustics, Speech and Signal Processing*, 34(5):1228–1244, October 1986.

[20] F. Marqués, S. Fioravanti, and P. Brigger. Coding of image partitions by morphological skeletons using overlapping structuring elements. In IEEE, editor, *1995 IEEE Workshop on Nonlinear Signal and Image Processing*, pages 250–253, Halkidiki, Greece, June 20-22 1995.

[21] F. Marqués, J. Sauleda, and A. Gasull. Shape and location coding for contour images. In *Picture Coding Symposium*, pages 18.6.1–18.6.2, Lausanne, Switzerland, March 1993.

[22] F. Marqués, V. Vera, and A. Gasull. A hierarchical image sequence model for segmentation: Application to object-based sequence coding. In *Proc. SPIE Visual Communication and Signal Processing-94 Conference*, pages 554–563, Chicago, USA, Oct 1994.

[23] F. Meyer and O. Ribes. Contour coding system in the hexagonal raster. In IEEE, editor, *1995 IEEE Workshop on Nonlinear Signal and Image Processing*, pages 274–277, Halkidiki, Greece, June 20-22 1995.

[24] T. Minami and K. Shinohara. Encoding of line drawings with multiple grid chain code. *IEEE, Transactions on Pattern Analyis and Machine Intelligence*, 8:265–276, March 1986.

[25] H.G. Musmann, M. Hőtter, and J. Ostermann. Object-oriented analysis-synthesis coding of moving images. *Signal Processing, Image Communication*, 1(2):117–138, October 1989.

[26] C. Oddou and A. Sirat. A region-based coding scheme for still image compression. In *Picture Coding Symposium*, pages 1.3.1–1.3.2, Lausanne, Switzerland, March 1993.

[27] M. Pardàs, P. Salembier, and B. González. Motion and region overlapping estimation for segmentation-based video coding. In *IEEE International Conference on Image Processing*, volume II, pages 428–432, Austin, Texas, November 1994.

[28] T. Pavlidis. Algorithms for shape analysis and waveforms. *IEEE Transactions Pattern Analysis and Machine Intelligence*, 2:301–312, July 1980.

[29] H. Peterson. *Image segmentation using human visual system properties with applications in image compression.* PhD thesis, School of Electrical Engineering, Purdue University, West Lafayette, Indiana, January 1990.

[30] P. Salembier. Motion compensated partition coding. In SPIE, editor, *Visual Communication and Image Processing'96*, volume 2727, Orlando, USA, March 1996.

[31] P. Salembier, L. Torres, F. Meyer, and C. Gu. Region-based video coding using mathematical morphology. *Proceedings of IEEE (Invited Paper)*, 83(6):843–857, June 1995.

[32] L.L. Schumaker. *Spline functions: basic theory.* Wiley-Interscience, New York, 1981.

[33] J. Serra. *Image Analysis and Mathematical Morphology.* Academic Press, 1982.

[34] C. Teh and R.T. Chin. On the detection of dominant points on digital curves. *IEEE Transactions on Pattern Analysis and Machine Intelligence*, PAMI-11(8):859–872, August 1989.

[35] P. Van Otterloo. *A contour-oriented approach for shape analysis.* Prentice Hall International (UK), 1991.

[36] P. Willemin, T. Reed, and M. Kunt. Image sequence coding by split and merge. *IEEE Transactions on Communications*, 39(12):1845–1855, December 1991.

5

REGION-ORIENTED TEXTURE CODING

Michael Gilge

Philips Communication Systems, 90411 Nuremberg, Germany

ABSTRACT

Region-oriented image representation offers several advantages over block-oriented schemes, for example better adaptation to the local image characteristics, or object motion compensation as opposed to block-wise motion compensation. For the task of region-oriented image coding, new algorithms are needed which work on arbitrarily shaped image regions or segments instead of the conventional rectangular image blocks. In this chapter we introduce the concept of a generalized transform coding algorithm: based on approximation theory the texture inside image regions is described by a weighted sum of basis functions, for example polynomials. Orthogonalization schemes can be used to obtain a set of basis functions which is orthogonal with respect to the shape of the region, resulting in a generalized region-oriented transform coder. Encoder and decoder structures are derived which do not necessitate the transmission of the basis functions for each segment. Finally we apply the new scheme to image sequence coding at low data rates, based on a segmentation of the motion compensated prediction error images.

1 INTRODUCTION

Image coding is a key technique in a wide variety of applications, ranging from medical imagery, remote sensing, image communication (e.g. video conference, picture phone), commercial products (e.g. digital VCRs) up to the printing industry. Today's products mostly use JPEG or MPEG based algorithms, due to the availability of VLSI chips and reasonable performance. The data compression factors obtained with these transform coding-based algorithms range from 5 to 15 for still images and go up to 100 or more for video coding. In spite the amount of research focused on new and improved image coding algorithms, no major breakthrough has been reported and a saturation point with respect to compression ratio seems to have been reached [46].

Only in the last few years have we seen the advent of so-called second generation techniques, which depart from the conventional waveform coding principle and introduce object- or region-oriented image processing. Figure 1 shows a typical block-based segmentation of an image, motivated by computational constraints and the absence of overhead requirements due to fixed block size, which does not require the transmission of shape information. Drawbacks of this approach stem from the independent treatment of each block, which results in the infamous blocking artifact, caused by discontinuities in the gray values between adjacent blocks. Also, correlation between the image content of adjacent blocks is not exploited, resulting in an upper limit on the obtainable compression ratio (e.g. large homogeneous image regions). Research in the past has concentrated on overcoming these limitations by introducing variable block size [74], overlapping blocks [20], or a combination of DPCM and transform coding [76], to name only a few approaches.

Figure 1 Conventional block-based image processing.

Figure 2 Region-oriented image coding through segmentation.

Second generation image coding techniques have departed from block-oriented techniques, using a region-oriented image segmentation as the basis for further processing. Depending on the application, regions may be formed with respect to motion (e.g. objects in the scene that exhibit similar motion parameters), or with respect to spatial properties (e.g. image regions with homogeneous texture). Another segmentation criterion is the prediction error signal after motion compensated prediction, which will be explained in a later section. Of the many more reasons for employing region-based techniques, two are mentioned here: image manipulation or combination of real-world images and computer graphics.

Second generation techniques are often termed *contour/texture* approaches in the literature; an example is depicted in Figure 2. The regions resulting after segmentation are statistically quasi-stationary and should therefore enable higher data compression ratios, while instationarities in the image are accounted for by the segment boundaries, which are separately coded. Because of this additional overhead, region-oriented image coding is not necessarily more efficient than traditional block-oriented coding. Only if powerful representation schemes for both, segment contours and the texture inside the regions are employed will further compression and better image quality be achieved.

The following contributions from the literature have considered the problem of describing image regions of arbitrary shape using functional approximation: Eden et al. [18] have investigated polynomial representations, but later on restrict themselves to rectangular regions for the sake of fast implementation and mathematical stability. Kocher and Leonardi [44] tackled the problem of arbitrary shape using the Gauss-Jordan algorithm to solve the equations, but they only consider polynomials up to the second order. In the comparison between block-wise DCT and segmented image coding by Biggar et al. [5], only zero and first order polynomials are used in the region representation. Other authors report on spline-functions for image approximation [9] [77] or polynomials, but again used on small rectangular blocks [29].

In addition to earlier papers by the author [22][23] recent work on region-oriented transform coding has been reported in the literature:

In contrast to most other work, Torres et al. [71] describe a combined segmentation and coding approach, which should perform favorably as compared to approaches treating both steps separately. Coding of textures inside the regions is accomplished through a combination of low-order polynomials and stochastic vector quantization (SVQ). This can be viewed as a form of subband coding: polynomials are suitable to represent the mean and low frequency grey value variations, while the synthesized codewords of the SVQ should account for good texture rendering. The approach is not fully region-oriented, however, as square codewords are used in the SVQ coding.

The paper by Chang et al. [10] compares several algorithms for region-oriented still image coding. The range of algorithms compared starts from conventional 2D-DCT coding, where pixels not contained in the regions are set to zero. Some improvements are obtained by filling squared blocks, which are only partly filled by the region to be coded, using mirroring techniques before 2D-DCT. A more involved iterative method attempts to approximate the segment by a few, non-orthogonal basis functions. However, perfect reconstruction can only be guaranteed at prohibitive computational expenses. An interesting aspect in this context is the integration of quantization into the iterative transformation process. The comparative study shows that all methods are outperformed by region-oriented transform coding, which is the focus of this chapter.

For the sake of low computational complexity, two-dimensional coding of arbitrarily shaped image regions is often given up and replaced by one-dimensional coding. Lavagetto et al. [47] scan an image region along a predefined path to convert the texture into a one-dimensional waveform. The obtained 1D-signal is in turn transform coded using orthogonal functions. Jensen et al. [42] and Sikora et al. [65] both use a one-dimensional DCT of variable length: In a first step, the pixel columns of an arbitrarily shaped segment are aligned at the top row and and transformed. The lenght of the 1D-DCT used for each column depends on the column size. In the second step, the procedure is repeated in the horizontal direction, aligning the rows of coefficients first on the left hand side, then again using variable length 1D-DCT to transform each row. Both procedures are computationally efficient, and are therefore attractive candidates for hardware implementations. However, the scanning or the shifting of pixel columns prior to transform coding will somewhat weaken the correlation between neighboring columns and will therefore limit the achievable coding gain. [42] contains an extensive comparison of different coding techniques for arbitrarily shaped image segments.

This chapter introduces a generalized transform coding scheme, which works on arbitrarily shaped image segments instead of rectangular image blocks. The approach contains the conventional block transform as a special case. Generalized moments are introduced in Section 2 as image properties which do include all transform classes as special cases (e.g. Fourier, DCT). The reconstruction from the moment representation leads to a least mean squares (LMS) approximation of the luminance function inside the region (=the texture). It will be shown that LMS approximation and transform coding are in fact equivalent. Shape-adapted orthogonalization of basis functions yields an uncoupled set of well posed normal equations. Practical encoder and decoder structures are introduced in Section 3. Section 4 gives experimental results using the algorithm for the coding of moving video. Section 5 summarizes the most significant results and discusses possible extensions.

2 DATA COMPRESSION BY APPROXIMATION

Assume an image patch with the luminance values given by the function $f(x,y)$. The region is a result of a previous segmentation procedure. To give an example, Figure 3 shows a 3-dimensional plot of a region's texture, obtained from a segmentation of the image shown in Figure 2. Only the luminance information is considered below. For color image coding the segmentation of the luminance component may also be used for the chrominance, due to the correlation between the color components. The coding principle used for the luminance is also applicable to the obtained chrominance segments. It should be noted, however, that the amplitudes of the chrominance samples may be negative for differential color spaces.

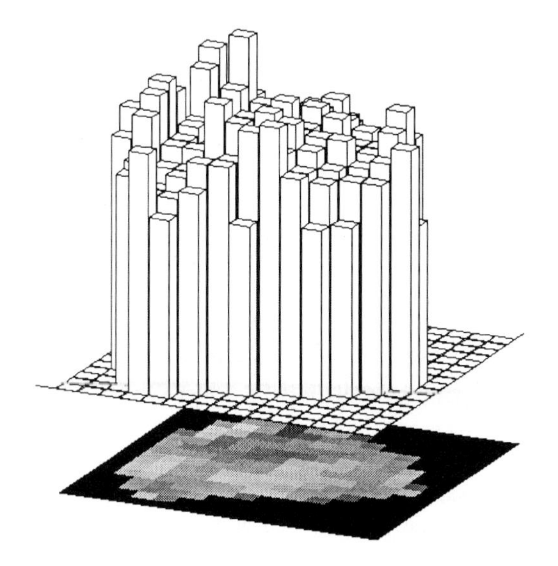

Figure 3 Plot of the luminance values of a region, extracted from Figure 2.

We are in search of a representation of $f(x,y)$, say $g(x,y)$, which approximates the given luminance values and describes them in a compact form, because we want to compress the amount of data. The compression ratio is defined as the number of bits needed to represent the given original pixel values $f(x,y)$ (usually 8 bits/pixel) divided by the number of bits needed to describe the approximation $g(x,y)$. In the case of arbitrarily shaped image segments, which is considered here, the number of bits to code the shape of the segment has to be added to the bit count for the texture representation. Contour coding has already been addressed in Chapter 4, therefore only texture coding will be considered below.

2.1 Generalized moments

A 2-dimensional probability density function of a 2-dimensional random variable is considered. Without loss of generalization any image function $f(x,y)$ restricted to a finite area can be considered such a density due to the non-negativity constraint of image pixels. The function $f(x,y)$ has non-zero values only in a finite part of the x-y plane, namely the region D corresponding to the shape of the segment. We can derive linear properties from the given function using the following operation:

$$m_{ij} = \int_{-\infty}^{\infty}\int_{-\infty}^{\infty} h_{ij}(x,y) \cdot f(x,y)\, dx\, dy = \iint_{D} h_{ij}(x,y) \cdot f(x,y)\, dx\, dy \qquad (5.1)$$

The measurements or properties m_{ij} are called *'generalized moments'*. According to the uniqueness theorem [34], the sequence of moments $\{m_{ij}\} = \{m_{00}, m_{01}, \ldots, m_{kl}\}$ is uniquely defined by the density $f(x,y)$, and conversely $f(x,y)$ is uniquely defined by its associated sequence of moments $\{m_{ij}\}$.

The specification of the measurement kernel $h_{ij}(x,y)$ depends on the application. The approach is general enough to show that different, commonly used measures all involve the same mathematics. For example, consider the kernel

$$h_{ij}(x,y) = e^{-j2\pi(u_i x + v_j y)}. \qquad (5.2)$$

We obtain m_{ij} as samples of the 2-dimensional Fourier transform of the image at the spatial frequency (u_i, v_j). With the choice

$$h_{ij}(x,y) = s(x_i - x, y_j - y) \qquad (5.3)$$

the measurements become samples of the convolution of the image with the point spread function h. Finally we take

$$h_{ij}(x,y) = x^i \cdot y^j \qquad (5.4)$$

and obtain m_{ij} as the geometric moments with the order of the moments denoted by $i+j$. The above definition has the form of a projection of the function $f(x,y)$ onto the monomial $x^i y^j$. This basis functional set $\{x^i y^j\}$ is complete, which is expressed by the Weierstrass approximation theorem [6], but not orthogonal. Orthogonal basis functional sets will be considered in Section 3.

Before we can employ the above derivations, we have to solve two main problems:

- Given a sequence of generalized moments, we need to find an inverse transform in order to retrieve $f(x,y)$ or at least an approximation of $f(x,y)$. In the case of the Fourier transform we know that an inverse exists. But does an inverse of the geometric moment representation exist?
- Assuming we found an inverse operation, for example using the geometric moments. The sequence of possible moments is not finite. How many and which of the moments do we need for an approximation satisfying a given error constraint?

It has been shown in [21] that no direct inverse exists for the case of a finite number of geometrical moments and a continuous function $f(x,y)$. Of course, for the discrete case, i.e. sample image function, Eq. (5.1) can be written in a form with the integrals replaced by summation over the defined area of the given segment

$$m_{ij} = \sum_k \sum_l x_k^i \, y_l^j \, f(x_k, y_l) \qquad x_k, y_l \in D. \tag{5.5}$$

or in matrix form

$$\begin{pmatrix} m_{00} \\ m_{10} \\ \vdots \\ m_{ij} \end{pmatrix} = \begin{pmatrix} 1 & 1 & \dots & 1 \\ x_1 & x_2 & \dots & x_M \\ \vdots & \vdots & \ddots & \vdots \\ x_1^i y_1^j & x_2^i y_2^j & \dots & x_M^i y_M^j \end{pmatrix} \cdot \begin{pmatrix} f(x_1,y_1) \\ f(x_2,y_2) \\ \vdots \\ f(x_M,y_M) \end{pmatrix} \tag{5.6}$$

$$\vec{m} = \Phi \cdot \vec{f} \tag{5.7}$$

If we additionally assume that we are to calculate exactly K moments for $K = M$ given image samples, the above equation has the form of a one-to-one transformation. We know that an inverse is given right away by

$$\vec{f} = \Phi^{-1} \cdot \vec{m} \tag{5.8}$$

with Φ^{-1} as the inverse matrix of Φ. Unfortunately, inverting the matrix Φ is a mathematically ill-posed problem. Besides, we want fewer moments than the given number of pixels inside the image segment to suffice as a description of $F(x, y)$. One possible solution is the method of *'moment matching'*, described in the following. This turns out to be equivalent to the method of least squares approximation, discussed in Section 2.2.

A finite set of K known moments m_{ij} is given and we use a model function $g(x, y)$ which may be expressed in parametric form with N unknown parameters a_n by

$$g(x_k, y_l) = a_1\varphi_1(x_k, y_l) + a_2\varphi_2(x_k, y_l) + \cdots + a_N\varphi_N(x_k, y_l). \tag{5.9}$$

The functions $\varphi_n(x, y)$ are called the basis functional set. For the task of approximation the set must be complete (Weierstarss approximation theorem). If we now require the moments M_{ij} of $g(x, y)$ to be identical to the given moments m_{ij}

$$\begin{aligned} m_{ij} &\stackrel{!}{=} M_{ij} \\ m_{ij} = \sum_k \sum_l h_{ij}(x_k, y_l) f(x_k, y_l) &\stackrel{!}{=} \sum_k \sum_l h_{ij}(x_k, y_l) g(x_k, y_l) \\ \sum_k \sum_l h_{ij}(x_k, y_l) f(x_k, y_l) &= \sum_k \sum_l h_{ij}(x_k, y_l) \sum_{n=1}^{N} a_n \varphi_n(x_k, y_l) \end{aligned} \tag{5.10}$$

we get a coupled set of K equations. If the number of parameters N is equal to the number of moments K, the set of equations has exactly one solution and we solve for the vector $\vec{a}^N$ (the superscript N denotes a vector consisting of N components). Note that the closeness of approximation is determined by the number of moments used. The sequence of given moments approximates $f(x, y)$ by the estimate $g(x, y)$ which follows from the uniqueness theorem:

$$g(x_k, y_l) \approx f(x_k, y_l) = \Phi \cdot \vec{a}^N \tag{5.11}$$

The equality holds in the discrete case for $K = N \rightarrow M$ or in the continuous case for $K = N \rightarrow \infty$.

Before we can attempt an application, we have to make three decisions:

- Select a measurement kernel h_{ij}.
- Select the approximation functions $\varphi_i(x, y)$.

- Determine the number of moments.

In the following we will use a geometrical moment generating measurement kernel $h_{ij}(x_k, y_l) = x_k^i \cdot y_l^j$ and a polynomial approximation function $g(x_k, y_l)$ which yields simple basis functions

$$\varphi_n(x_k, y_l) = x_k^{k(n)} \cdot y_l^{l(n)}, \quad k + l \leq N. \tag{5.12}$$

The justification for the choice of geometrical moments and polynomial basis functions with regard to image representation is motivated by the following reasons:

- Under certain constraints all functions can be developed as polynomials, e.g. Taylor series. We therefore have a kind of generalized tool for describing textures.
- Images predominantly consist of slowly varying surfaces which are well represented by polynomials. The subjective quality of the reconstructed images is pleasant to the human eye.
- Mathematically, polynomials have very simple expressions.

In order to give an example we consider only the moments m_{ij} up to first order, making it a total of three moments ($K = 3$), namely $\vec{m} = \{m_{00}, m_{01}, m_{10}\}$:

$$\begin{aligned} m_{ij} = M_{ij} &= \sum_k \sum_l h_{ij}(x_k, y_l)\ g(x_k, y_l) \\ m_{ij} &= \sum_k \sum_l h_{ij}(x_k, y_l) \sum_{n=1}^{N} a_n \varphi_n(x_k, y_l). \end{aligned} \tag{5.13}$$

Using $\varphi_1 = 1$, $\varphi_2 = y_l$ and $\varphi_3 = x_k$ we can write

$$\begin{aligned} m_{ij} &= \sum_k \sum_l x_k^i y_l^j (a_1 + a_2 y_l + a_3 x_k), \\ m_{00} &= \sum_k \sum_l (a_1 + a_2 y_l + a_3 x_k), \\ m_{01} &= \sum_k \sum_l (a_1 y_l + a_2 y_l^2 + a_3 x_k y_l), \\ m_{10} &= \sum_k \sum_l (a_1 x_k + a_2 x_k y_l + a_3 x_k^2). \end{aligned} \tag{5.14}$$

The summation is over the area of the given segment (including M pixels). In the following section it will be shown that the above approach is identical to least mean square approximation if the basis functions $\phi_n(x, y)$ agree with the measurement kernel h_{ij}.

2.2 Least squares approximation and normal equations

Two kinds of errors are connected with an approximation of the original image region:

- *Measurement errors:* Taking the samples (pixels) from the original continuous luminance function can be thought of as a measurement process. The values obtained differ from the true brightness values by the measurement error, e.g. due to camera noise. Denoting the true but unknown gray values by $s(x_k, y_l)$ and the error term by $e(x_k, y_l)$, the following equation holds:
$$f(x_k, y_l) = s(x_k, y_l) + e(x_k, y_l) \tag{5.15}$$
- *Modelling errors:* The approximation tries to fit the measurements to a selected model. Depending on the physical data generating process, the selected model is more or less suitable for the description. Only in rare cases do we know the physics behind the data generating process which dictates the choice of the model. In the case of modelling image intensities, little can be assumed about the nature of $s(x, y)$: the function is positive and bounded. Considering the human observer, we can also assume $s(x, y)$ to be band limited, otherwise it would contain information that a human observer cannot perceive. Summarizing, it can be said that the approximation error is dependent on the model used.

In order to find the 'best' approximation for a given set of data, we have to select not only a model but we also have to express 'best' in a defined way. For an approximation $g(x, y)$ of $f(x, y)$ in the least mean square sense we have to use the Euclidean distance, also called *'sum of the squared errors'*. Considering the discrete case, the Euclidean distance is computed as

$$||g - f||^2 = \sum_k \sum_l \left(g(x_k, y_l) - f(x_k, y_l)\right)^2. \tag{5.16}$$

The least squares approximation was introduced by Gauss as early as 1795 [67] and offers the following advantages [61]:

- Easy to implement
- Does not require an iterative solution
- Filters zero mean, finite variance noise in an unbiased manner
- If the measurement error $e(x_k, y_l)$ has Gaussian distribution and if the measured data is considered to be a realization of a stochastic process, then the estimate $g(x_k, y_l)$ obtained using the least squares criterion yields the minimum variance unbiased estimate among all unbiased estimators. Furthermore the estimate $g(x_k, y_l)$ is the *Maximum Likelihood* estimate.

The approximating function $g(x_k, y_l)$ can be written in parameterized form as

$$g(x_k, y_l) = a_1\varphi_1(x_k, y_l) + a_2\varphi_2(x_k, y_l) + \cdots + a_N\varphi_N(x_k, y_l), \qquad (5.17)$$

with the basis functions $\varphi_1,\ \varphi_2,\ \cdots\varphi_N$ given through the selected model, e.g. polynomials, and $a_1, a_2, \cdots a_N$ being the parameters which are to be determined in a way that Eq. (5.16) will be a minimum:

$$E = ||g - f||^2 = \sum_k \sum_l \Big(\sum_{n=1}^{N} a_n\varphi_n(x_k, y_l) - f(x_k, y_l)\Big)^2 \qquad \Longrightarrow MIN. \quad (5.18)$$

The parameter vector $\vec{a}^N$, corresponding to a minimum squared error, is commonly determined by differentiating Eq. (5.18) with respect to every component a_q of $\vec{a}^N$ and requiring it to be zero for the minimum:

$$\frac{\partial E}{\partial a_q} = 2\sum_k \sum_l \Big(\sum_{n=1}^{N} a_n\varphi_n(x_k, y_l) - f(x_k, y_l)\Big)\cdot\varphi_q(x_k, y_l) \overset{!}{=} 0 \qquad q = 1, \cdots, N \qquad (5.19)$$

$$\Rightarrow \sum_{n=1}^{N} a_n \sum_k \sum_l \varphi_n(x_k, y_l)\varphi_q(x_k, y_l) \overset{!}{=} \sum_k \sum_l f(x_k, y_l)\varphi_q(x_k, y_l) \qquad q = 1, \cdots, N. \qquad (5.20)$$

Comparing the *normal equations* above with Eq. (5.10), and using the measurement kernel h_{ij} as $\varphi_q(x, y)$, it can be seen that the method of moment matching is equivalent to the Gaussian method of least squares approximation. This proves our earlier statement.

Equations (5.20) are called the normal equations because in an N-dimensional Euclidean space the smallest error vector $e(x, y) = g(x, y) - f(x, y)$ is perpendicular (normal) to the hyperplane spanned by the basis functions $\varphi_q(x, y)$, see Figure 4. The maximum dimension of the Euclidean space is given by the number of pixels in the current segment, M in our case.

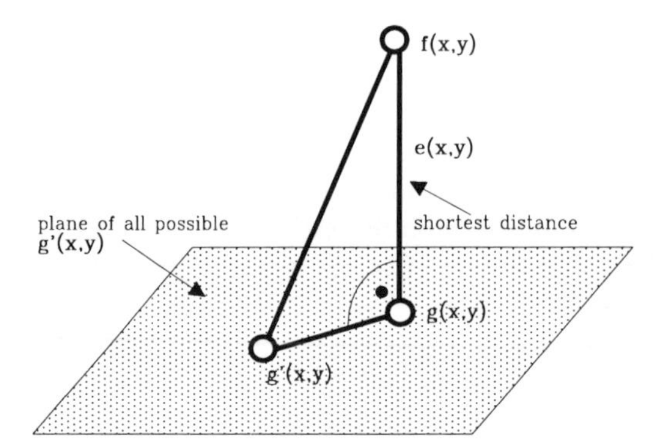

Figure 4 Relationship between a given function f, approximation function g, and error e.

The N equations can be solved for the N values of $\vec{a}^N$ which minimize the approximation error. With regard to the number of unknowns N and the number of given pixels M we can distinguish between three cases:

- M < N: We have fewer data points than unknowns, which cannot be solved for. The system is said to be underdetermined.
- M = N: We have as many data points (pixels) as parameters and the system is guaranteed to have exactly one solution if the columns of Φ are linearly independent. The solution can either be viewed as a *transformation*: mapping the data values onto the basis functional set given by the matrix Φ yielding the parameters $\vec{a}^N$, which are often called coefficients. Another interpretation is to regard the analytical function $g(x, y)$ as an *interpolation* of $f(x_k, y_l)$. Both functions agree on the nodes of the sampling grid, but $g(x, y)$ also gives intermediate values.
- M > N: Now we have more data points than parameters, which is what we want for data compression (unless the representation of the fewer parameters requires more bits than the given pixels, e.g. floating point vs. 8 bit integer). But the distance between g and f cannot be zero for all M sampling positions. This follows directly when attempting to pass a straight line, determined by two parameters, through more than two points in the plane. The system of linear equations is said to be overdetermined. It can

be solved by distributing the fitting error in a way that minimizes the sum of squared errors.

This last case for $M > N$ in the context of image data compression offers the following potential: usually $f(x, y)$ is corrupted by measurement errors as stated above, therefore an exact reproduction of $f(x, y)$ requires not only the information but also the noise to be coded. A least squares approximation, on the other hand, can reduce stochastic errors. The function g may also be selected with respect to the image content. For example, polynomials are well suited for smooth luminance transitions and avoid the oscillatory effect associated with periodic functions, e.g. the Fourier transform. The analytic nature of the approximation simplifies geometrical image operations, for example affine transformations or mathematical operations (differentiation, integration).

On the negative side, the set of equations derived above is coupled and mathematically not necessarily well posed. A solution for $\vec{a}^N$ can be obtained via the Gauss-Jordan algorithm, which is computationally expensive. The coupled nature of the normal equations also prohibits a successive improvement of the reconstruction quality by the addition of more coefficients. In this case all equations have to be solved again from the beginning.

All these problems can be overcome for an orthogonal functional basis. Orthogonal basis functions satisfy the following equation:

$$\sum_k \sum_l \varphi_n(x_k, y_l) \cdot \varphi_q(x_k, y_l) = 0 \qquad \text{for} \quad n \neq q \tag{5.21}$$

The notion of orthogonality greatly simplifies the solution of Eq. (5.20). Now the system of equations is decoupled, yielding

$$\sum_{n=1}^{N} a_n \cdot \sum_k \sum_l \varphi_n(x_k, y_l)\varphi_q(x_k, y_l) = \sum_k \sum_l f(x_k, y_l)\varphi_q(x_k, y_l) \quad q = 1, \cdots, N \tag{5.22}$$

$$a_q \cdot \sum_k \sum_l \varphi_q(x_k, y_l)\varphi_q(x_k, y_l) = \sum_k \sum_l f(x_k, y_l)\varphi_q(x_k, y_l) \quad q = 1, \cdots, N \tag{5.23}$$

$$\Longrightarrow a_q = \frac{\sum_k \sum_l f(x_k, y_l)\varphi_q(x_k, y_l)}{\sum_k \sum_l \varphi_q(x_k, y_l)\varphi_q(x_k, y_l)} \qquad q = 1, \cdots, N. \tag{5.24}$$

The basis functions are said to be orthonormal if they satisfy

$$\sum_k \sum_l \varphi_q(x_k, y_l)\varphi_q(x_k, y_l) = 1. \tag{5.25}$$

Then we can simplify Eq. (5.24) even further

$$\Longrightarrow a_q = \sum_k \sum_l f(x_k, y_l)\varphi_q(x_k, y_l) \quad q = 1, \cdots, N. \tag{5.26}$$

Now the system of equations can be solved easily for each of the approximation coefficients by just mapping the given image onto the respective basis function $\varphi_q(x, y)$. Including more coefficients this time does *not* necessitate the recalculation of previously computed coefficients. This is useful for progressive image transmission, e.g. database browsing. However, the main advantage connected with the orthogonal basis functions is the mathematical stability and simplicity of the approach: no matrix inversion or iterative solution is required and we get a well conditioned system of equations.

Summarizing, it has been shown that, starting from a generalized moment representation, which includes many well-known techniques as special cases (e.g. Fourier analysis), we can obtain a reconstruction of the original function by the method of moment matching. This method conveniently turns out to be equivalent to a least squares approximation. The cumbersome mathematics involved in finding a solution for the normal equations can be completely avoided by using orthogonal basis functions. Taking as many approximation coefficients as given pixels in the image segment results in an error-free reconstruction. For a smaller set of coefficients an approximation of the original luminance of the image can be recovered. The compression ratio depends on the basis functions used. In the following section we will investigate how to obtain orthogonal basis functions with respect to image segments of arbitrary shape.

3 ORTHOGONAL BASIS FUNCTIONS FOR ARBITRARILY SHAPED REGIONS

We have already seen that functions are orthogonal if they satisfy Eq. (5.21). In order to account for a non-rectangular shape of a given image segment, this equation can be extended to regard only coordinate pairs (x_i, y_j) inside the segment boundaries. Formally this may be regarded as a binary weight function $w(x_i, y_j)$ which is zero outside the given segment and equal to 1 inside (the binary weight function may be extended to multiple levels, yielding a spatial prioritization, e.g. at the segment boundaries). The summation is still carried out over the circumscribing rectangle as in Eq. (5.21). But now the functions f and g are orthogonal with respect to the weight function $w(x_i, y_j)$:

$$\sum_k \sum_l w(x_k, y_l)\varphi_n(x_k, y_l)\varphi_q(x_k, y_l) = 0 \qquad \text{for} \quad n \neq q \tag{5.27}$$

A sample image segment, the sampling grid, and the weight function are depicted in Figure 5 to explain the introduction of shape to the principle of orthogonality. It is clear that two given functions cannot be orthogonal with respect to several differently shape segments. Therefore it is necessary to find a specific set of orthogonal functions for each particular shape. Rectangular image blocks are contained in the treatment here as a special case without loss of generality, i.e. the discrete cosine transform is a special case of the more general method introduced here.

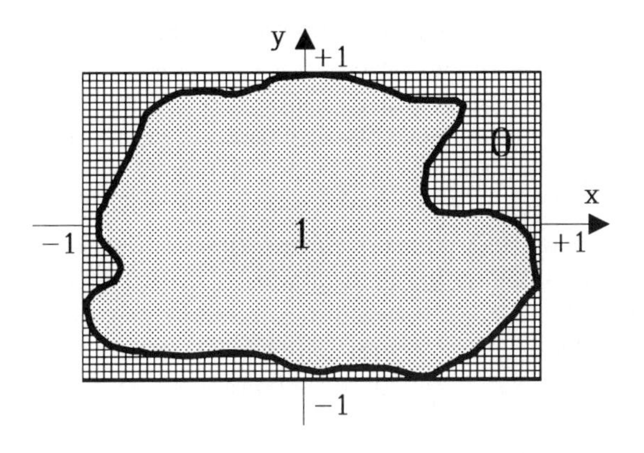

Figure 5 Example of a window function for shape-adapted orthogonal functions.

The following theorem guarantees that we can actually find an orthogonal set of basis functions for a given shape:

Theorem: *If the set of functions $\vec{u}_0, \vec{u}_1, \ldots, \vec{u}_n$ is linearly independent in an n-dimensional subspace A_n, then there exists a set of functions $\vec{q}_0, \vec{q}_1, \ldots, \vec{q}_n$ which are orthogonal with respect to the same subspace and moreover the functions $\vec{q}_n$ are linear combinations of the functions $\vec{u}_0, \vec{u}_1, \ldots, \vec{u}_n$.*

The proof of the theorem can be found in the literature [75]. The Appendix gives a review of two algorithms which generate orthogonal basis functions for a given shape and given linear independent approximation functions.

Orthogonalization example

Orthogonalization of a given set of basis functions may also be necessary in the conventional block-oriented mode of operation, if the basis functions used are not orthogonal from the beginning. Considering a rectangular area of support for the moment and using the lowest order polynomials $1, x, y, x^2, xy, y^2, \ldots$ which are not orthogonal on any area of support, the orthogonalization process yields the well known Gram-Schmidt polynomials. These polynomials are discrete orthogonal versions of the Legendre polynomials, which are only orthogonal in the continuous case. Cosine functions are special in that they are

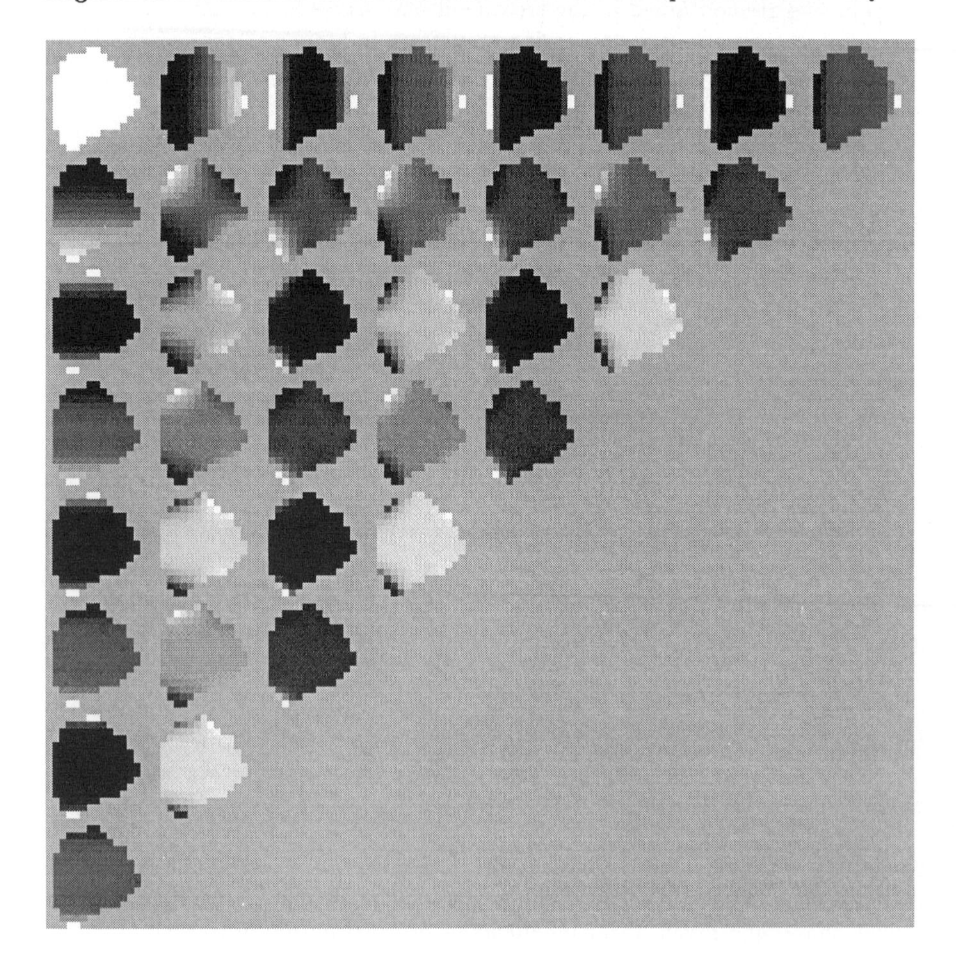

Figure 6 Basis images for non-orthogonal polynomials.

orthogonal in the continuous as well as in the discrete case; however, only on rectangular areas of support again.

Considering the shape of the segment already depicted in Figure 3 and using the non-orthogonal polynomials $x^i \cdot y^j$ as linear independent starting base, we obtain the set of basis functions shown in Figure 6. The polynomials have been defined on a square circumscribing the shape. The functional values of the

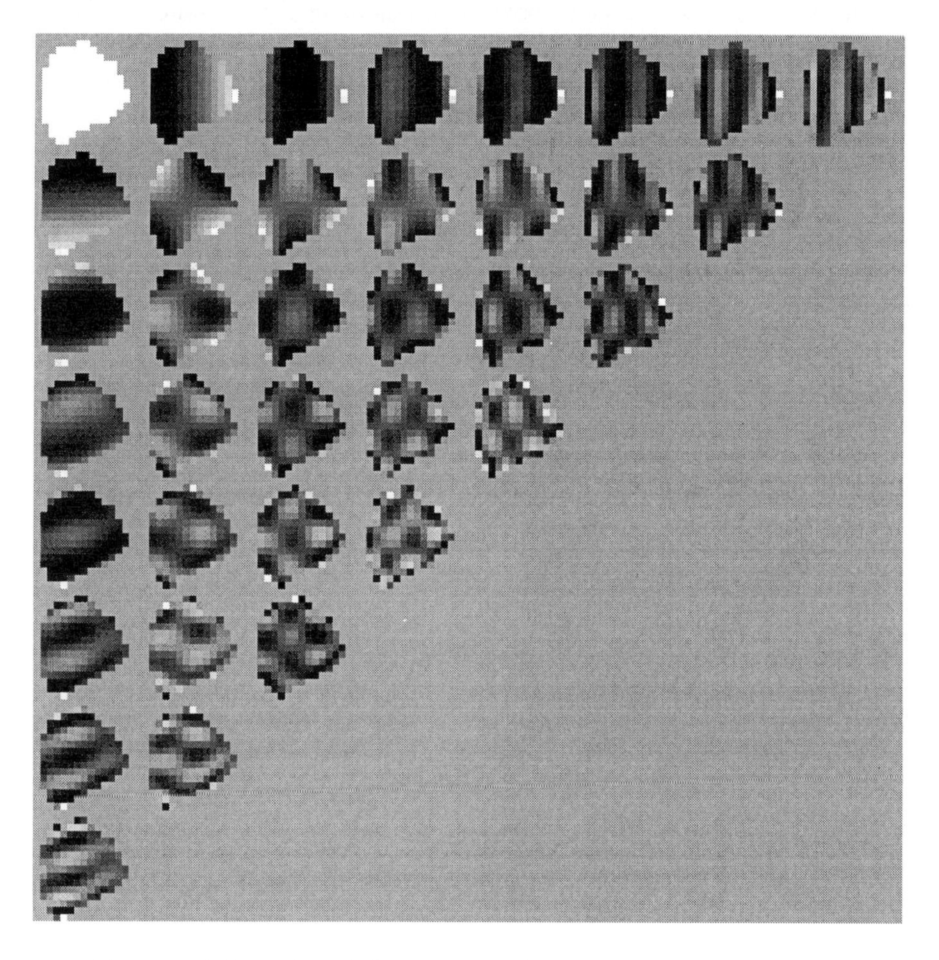

Figure 7 Polynomial basis images after orthogonalization with respect to the given shape.

polynomials are in the range [-1, +1], in the vertical as well as the horizontal directions. The lowest order (zero) basis function is in the upper left corner of Figure 6, with increasing order of the polynomials in the direction of the respective coordinate axes. Adjacent to the zero order basis function are the ramp functions $\varphi_{10} = x$ and $\varphi_{01} = y$. Most of the higher order polynomials exhibit a similar structure, because most of the variations are to be found towards the boundaries for $|x|, |y| \rightarrow 1$.

The problems connected with polynomial approximation, as reported in the literature, immediately become obvious in regard to the structure of the basis functions. It is clear that higher order polynomials do not contribute to an increased quality image reconstruction, due to the similarity of the higher order polynomials to the lower order ones. Also, textures inside the segments are hard to represent due to the flat response in the center of the segment. These problems have prompted a finer segmentation into tiny segments, which could be represented using non-orthogonal and low-order polynomials. This approach, however, is in contrast to the original objective of representing large segments as long as they exhibit homogeneous texture. Note that the size of the segments directly affects the overall compression ratio obtained, caused by the bit rate needed for contour coding.

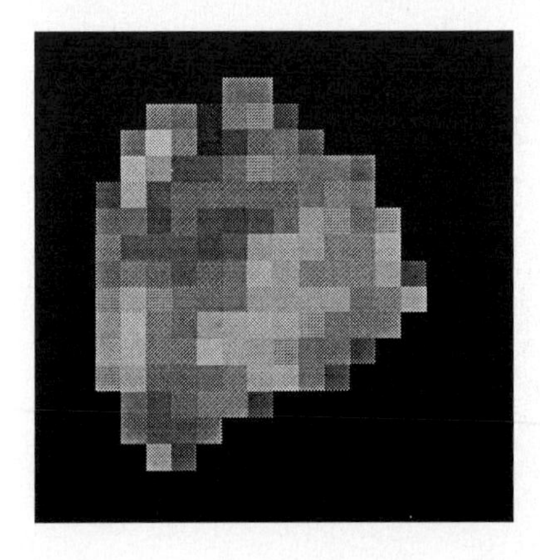

Figure 8 Original segment.

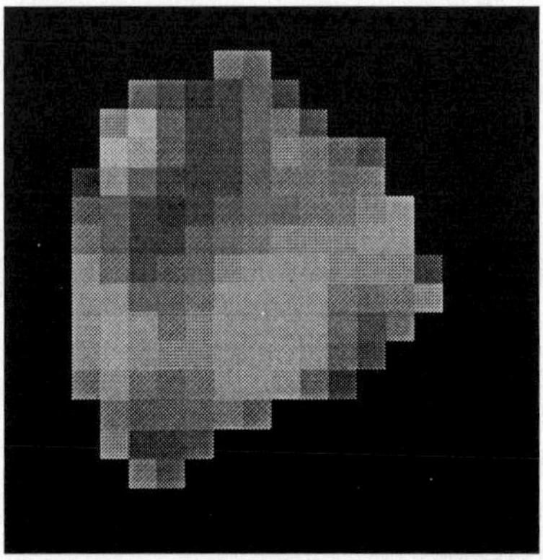

Figure 9 Reconstructed segment using the orthogonal basis functions depicted in Figure 7.

After employing an orthogonalization method we obtain the orthogonal polynomials depicted in Figure 7. The different structures represented by the basis functions can clearly be seen and immediately suggest a good representation of textured image regions.

A sample reconstruction using the orthogonal basis function is shown in Figure 9. Figure 8 shows the original segment for comparison. The segment consists of 132 pixels, while the reconstruction uses polynomials up to order 9, totalling 45 coefficients. The figure clearly shows the noise reducing effect of the approximation. At the same time, the features of the segment are retained. This would not have been possible using non-orthogonal polynomials up to second order, a popular approach often found in the literature.

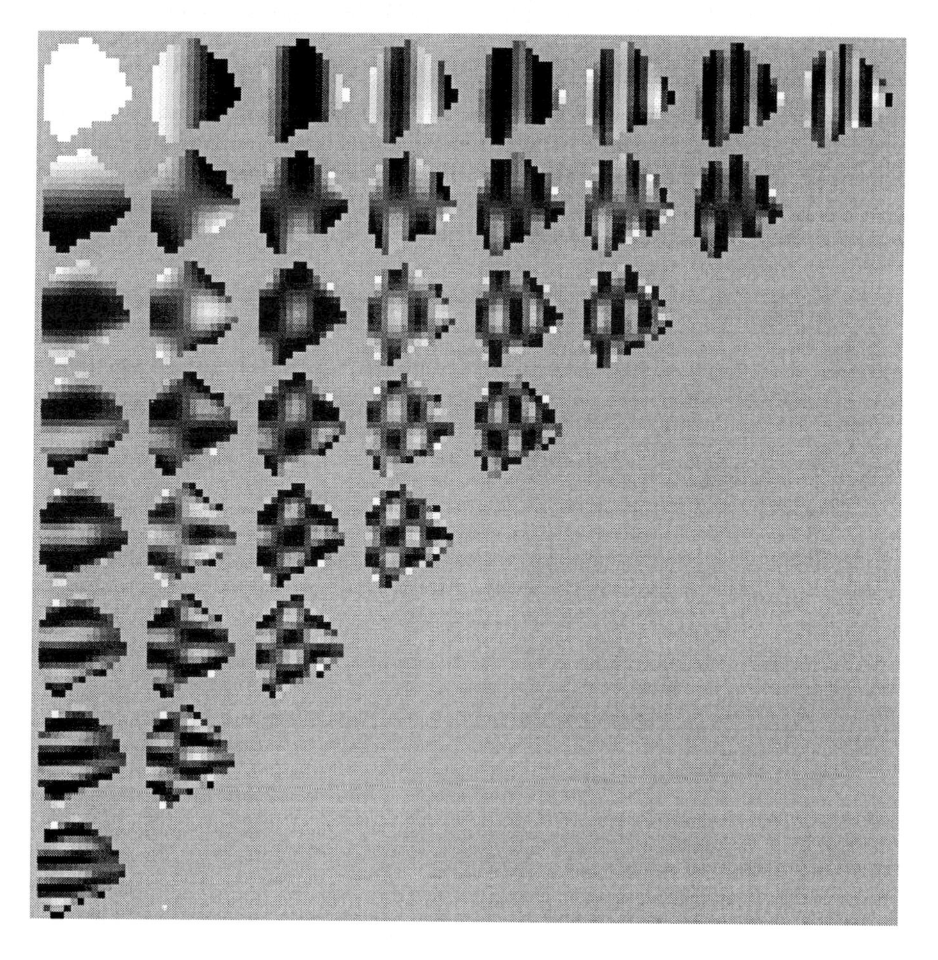

Figure 10 Shape adapted cosine basis functions yielding a shape adapted DCT.

The discrete cosine transform (DCT) is renowned for its good energy compacting capabilities. It has also been shown that the DCT approaches the decorrelation properties of the optimum Karhunen-Loéve transform (KLT) for a specific class of image functions [36][73]. These good properties of the DCT may be employed for region-oriented texture coding by orthogonalization of the cosine basis functions with respect to shape. Using the same shape as above, Figure 10 shows the result of orthogonalization. Comparison with Figure 9 verifies that the orthogonal polynomials are similar to the cosine basis functions. A more elaborate comparative analysis of the performance of different basis functions, including Walsh functions and splines, indeed proves to be marginally favorable for the orthogonal polynomials.

Summarizing, it has been shown that the introduction of orthogonal basis functions, which are orthogonal to the specific shape of a given segment, gives a solution to the approximation problem for arbitrarily shaped regions. The approach is characterized by mathematically stable and simple operations, i.e. no iterations and no matrix inversion. The set of approximation coefficients is finite, unlike the infinite set of non-orthogonal moments. For a progressive or improved reconstruction, higher order coefficients can be added *without* recalculating any previously computed coefficients. This property is required for hierarchical representation. As the orthogonalization process is not limited to polynomials, any linear set of independent functions can be used for the approximation. Examples of basis functions suitable for the task of image data compression are polynomials and cosine functions. These functions have been used for a sample orthogonalization with respect to a real world image segment.

The following section introduces encoder and decoder structures which implement a complete image coding system using region-oriented texture coding.

4 REGION-ORIENTED TRANSFORM CODING

The first step in region-oriented transform coding is segmentation. Clearly, the segmentation is connected to the coding process, but also depends on the application. Therefore both steps – segmentation and coding – should be considered as a combined optimization problem [44]. For example, in the case of still image coding a segmentation into quasi-stationary image regions is adequate. One criterion for a 'good' segmentation – bearing in mind the achievable compression ratio as a goal – is the codability of a segment: only if the aggregate data rate needed for contour and texture coding is minimal. If the segmentation process does not take the capabilities of the texture coding process into account, over- or undersegmentation of the image will result. Oversegmentation results in many tiny segments of low frequency content, caused by breaking up even, homogeneous textures into subsegments. The data rate is largely dominated by the contour code. If the segmentation process overestimates the structure describing capabilities of the texture coder the result is undersegmentation: few large segments contain residual discontinuities, causing an excessive data rate for the texture code. In order to concentrate on the coding aspects, the following discussion of region-oriented transform coding will be based on a given segmentation, which may be assumed to be the outcome of such an optimization process.

The next paragraph introduces the principle operation for region-oriented texture coding as it applies to still and moving image coding. In the context of video coding, the first frame and intermediate anchor frames (I-frames in MPEG schemes) are coded in intraframe mode, e.g. as still images. The scheme may also be used for coding of prediction error images, which has been demonstrated in the paper by [59]. Parallel to conventional block oriented coding, transform coefficients are quantized before transmission. An experimental still image coding example will illustrate the effect of quantization. An example for region-oriented video coding is discussed in Section 5.

4.1 Encoder and decoder structure

A set of orthogonal basis functions can only be orthogonal with respect to the considered shape and not to several shapes at the same time. This would necessitate the transmission of the basis functions used in addition to the contour code and the approximation coefficients in order to enable a reconstruction at the receiving end. This additional overhead, however, would certainly nullify all compression gains. In the following a scheme is introduced which does not require the transmission of the used basis functions. Figure 11 shows a general block diagram of the proposed coding scheme, which may be used for still and moving video coding. Starting with a given segmentation the coder can be structured into a contour and a texture coder.

The *contour* coder has to find an efficient representation for the shape of the current segment to be coded. The representation may either be errorless, e.g. using run-length codes, or approximate. Approximately 1.2 bits/contour point are usually required for error-free coding [17]. Approximate coding is attractive from the data compression perspective; however, segment overlap and gaps need to be resolved. For approximate contour coding, the orthogonalization

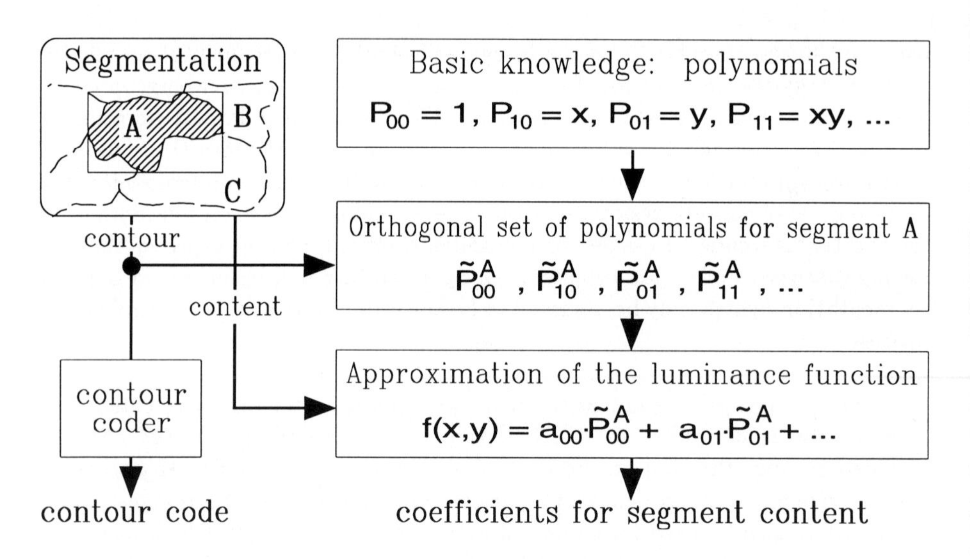

Figure 11 Structure of a region-oriented image encoder based on orthogonal polynomials for texture coding.

process of the basis function may be based on the reconstructed shapes inside the encoder. Investigations have also shown that shape changes due to approximate contour coding have only a negligible effect on the orthogonalized basis functions. See Chapter 4 for a detailed discussion of contour coding.

The *texture* coder operates segment by segment; the current segment under coding, for example, is labeled A in Figure 11. Starting with a set of basis functions P_{ij}, the so-called 'basic knowledge', an orthogonal set of basis functions $\tilde{P}_{ij}^{A}$ with respect to the shape of the segment is generated. Finally the orthogonal basis functions produced are used for an approximation of the texture inside the segment. The approximation coefficients A_{ij} are transmitted to the receiver, together with the contour information.

A block diagram of the decoder is depicted in Figure 12. The shape of the current segment is recovered first. It is easy to see why the transmission of the orthogonal basis functions used is not required: if the receiver uses the same 'basic-knowledge', e.g. the same polynomials, as a linear independent base, the orthogonal basis can be regenerated at the receiver. The coded segment shape contains the information on how to derive the basis functions. Finally, the texture inside the segment is reconstructed by a weighted summation of the orthogonal basis functions, with the approximation coefficients being the weighting factors.

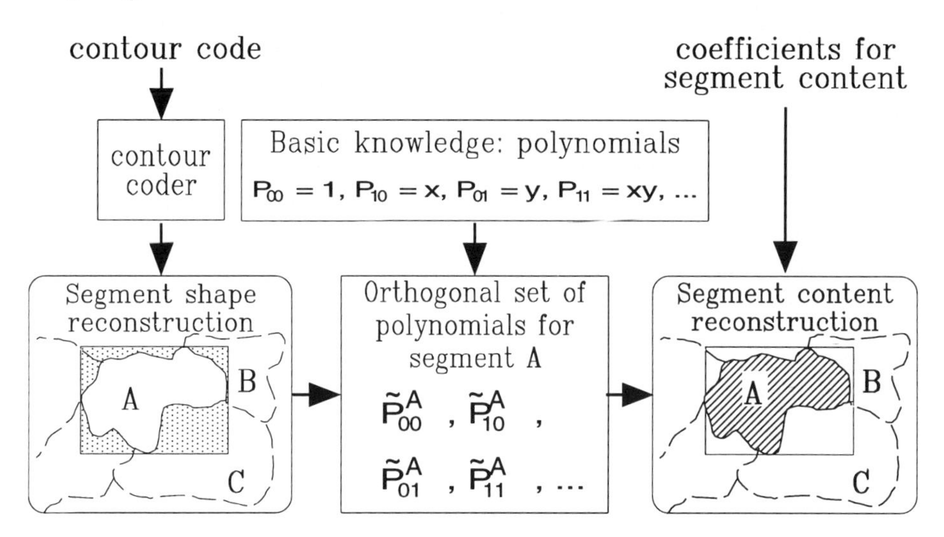

Figure 12 Structure of a region-oriented image decoder. Note that orthogonal polynomials used in the encoding are recovered using the shape information.

Summarizing, the coding and decoding process consist of three steps:

- *Encoder:* Description of the current segments contour $\Rightarrow$ generation of a set of orthogonal basis functions for the current segment $\Rightarrow$ approximation of the current segment's texture.
- *Transmission:* Contour code and approximation coefficients
- *Decoder:* Reconstruct the shape of the segment $\Rightarrow$ generate a set of orthogonal basis functions for the current segment $\Rightarrow$ recover an approximation of current segment's texture.

The basis set of linear independent functions, the 'basic-knowledge' of Figures 11 and 12, is certainly not limited to polynomials. As already mentioned in Section 2, the only constraint limiting the range of possible functional sets is the requirement of linear independence, which is not a serious limitation. This opens up the possibility of using a wide variety of basis functions as a starting point for the orthogonalization. Even more interesting might be the possibility of adapting the used set to the properties of the current segment under coding: smooth luminance transitions might be described favorably by polynomials, while Walsh functions might be better suited for areas containing text or binary structures. Last not least, cosine basis functions, resembling a shape adapted DCT, might be best for the representation of highly structured textures.

Another possible adaptation parameter is the quality of the approximation, expressed in terms of the number of approximation coefficients taken into consideration. Most polynomial representation schemes in the literature use a fixed order of approximation, e.g. Eden [18] uses polynomials up to the second order. As shown in the previous section this is due to the use of non-orthogonal functions, which necessitates the recalculation of all previously determined coefficients, if only a single additional coefficient is to be included. The situation changes for the orthogonal scheme used here: thanks to the decoupled normal equations, the quality of the approximation may be successively improved, coefficient by coefficient.

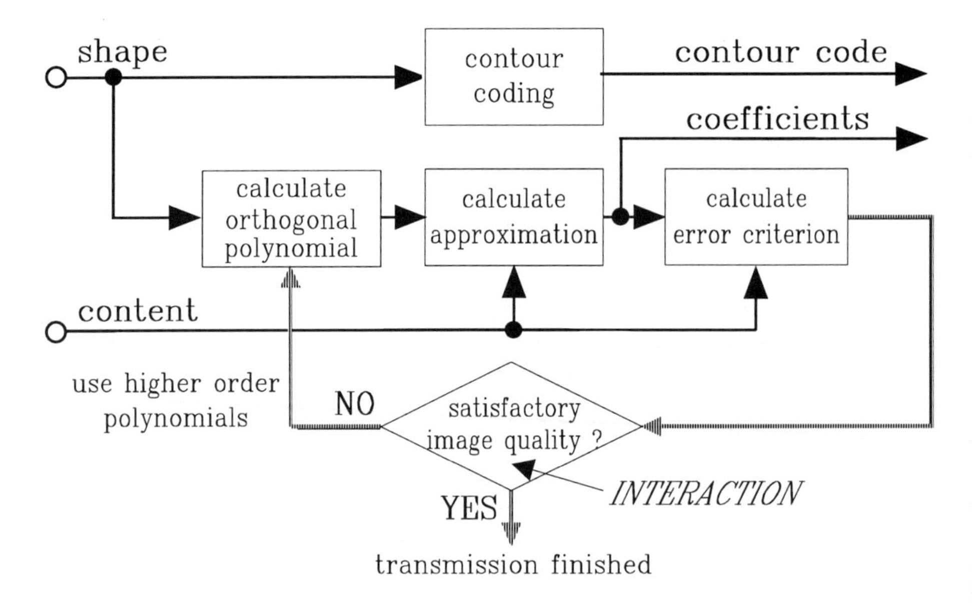

Figure 13 Structure of a hierarchical image encoder for progressive image transmission or interactive services.

This is also useful for hierarchical schemes, e.g. progressive image transmission, where coarse picture information is transmitted first. The user will see an image of contours with only mean luminance values filled in the segments. Such a scheme may be valuable for database browsing or low data rate transmission links. With the inclusion of additional coefficients, textures will begin to appear inside the segments. Figure 13 shows the block diagram for such a hierarchical image encoder. The approximation error is monitored in the encoder for each step of the hierarchy by a comparison of the current reconstruction with the original image. Coefficients may be selected automatically by an overall quality objective that has to be met. Alternatively, user interaction could prompt the calculation of additional coefficients, e.g. the user selects an image area where more detail is wanted. Hierarchical region-oriented transform coding is especially facilitated by the orthogonalization scheme of Schmidt (see Appendix) which allows for an orthogonalization of the respective next basis function, as needed. In contrast, the scheme by Householder computes all orthogonal basis functions at once.

Fast algorithms have been developed [56] in order to overcome the computational requirements of the orthogonalization.

4.2 Quantization of coefficients

A transform is a special kind of approximation, where all approximation coefficients are calculated and therefore an exact reconstruction is possible (in the discrete case only). So far, a transform does not yield any data compression, but it is fully reversible. Only in connection with a quantization of the coefficients is an irreversible reduction in the amount of data obtained. Usually not all coefficients are retained. Block-oriented transform coding schemes are mostly zonal or threshold coder [78][39], where the position or the amplitude determines whether a coefficient is retained or not. For the region-oriented scheme introduced above, only a subset of all possible coefficients is calculated to begin with.

The data rate required for the description of the texture inside a given segment is the product of the number of approximation coefficients calculated and the number of bits allotted to each coefficient. Quantization of the coefficients will reduce the overall data rate but will result in an increased reconstruction error. The relationship between reconstruction error and the required accuracy of the coefficient quantization is discussed in this section. Note that the reconstruction quality is a combination of the errors caused by the quantization of the coefficients and the error caused by the restriction to a finite number of approximation coefficients.

In order to illustrate the quantization task, look at the image depicted in Figure 2. It has been segmented into 128 regions, indicated by the overlayed contours. Using a total of 3209 coefficients for an approximation with orthogonal polynomials the reconstruction depicted in Figure 14 is obtained. The coefficients

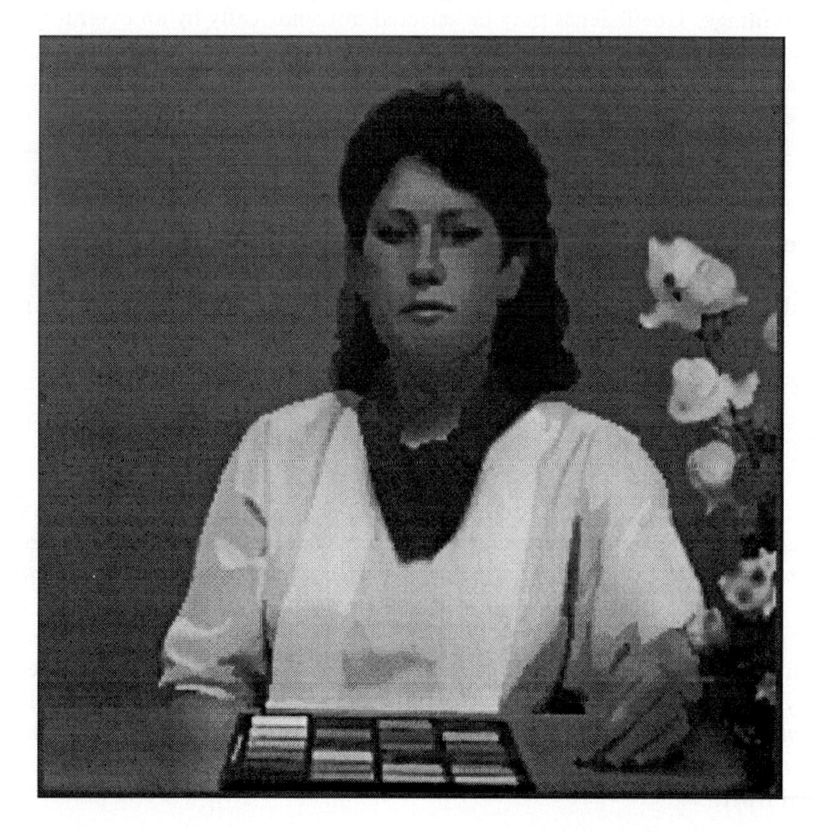

Figure 14 Reconstructed image using a total of 3209 coefficients *without* quantization. 128 segments were used based on the segmentation depicted in Figure 2.

have not been quantized. Any degradation in image quality is attributable to the limited number of coefficients used in the recons. The order of the used polynomials used depends on the segment: the large background area has only been approximated using polynomials up to second order, while the maximum order used for any segment included polynomials up to an order of 9. A close-up comparison of the facial region, which happened to be segmented into a single larger segment, is shown in the following Figures 15 and 16. Figure 15 just introduces the original image, without any coding, at the same enlargement as

of bits per coefficient used. At 8 bits/coefficient the reconstruction quality remains mostly degraded by the limited number of approximation coefficients. Only for an expenditure of fewer bits per coefficient will the coarseness of the quantization have a significant influence on the quality of the reconstructed image. The result of this quantization scheme using 8 bits/coefficient on the reconstruction quality for the sample segment is depicted in Figure 22. This example verifies that at 8 bits/coefficient the quality remains almost unchanged from the reconstruction without quantization.

Summarizing, in spite of the wide dynamic range of the coefficients and a necessary fine resolution for small coefficient amplitudes, a quantization scheme has been designed, using at most 8 bits/coefficient. In combination with a successive approximation of a given segment, basis function by basis function, a further refinement is possible: after computing an additional basis function φ_q and the corresponding approximation coefficient a_q, the reconstruction quality including the new basis function is compared to the previous one. Only if the reconstruction is improved by a given threshold will the coefficient a_q be quantized, coded, and transmitted.

5 REGION-ORIENTED VIDEO CODING

When coding moving images, the addition of the time domain further complicates the segmentation process. In general the performance of a video coding system depends to a fair degree on the ability to compensate motion in the scene (motion compensated prediction). A segmentation of image sequences without consideration of the motion compensation aspect is likely to be suboptimal. Segmentation and motion compensation have to be considered a mutually dependent problem [59][31]: boundaries of moving object have to be known in order to compensate for their motion. However, the detection of object boundaries requires information about object motion. Considering a static object for the moment, structures in the image may be attributable either to textures or to object boundaries. The segmentation algorithm, lacking information about the object shape, will produce segments in order to accommodate texture coding, but the segment borders generated will in general not coincide with object boundaries, unless the object clearly stands out against the background. As soon as the object starts moving, knowledge about the shape of the moving object may be gathered in order to build up an object boundary description. Now motion compensation is facilitated using this shape description and the segmentation should be adapted accordingly in order to account for recognized moving objects. Chapter 3 is dedicated to image sequence segmentation. Below we introduce an alternative to the computationally expensive image sequence segmentation prior to coding, which is based on a segmentation of the prediction error images.

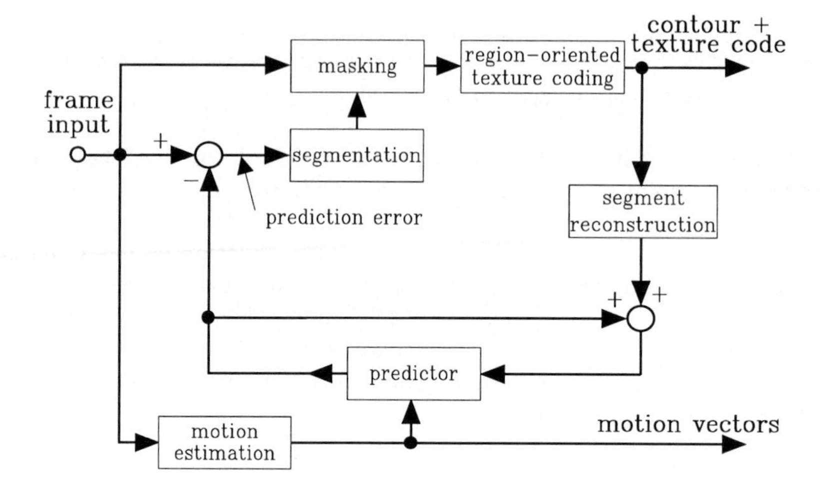

Figure 23 Block diagram of the proposed image sequence coding scheme based on conventional motion estimation in combination with segmented prediction error image coding.

Block-oriented motion compensation, for example by block matching, has been thoroughly investigated in the past and appeals because of its low computational complexity. Figure 23 shows the block diagram of the proposed region-oriented image sequence coding scheme. Motion estimation is performed conventionally, using hierarchical block matching. Motion information is transmitted to the receiver using block-oriented motion vectors. Encoding of the motion vectors follows the MPEG scheme of variable length coding. The motion compensated prediction image is then subtracted from the actual incoming new image, yielding the prediction error image. In order to illustrate the operation of the coding scheme, Figure 24 shows the effect of motion compensation. Figure 24(a) and 24(b) show two adjacent original frames of a test sequence. The difference between both frames is caused by motion in the scene, as depicted by the difference shown in Figure 24(c), where medium gray indicates zero difference. With motion compensation most of the differences attributable to scene motion can be accounted for and the resulting prediction error image is depicted in Figure 24(d).

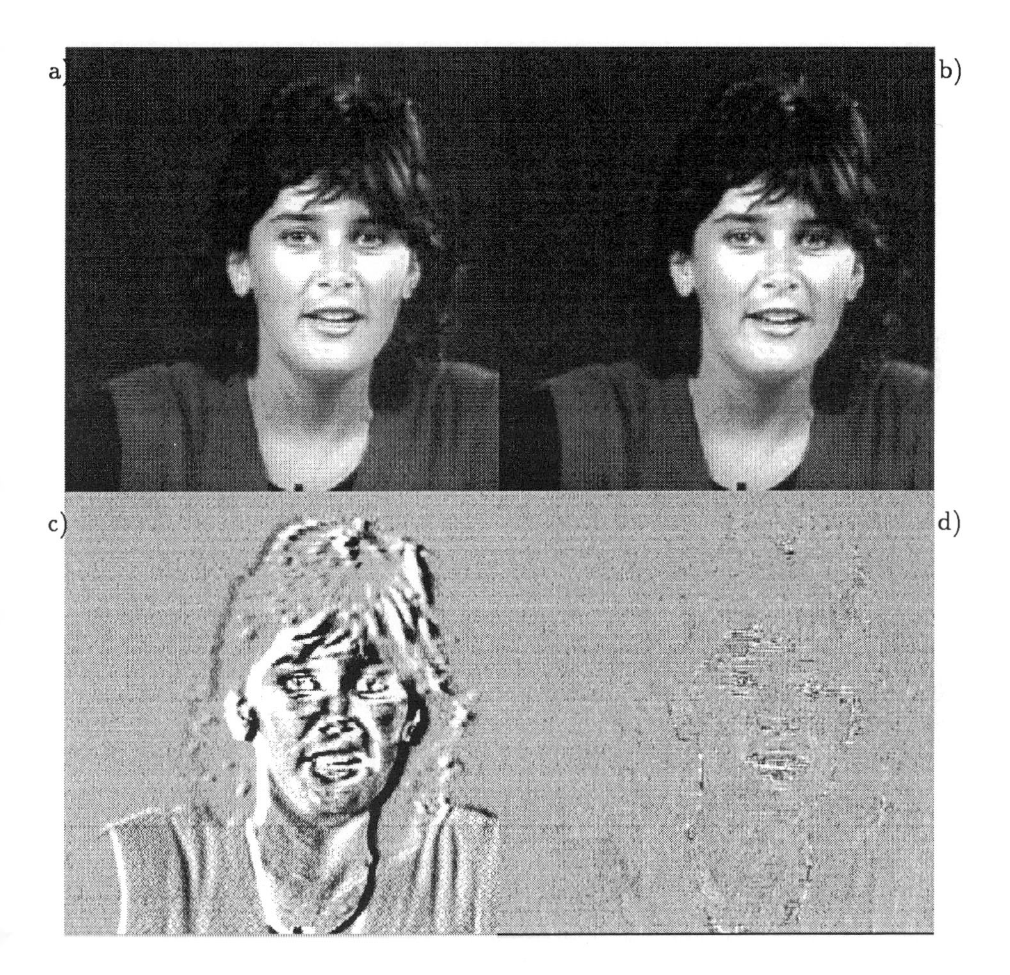

Figure 24 Test sequence "Miss America": a) original frame 80, b) original frame 83, c) difference between original frames (amplified four times, medium gray indicates zero difference), d) difference after motion compensation = prediction error image (amplified four times).

5.1 Prediction error image segmentation

The visible structures in Figure 24(d) indicate that motion compensation failed only in a finite number of small regions. This observation makes prediction error images a candidate for segmentation: only areas with high prediction error values need to be coded. The conventional block oriented approach is very coarse in this respect, because relevant structures are often only a few pixels wide. Block-wise coding will therefore result in an unnecessary coding of areas where motion compensation did in fact work satisfactorily. To make things even worse, block-based coding might introduce error in these successfully compensated regions due to the quantization of the spectral coefficients. One possible

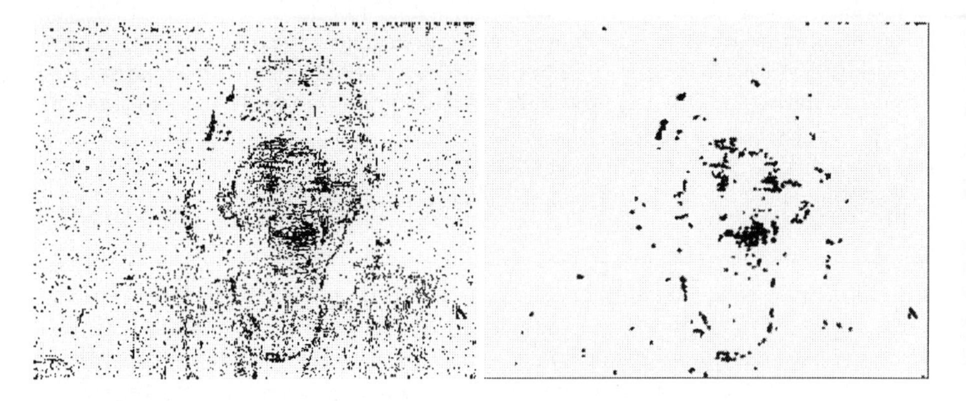

Figure 25 Result of the thresholding operation with a value of *pm* 5 on the prediction error image.

Figure 26 Thresholded prediction error image after median filtering. Small segments may be removed in a further step.

segmentation method for prediction error images consists of a simple thresholding operation followed by a filtering step. The thresholding operation on the prediction error image will filter all pixels where motion compensation worked well. The result obtained with a threshold value of $\pm$ 5 is depicted in Figure 25. All pixels shown in black exceed this error amplitude. In addition to regions corresponding to moving parts in the image, many other pixels appear scattered over the image, which do not seem to be connected to motion in the scene but are rather caused by noise. A filtering step eliminates these noise induced pixels exceeding the threshold. A modified median filtering using a 3 x 3 neighborhood is employed to retain only those points which have at least 5 super-threshold points in their direct surrounding. Finally, all small segments including no more than 8 pixels are discarded. The result of this simple approach is a segmentation yielding only larger areas where motion compensation

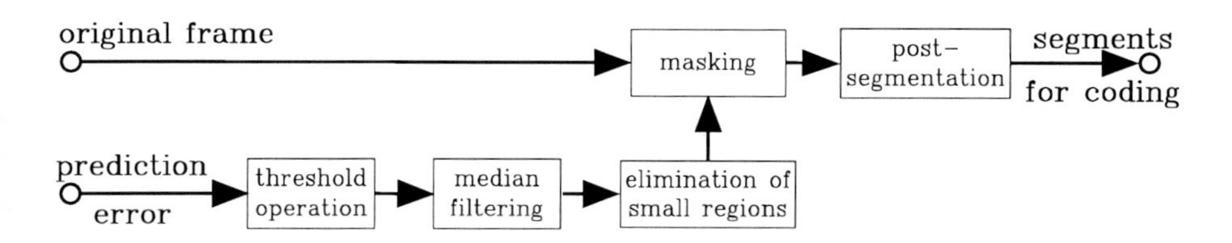

Figure 27 Block diagram of the prediction error image segmenter, comprising a thresholding operation followed by a filtering step. Retained regions are used to mask the corresponding regions in the original frame for region-oriented texture coding and replenishment.

failed and replenishment is required. A block diagram of the thresholding and filtering steps is depicted in Figure 27. The simulation result after filtering is shown in Figure 26. The outline procedure may be refined by an adaptation of the threshold value, the required neighborhood for sustained pixels and the size of the smallest retained segment to the respective available data rate. Inside a coder architecture working over constant rate communication links, buffer fullness is usually used to control these values.

Statistical investigation of motion-compensated prediction error images reveals a low correlation between adjacent pixels and a high-pass characteristic of the error image [24]. A direct approximation of the prediction error segments is thus not promising, especially if only low order polynomials are used. Instead, the prediction error segments are used to mask the corresponding regions in the original image frame. In image sequence coding terms, all coding is done in intraframe mode, using conditional replenishment. The prediction error image serves as an indicator, where prediction failed and replenishment is required. Coding the original image region instead of the prediction error image seems like a step back, since most of the information is already at the receiver. It turns out, however, that in most cases coding of the errors requires a higher bit rate as compared to coding the original image information.

Motion estimation tends to fail around the borders of moved objects, as can be seen from the prediction error images, which will cause both sides along the edge to appear in the prediction error image. Therefore, after masking the original image frame, the resulting segments may contain discontinuities. To overcome this situation, larger segments containing more than 40 pixels are subjected to a post-segmentation process. Based on known region-growing or split and merge techniques, for example, these few larger segments are eventually broken up into two smaller ones in case the overall segment contains a luminance edge.

The following paragraph will illustrate the obtained coding resulting from the introduced scheme.

5.2 Experimental results

A complete image sequence coder according to Figure 23 has been simulated for operation at low data rates. The hierarchical motion estimation works with a block size of 16×16 pels. The common intermediate image format (CIF) at 352×288 pixels is coded at 15 frames per second. Figure 28 shows the original

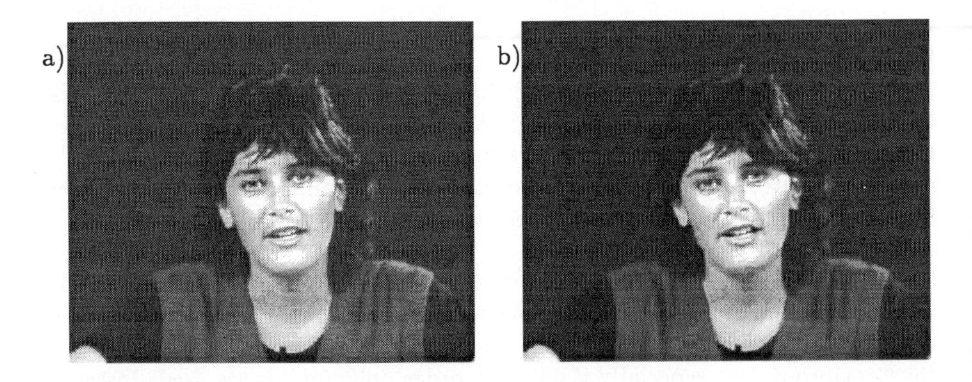

Figure 28 Test sequence "Miss America": a) original frame b) reconstructed frame #150 after region-oriented prediction error image coding at 64 kbit/s.

Item	**frame 80**	**frame 100**
Number of segments	186	142
Number of contour pixels	1969	1894
Number of approximation coefficients	1107	891
Bitrate for motion vectors	1814 bits	2140 bits
Bitrate for contours	2362 bits	2272 bits
Bitrate for texture	8856 bits	7128 bits
Total bitrate	13032 bits	11540 bits
Signal to noise ratio (SNR)	35.7 dB	35.6 dB

Table 1 Accounting information for coding of frame number 80 and 100.

frame number 80 and 100 in the upper half and the corresponding coding results in the lower half. Table 1 gives the bit rate accounting information for the two frames. Note that the contour coding has been simulated at a rate of 1.2 bits per contour point [17] and included in the total bit rate for each frame.

6 CONCLUSIONS

A texture coding scheme for arbitrarily shaped image segments has been introduced. Starting from a generic moment representation it could be shown how special transforms such as Fourier are derived as special cases. Furthermore the connection between moment representation, transform and approximation techniques has been derived. The problems associated with the important method of least squares approximation, namely coupled and eventually ill-posed normal equations, which do not allow for a hierarchical description, could be overcome by the introduction of orthogonal basis functions. In order to solve the original problem to provide a description for arbitrarily shaped image regions, a window function was introduced. This led to the definition and construction of basis functions, which are orthogonal with respect to a particular shape. Two orthogonalization schemes are described in the following Appendix.

Based on the theoretical results, an encoder and decoder structure has been introduced which implements shape-adapted transform coding without the need to transmit the used set of basis functions for each segment. The system enables progressive image transmission and/or user interaction. For all practical implementations the approximation coefficients need to be quantized. A combination of a uniform and a non-uniform quantizer has been designed, which gives almost imperceptible reconstruction errors at a rate of 8 bits/coefficient.

The important field of image sequence coding has been addressed, combining conventional, economic block-matching motion estimation with second generation coding techniques for the prediction error image. Prediction error images indicate the spatial success of the motion compensation process. Significant non-zero values show up in areas where motion compensation failed and that need to be be replenished. This has been accomplished by thresholding and filtering the prediction error image, yielding segments which were used to mask the corresponding areas of the original frame. These segments were in turn coded using region-oriented texture coding. Experimental results demonstrated the viability of the concept for the application of low bitrate image sequence coding.

Also for the emerging field of very low bit rate coding, targetting rates as low as 16 kbit/s, the introduced region-oriented approach should be useful: at these low data rates, coding must be restricted to relevant picture information, e.g. facial regions or moving objects. In order to avoid blocky descriptions of these image features, powerful region-oriented schemes are needed. The computational burden of the segment-wise orthogonalization of the basis functions can be lowered with fast algorithms [56].

Up to now, the overall compressed data rate for most video coding algorithms in use today is the sum of bits needed for motion compensation and prediction error coding. It is up to the designer to split the allowable data rate between the coding of motion and the coding of prediction errors in order to arrive at the best possible overall coding performance in a given situation. This becomes ever more important at lower data rates. For a successful region-oriented video coding at very low bit rates, a joint treatment of segmentation and coding is essential in order to arrive at the global minimum.

APPENDIX A

A.1 ORTHOGONALIZATION SCHEME OF SCHMIDT

Before giving the general algorithm, the problem of finding a single orthogonal basis function with respect to a given function is considered. Starting from a linear independent set of functions $\vec{u}_0, \vec{u}_1, ..., \vec{u}_n$ we consider a new function

$$\vec{q_1} = \vec{u}_1 - r_{01} \cdot \vec{n}_0 \tag{A.1}$$

with

$$\vec{n}_0 = \frac{\vec{q}_0}{||\vec{q}_0||} \quad \text{and} \quad \vec{q}_0 = \vec{u}_0. \tag{A.2}$$

By choosing

$$r_{01} = (\vec{u}_1, \vec{n}_0) \Longrightarrow \vec{q}_1 = \vec{u}_1 - (\vec{u}_1, \vec{n}_0) \cdot \vec{n}_0 \tag{A.3}$$

we get a vector space $\vec{q}_1$ which is orthogonal to the vector space $\vec{q}_0$. The term $r_{01} \cdot \vec{q}_0$ yields the projection of $\vec{u}_1$ onto $\vec{q}_0$. The difference of $\vec{u}_1$ and its projection has to be normal to $\vec{q}_0$ [1]. Testing the orthogonality shows

$$(\vec{q}_1, \vec{n}_0) = \left((\vec{u}_1 - r_{01} \cdot \vec{n}_0), \vec{n}_0\right) = (\vec{u}_1, \vec{n}_0) - r_{01} \cdot (\vec{n}_0, \vec{n}_0) = 0. \tag{A.4}$$

In general we can write

$$\vec{q}_k = \vec{u}_k - r_{0k} \cdot \vec{q}_0 - r_{1k} \cdot \vec{n}_1 - r_{2k} \cdot \vec{n}_2 - \ldots - r_{(k-1)k} \cdot \vec{n}_{k-1} \tag{A.5}$$

with

$$r_{1k} = (\vec{u}_k, \vec{n}_1) \quad \text{and} \quad \vec{n}_1 = \frac{\vec{q}_1}{||\vec{q}_1||}. \tag{A.6}$$

Using matrix notation with all the functional values of $\vec{u}_k$ or $\vec{q}_k$ in the kth column of the following matrices respectively, we get

$$\mathbf{U} = \mathbf{Q} \cdot \mathbf{R} \tag{A.7}$$

where

$$\mathbf{U} = \{\vec{u}_0, \vec{u}_1, \ldots, \vec{u}_n\} \tag{A.8}$$

is the matrix of the given basis functions, and

$$\mathbf{Q} = \{\vec{q}_0, \vec{q}_1, \ldots, \vec{q}_n\} \tag{A.9}$$

is the matrix of the resulting orthogonal basis functions. $\mathbf{R}$ is an upper triangular matrix consisting of the orthogonalization coefficients:

$$\mathbf{R} = \begin{pmatrix} \frac{1}{||\vec{q}_0||} & r_{01} & r_{02} & \cdots & r_{0n} \\ 0 & \frac{1}{||\vec{q}_1||} & r_{12} & \cdots & r_{1n} \\ 0 & 0 & \frac{1}{||\vec{q}_2||} & \cdots & r_{2n} \\ \vdots & \vdots & \vdots & \ddots & \vdots \\ 0 & 0 & 0 & \cdots & \frac{1}{||\vec{q}_n||} \end{pmatrix} \tag{A.10}$$

The above algorithm gives orthogonal basis functions step by step, starting from two given, linear independent functions. Successively, orthogonal functions can be calculated, until $\mathbf{U}$ and $\mathbf{Q}$ are quadratic. When this happens, the number of functions equals the number of given data points. Any further function has to be linearly dependent on the previous functions.

A.2 ORTHOGONALIZATION SCHEME OF HOUSEHOLDER

With the matrix $\mathbf{R}$ as defined in Eq. (A.10) and an $n \times n-$matrix $\mathbf{U}$, consisting of n linear independent basis functions $\vec{u}_k$ each given at the n nodes of the sampling grid, Householder's scheme uses the factorization $\mathbf{PU} = \mathbf{R}$, [32]. The matrix $\mathbf{P}$ is the desired $n \times n-$matrix of the orthogonal base. In contrast to Schmidt's scheme the orthogonal functions are now along the rows of the matrix $\mathbf{P}$ instead of along the columns.

In order to obtain $\mathbf{P}$ we start with a $n \times n-$matrix $\mathbf{P}_1$:

$$\mathbf{P}_1 = \mathbf{E} - 2\vec{w}\vec{w}^{\,\mathrm{T}}, \tag{A.11}$$

with the identity matrix $\mathbf{E}$ and the normalized column vector $\vec{w}$: $\vec{w}^{\,\mathrm{T}}\vec{w} = 1$. $\mathbf{P}_1$ is a symmetrical and orthogonal matrix as can be seen by evaluating the following two equations

$$\begin{aligned} \mathbf{P}_1^{\mathrm{T}} &= \mathbf{E}^{\mathrm{T}} - \left(2 \cdot \vec{w} \cdot \vec{w}^{\,\mathrm{T}}\right)^{\mathrm{T}} \\ &= \mathbf{E} - 2 \cdot \left(\vec{w}^{\,\mathrm{T}}\right)^{\mathrm{T}} \cdot \vec{w}^{\,\mathrm{T}} \\ &= \mathbf{E} - 2 \cdot \vec{w} \cdot \vec{w}^{\,\mathrm{T}} = \mathbf{P}_1, \end{aligned} \tag{A.12}$$

$$\begin{aligned} \mathbf{P}_1^{\mathrm{T}} \cdot \mathbf{P}_1 = \mathbf{P}_1 \cdot \mathbf{P}_1 &= (\mathbf{E} - 2 \cdot \vec{w}\vec{w}^{\,\mathrm{T}}) \cdot (\mathbf{E} - 2 \cdot \vec{w}\vec{w}^{\,\mathrm{T}}) \\ &= \mathbf{E} - 2 \cdot \vec{w}\vec{w}^{\,\mathrm{T}} - 2 \cdot \vec{w}\vec{w}^{\,\mathrm{T}} + 4 \cdot \vec{w}\vec{w}^{\,\mathrm{T}} \\ &= \mathbf{E}. \end{aligned} \tag{A.13}$$

A particular vector $\vec{w}$, which will be derived later, yields a matrix $\mathbf{P}_1$, that maps the first column of the matrix $\mathbf{U}$ into a column vector with just one nonzero element in the first position. With the first column of $\mathbf{U}$ written as vector $\vec{x}$ and the first column of the identity matrix as $\vec{e}_1$ it follows:

$$k \cdot \vec{e}_1 = \mathbf{P}_1 \cdot \vec{x} \quad \Leftrightarrow \quad \begin{pmatrix} k \\ 0 \\ 0 \\ \vdots \\ 0 \end{pmatrix} = \mathbf{P}_1 \cdot \begin{pmatrix} x_1 \\ x_2 \\ x_3 \\ \vdots \\ x_n \end{pmatrix} \tag{A.14}$$

Applying $\mathbf{P}_1$ to $\mathbf{U}^{(0)} = \mathbf{U}$ yields a new matrix $\mathbf{U}^{(1)}$:

$$\mathbf{U}^{(1)} = \mathbf{P}_1 \cdot \mathbf{U}^{(0)} \tag{A.15}$$

The first line of this matrix $\mathbf{U}^{(1)}$ is identical to the first line of the wanted matrix $\mathbf{R}$. The first column contains zeroes below the first element. In a second step, constructing a $(n-1)\times(n-1)$–matrix $\mathbf{P}_2'$ which is smaller by one row and one column. This matrix is used in Eq. (A.14) instead of $\mathbf{P}_1$ with the vector $\vec{x}$, which now contains the lower $(n-1)$ elements of the second column of the matrix $\mathbf{U}^{(1)}$. Finally $\mathbf{P}_2'$ is fitted with the identity matrix to yield an $n\times n$–matrix $\mathbf{P}_2$:

$$\mathbf{P}_2 = \begin{pmatrix} \mathbf{E} & 0 \\ 0 & \mathbf{P}_2' \end{pmatrix} \tag{A.16}$$

Now the product with $\mathbf{U}^{(1)}$ yields a matrix where the first two rows are identical to $\mathbf{R}$:

$$\mathbf{U}^{(2)} = \mathbf{P}_2 \cdot \mathbf{U}^{(1)} \tag{A.17}$$

After $n-1$ steps the matrix $\mathbf{U}^{(n-1)}$ is identical to the desired matrix $\mathbf{R}$:

$$\mathbf{P}_{n-1}\cdot\mathbf{P}_{n-2}\cdot\ldots\cdot\mathbf{P}_1\cdot\mathbf{U} = \mathbf{R} \tag{A.18}$$

All matrices $\mathbf{P}_k$ are orthogonal, therefore the product matrix $\mathbf{P}$ is also orthogonal:

$$\mathbf{P} = \mathbf{P}_{n-1}\cdot\mathbf{P}_{n-2}\cdot\ldots\cdot\mathbf{P}_1 \quad \text{with} \quad \mathbf{P}^{\mathrm{T}}\cdot\mathbf{P} = \mathbf{E} \tag{A.19}$$

Comparing the two algorithms, the matrix $\mathbf{P}^{\mathrm{T}}$ is identical to the matrix $\mathbf{Q}$ of Schmidt's algorithm:

$$\mathbf{U} = \mathbf{P}^{\mathrm{T}}\cdot\mathbf{R} \tag{A.20}$$

A vector $\vec{w}$ remains to be found that satisfies the equation $k\cdot\vec{e}_1 = \mathbf{P}_k'\cdot\vec{x}$. This vector is given by

$$\vec{w} = \frac{\vec{x} - k\cdot\vec{e}_1}{||\vec{x} - k\cdot\vec{e}_1||} \tag{A.21}$$

where

$$k = -||\vec{x}|| = -\sqrt{(\vec{x},\vec{x})}. \tag{A.22}$$

Proof:

$$\begin{aligned} k\cdot\vec{e}_1 &= \mathbf{P}_k'\cdot\vec{x} \\ &= (\mathbf{E} - 2\cdot\vec{w}\cdot\vec{w}^{\,\mathrm{T}})\cdot\vec{x} \\ &= \vec{x} - 2\cdot\vec{w}\cdot\vec{w}^{\,\mathrm{T}}\cdot\vec{x} \\ &= \vec{x} - 2\cdot\frac{(\vec{x} - k\cdot\vec{e}_1)\cdot(\vec{x} - k\cdot\vec{e}_1)^{\mathrm{T}}\cdot\vec{x}}{||\vec{x} - k\cdot\vec{e}_1||^2} \end{aligned} \tag{A.23}$$

with

$$||\vec{x} - k\cdot\vec{e}_1||^2 = (\vec{x} - k\cdot\vec{e}_1)^{\mathrm{T}}\cdot(\vec{x} - k\cdot\vec{e}_1)$$

$$
\begin{aligned}
&= \vec{x}^{\,\mathrm{T}} \cdot \vec{x} - k \cdot \vec{x}^{\,\mathrm{T}} \cdot \vec{e}_1 - k \cdot \vec{e}_1^{\,\mathrm{T}} \cdot \vec{x} + k^2 \\
&= k^2 - 2 \cdot k \cdot x_1 + k^2 = 2 \cdot (k^2 - k \cdot x_1) \qquad \text{(A.24)}
\end{aligned}
$$

and

$$
\begin{aligned}
(\vec{x} - k \cdot \vec{e}_1) \cdot (\vec{x} - k \cdot \vec{e}_1)^{\mathrm{T}} \cdot x &= (\vec{x} - k \cdot \vec{e}_1) \cdot (\vec{x}^{\,\mathrm{T}} \cdot \vec{x} - k \cdot \vec{e}_1^{\,\mathrm{T}} \cdot \vec{x}) \\
&= (\vec{x} - k \cdot \vec{e}_1) \cdot (k^2 - k \cdot x_1) \qquad \text{(A.25)}
\end{aligned}
$$

Substituting Eqs. (A.24) and (A.25) into Eq. (A.23) yields:

$$
k \cdot \vec{e}_1 = \vec{x} - (\vec{x} - k \cdot \vec{e}_1) \qquad \text{(A.26)}
$$

Householder's algorithm differs from Schmidt's scheme in two respects: Householder computes all orthogonal functions at once while Schmidt allows for a successive orthogonalization of single functions. Householder's scheme is mathematically better posed because all computations are done using orthogonal matrices. If almost linearly dependent functions $\vec{u}_k$ are input into Schmidt's algorithm, the resulting function $\vec{q}_k$ may not be orthogonal due to the accumulation of rounding errors. In the same case Householder's algorithm is more likely to produce functions which are sufficiently orthogonal. Neglecting rounding effects and a scaling factor, both schemes do produce the same set of orthogonal functions.

REFERENCES

[1] N.I.Achieser, *Vorlesungen über Approximationstheorie*, Akademie, Berlin, 2nd Ed., 1967.

[2] N.Ahmed, T.Natarajan, and K.R.Rao, "Discrete cosine transform," IEEE Trans. Computers, no. 1, January 1974, pp. 90-93.

[3] N.Ahmed, K.R.Rao, *Orthogonal transforms for digital signal processing*, Springer, Berlin Heidelberg New York, 1975.

[4] H.C.Andrews and C.L.Patterson, "Singular value decompositions and digital image processing," IEEE Trans. ASSP, vol. 24, no. 1, February 1976, pp. 26-53.

[5] M.J.Biggar, O.J.Morris and A.G.Constantinides, "Segmented-image coding: performance comparison with the discrete cosine transform," IEE Proceedings, vol. 135, no. 2, April 1988, pp. 121-132.

[6] Å.Björck and G.Dahlquist, *Numerische Methoden*, Oldenbourg, München Wien, 2nd Ed., 1979.

[7] F.W.Campbell and J.G.Robson, "Application of Fourier analysis to the visibility of gratings," Journal of Physiology, vol. 197, 1968, pp. 551-566.

[8] G.L.Cash and M.Hatamian, "Optical character recognition by the method of moments," Computer Vision, Graphics, and Image Processing, vol. 39, 1987, pp. 291-310.

[9] A.K.Chan and C.K.Chui and K.B.Chan, "Image reconstruction by bivariate quadratic splines," IEEE Trans. ASSP, vol. 36, no. 9, September 1988, pp. 1525-1529.

[10] S.-F. Chang and D.G. Messerschmitt, "Transform Coding of arbitrarily-shaped image segments", ACM Multimedia, no. 6, 1993, pp. 83-90.

[11] L.Chen and J.-G.Leu, "Polygonal approximation of 2-D shapes by line growing", Proceedings Phoenix Conf. on Computers and Communication, Scottsdale, March 1986, pp. 491-495.

[12] E.W.Cheney, *Introduction to Approximation Theory*, McGraw-Hill, New York, 1966.

[13] R.Courant and D.Hilbert, *Methoden der mathematischen Physik I und II*, Springer, Berlin Heidelberg New York, 3rd Ed., 1968.

[14] W.B.Davenport and W.L.Root, *An Introduction to the Theory of Random Signals and Noise*, McGraw-Hill, New York, 1958.

[15] L.D.Davisson, "Rate-distortion theory and application", Proceedings IEEE, vol. 60, July 1972, pp. 800-808.

[16] S.A.Dudani, K.J.Breeding, and R.B.McGeeh, "Aircraft identification by moment invariants", IEEE Trans. Computers, vol. 26, no. 1, January 1977, pp. 39-47.

[17] M.Eden and M.Kocher, "On the performance of a contour coding algorithm in the context of image coding - Part I: Contour segment coding", Signal Processing, vol. 8, no. 4, July 1985, pp. 381-386.

[18] M.Eden, M.Unser, and R.Leonardi, "Polynomial representation of pictures", Signal Processing, vol. 10, no. 4, June 1986, pp. 385-393.

[19] J.D.Eggerton and M.D.Srinath, "A visually weighted quantization scheme for image bandwidth compression at low data rates", IEEE Trans. Communications, vol. 34, no. 8, August 1986, pp. 840-847.

[20] P.M.Farelle and A.K.Jain, "Recursive block coding - a new approach to transform coding", IEEE Trans. COM, vol. 34, no. 2, February 1986, pp. 161-179.

[21] M.Gilge, "Coding of arbitrarily shaped image segments using moment theory", Proceedings European Signal Processing Conference EUSIPCO'88, Grenoble, September 1988, pp. 855-858.

[22] M.Gilge, T.Engelhardt, and R.Mehlan, "Coding of arbitrarily shaped image segments based on a generalized orthogonal transform", Visual Communication, Special Issue of Signal Processing on 64 kbit/s Coding of Moving Video, vol. 1, no. 2, October 1989, pp. 153-180.

[23] M.Gilge, "Region oriented transform coding (ROTC) of images", International Conference on Acoustics, Speech, and Signal Processing ICASSP-90, Albuquerque, USA, April 1990.

[24] B.Girod, "The efficiency of motion-compensating prediction for hybrid coding of video sequences", IEEE Journal SAC, vol. 5, no. 7, August 1987, pp. 1140-1154.

[25] R.C.Gonzales and P.Wintz, *Digital Image Processing*, Addison-Wesley, Reading, Massachusetts, 2nd Ed., 1987.

[26] M.Guglielmo, "An analysis of error behavior in the implementation of 2-D orthogonal transformations", IEEE Trans. COM, vol. 34, no. 9, September 1986, pp. 973-975.

[27] L.Gupta and M.D.Srinath, "Contour classification using invariant moments", Phoenix Conf. Comp. and Commun., March 1986, Cat.-No. 86CH2371-3, pp. 482-486.

[28] A.Habibi and P.A.Wintz, "Image coding by linear transformation and block quantization", IEEE Trans. COM, vol. 19, no. 1, February 1971, pp. 50-62.

[29] R.M.Haralick and L.Watson, "A facet model for image data", Computer Graphics and Image Processing, vol. 15, 1981, pp. 113-129.

[30] H.F.Harmuth, *Transmission of Information by Orthogonal Functions*, Springer, Berlin Heidelberg New York, 1969.

[31] M.Hötter and R.Thoma, "Image segmentation based on object oriented mapping parameter estimation", Signal Processing, vol. 15, no. 3, October 1988, pp. 315-334.

[32] A.S.Householder, *The Theory of Matrices in Numerical Analysis*, Blaisdell Publishing Company, New York, 1964.

[33] T.C.Hsia, "A note on invariant moments in image processing", IEEE Trans. SMC, vol. 11, no. 12, December 1981, pp. 831-834 .

[34] M.-K.Hu, "Visual pattern recognition by moment invariants", IRE Trans. IT, February 1962, pp. 179-187.

[35] B.R.Hunt and O.Kübler, "Karhunen-Loève multispectral image restoration, part I: theory", IEEE Trans. ASSP, vol. 32, no. 3, June 1984, pp. 592-599.

[36] A.K.Jain, "A fast Karhunen-Loève transform for a class of stochastic processes", IEEE Trans. COM, vol. 24, no. 9, 1976, pp. 1023-1029.

[37] A.K.Jain, "A sinusoidal family of unitary transforms", IEEE Trans. PAMI, vol. 1, no. 4, 1979, pp. 356-365.

[38] A.K.Jain, "Advances in mathematical models for image processing", Proceedings IEEE, vol. 69, no. 5, May 1981, pp. 502-528.

[39] A.K.Jain, "Image data compression: A review", Proceedings IEEE, vol. 69, no. 3, March 1981, pp.349-389.

[40] N.S.Jayant and P.Noll, *Digital Coding of Waveforms*, Prentice-Hall, Englewood Cliffs, 1984.

[41] E.T.Jaynes, "Prior probabilities", IEEE Trans. SSC, vol. 4, no. 3, September 1968, pp. 227-241.

[42] E.Jensen, K.Rijkse, I.Lagendijk, and P.van Beek, "Coding of arbitrarily shaped image segments", Proc. Workshop on image analysis and synthesis in image coding, Berlin, October 1994, paper E2.

[43] M.Kavehrad and M.Joseph, "Maximum entropy and the method of moments in performance evaluation of digital communication systems", IEEE Trans. COM, vol. 34, no. 12, December 1986, pp. 1183-1189.

[44] M.Kocher and R.Leonardi, "Adaptive region growing technique using polynomial functions for image approximation", Signal Processing, vol. 11, no. 1, July 1986, pp. 47-60.

[45] V.A.Kotel'nikov, *The theory of optimum noise immunity*, McGraw-Hill, New York Toronto London, 1959.

[46] M.Kunt, A.Ikonomopoulos, and M.Kocher, "Second-generation image-coding techniques", Proceedings IEEE, vol. 73, no. 4, April 1985, pp. 549-574.

[47] F.Lavagetto and F.Cocorullo, "texture Approximation through Discrete Legendre Polynomials", COST 211ter European workshop on new techniques for coding of video signals at very low bitrates, University of Hannover, Dec. 1993, paper 2.4.

[48] J.S.Lim and N.A.Malik, "A new algorithm for two-dimensional maximum entropy power spectrum estimation", IEEE Trans. ASSP, vol. 29, no. 3, June 1981, pp. 401-413.

[49] S.P.Lloyd, "Least squares quantization in PCM", IEEE Trans. Information Theory, vol. 28, 1982, pp. 129-137.

[50] S.Maitra, "Moment invariants", Proceedings IEEE, vol. 67, no. 4, April 1979, pp. 697-699.

[51] J.Makhoul, "Linear prediction. A tutorial review", Proceedings IEEE, vol. 63, no. 4, April 1975, pp. 561-580.

[52] P.Max, "Quantizing for minimum distortion", IEEE Trans. IRE, IT-6, 1960, pp. 7-12.

[53] H.G.Musmann, P.Pirsch, and H.-J.Grallert, "Advances in picture coding", Proceedings IEEE, vol. 73, no. 4, April 1985, pp. 523-548.

[54] D.P.O'Leary and S.Peleg, "Digital image compression by outer product expansion", IEEE Trans. COM, vol. 31, no. 3, March 1983, pp. 441-444.

[55] K.Paton, "Picture description using legendre polynomials", Computer Graphics and Image Processing, vol. 4, 1975, pp. 40-54.

[56] W.Philips, "A fast algorithm for the generation of orthogonal base functions on an arbitrarily shaped region", Proceedings of ICASSP 92, San Francisco, March 1992, vol. III, pp. 421-424.

[57] W.K.Pratt, *Digital Image Processing*, John Wiley, New York, 1978.

[58] A.Rosenfeld and A.C.Kak, *Digital Picture Processing*, Volume 2, Academic Press, New York, 2. Ed., 1982.

[59] P.Salembier, L.Torres, F. Meyer, C.Gu, "Region-based video coding using mathematical morphology", Proceedings of the IEEE, Special Issue on Digital TV, June 1995, vol. 83, no.6, pp. 843 - 857.

[60] R.Sauer and I.Szabó (Ed.), *Mathematische Hilfsmittel des Ingenieurs*, part III, Springer, Berlin Heidelberg New York, 1968.

[61] M.Schlatter and J.Eichler, "An introduction to the Gaussian least squares approximation and its application in signal processing and system modeling", Signal Processing, vol. 1, no. 3, July 1979, pp. 211-225.

[62] K.Shibata, "Waveform analysis of image signals by orthogonal transformation", Proceedings Walsh Functions, 1972, pp. 210-215.

[63] J.E.Shore, "Minimum cross-entropy spectral analysis", IEEE Trans. ASSP, vol. 29, no. 2, April 1981, pp. 230-237.

[64] J.E.Shore and R.W.Johnson, "Properties of cross-entropy minimization", IEEE Trans. IT, vol. 27, no. 4, July 1981, pp. 472-482.

[65] T. Sikora and B. Makai, "Shape-Adaptive DCT for Generic Coding of Video", IEEE Trans. Circuits and Systems for Video Technology, vol. 5, no.1, February 1995, pp. 59-62.

[66] A.Sommerfeld, *Partial Differential Equations in Physics*, Academic Press, New York, 1949.

[67] H.W.Sorenson, "Least Squares Estimation: From Gauss to Kalman", IEEE Spectrum, New York, July 1970.

[68] M.R.Teague, "Image analysis via the general theory of moments", Journal Optc. Soc. Am., vol. 70, no. 8, August 1980, pp. 920-930.

[69] C.-H.Teh and R.T.Chin, "On digital approximation of moment invariants", Computer Vision, Graphics, and Image Processing, vol. 33, 1986, pp. 318-326.

[70] C.-H.Teh and R.T.Chin, "On image analysis by the method of moments", IEEE Trans. PAMI, vol. 10, no. 4, July 1988, pp. 496-513.

[71] L. Torres, J.R. Casas, and S. de Diego, "Segmentation-based coding of textures using stochastic vector quantization", Proceedings of ICASSP 94, Adelaide, Australia, April 1994, pp. V-597 - V-600.

[72] N.S.Tzannes and P.A.Jonnard, "Maximum entropy reconstruction of moment coded images", Optical Engineering, vol.26, no.10, October 1987, pp. 1077-1083.

[73] M.Unser, "On the approximation of the discrete Karhunen-Loève transform for stationary processes", Signal Processing, vol. 7, 1984, pp. 231-249.

[74] J.Vaisey and A.Gersho, "Variable Rate Image Coding using quad-trees and vector quantization", Proceedings European Signal Processing Conference EUSIPCO'88, Grenoble, September 1988, pp. 1133-1136.

[75] J.L.Walsh, *Interpolation and approximation by rational functions in the complex domain*, American Mathematical Society, Providence, Rhode Island, 4th Ed., 1965.

[76] L.T.Watson, R.M.Haralick, and O.A.Zuniga, "Constrained transform coding and surface fitting", IEEE Trans. COM, vol. 31, no. 5, May 1983, pp. 717-726.

[77] L.T.Watson, T.J.Laffey, and R.M.Haralick, "Topographic classification of digital image intensity surfaces using generalized splines and the discrete cosine transformation", Computer Vision, Graphics, and Image Processing, vol. 29, 1985, pp. 143-167.

[78] S.Yuan and K.-B.Yu, "Zonal sampling and bit allocation of HT coefficients in image data compression", IEEE Trans. Communications, vol. 34, no. 12, December 1986, pp. 1246-1251.

6

SEGMENTATION-BASED MOTION ESTIMATION FOR SECOND GENERATION VIDEO CODING TECHNIQUES

Frédéric Dufaux* and Fabrice Moscheni**

** The Media Laboratory,*
Massachusetts Institute of Technology,
Cambridge, Massachusetts 02139, USA

*** Signal Processing Laboratory,*
Swiss Federal Institute of Technology,
1015 Lausanne, Switzerland

ABSTRACT

This chapter addresses the development of a new approach to motion estimation which rely on a semantic representation of the scene in terms of objects. More specifically, motion models, segmentation-based motion estimation, spatio-temporal segmentation and tracking are more thoroughly discussed. Simulation results are presented to show the efficiency of the new approach.

1 INTRODUCTION

One of the most challenging problems in video coding lies in an efficient reduction of temporal redundancy in image sequences. Motion estimation and compensation techniques have already shown their efficiency in reducing temporal redundancies. These techniques are clearly the key to reaching very high performances in future image sequence coding applications. This chapter addresses the problem of motion estimation in the context of second generation coding techniques.

The coding techniques of the first generation rely on a representation of the image sequence as a stochastic process. In an attempt to reduce the statistical redundancies in the data, pixels or small blocks of pixels are coded. Well known examples relying on this principle are the video coding schemes based on the Discrete Cosine Transform (DCT) such as the recent standards MPEG-1 [1, 2], MPEG-2 [3, 4], H.261 [5] and H.263 [6]. In this context, block matching motion estimation techniques [7, 8, 9, 10] are very well-suited and are the most widely used to reduce temporal redundancy. Basically, these techniques partition the image into small blocks and assign a translational displacement to each of these blocks by minimizing a disparity measure. This motion information is then exploited for efficient interframe predictive coding. The efficiency of these techniques resides, in a large part, in their ability to provide an accurate temporal prediction by directly minimizing the Displaced Frame Difference (DFD). Furthermore, thanks to their block-based nature, they are well adapted to block-based coding schemes and require low overhead information. Finally, they are easily implementable in hardware.

Although the above first generation coding techniques have reached good performances, they suffer from a severe drawback: the image is represented in terms of pixels or blocks of pixels which do not carry any semantic meaning. Consequently, visually annoying artifacts occur at high compression ratios, e.g. block artifacts in DCT-based schemes. Therefore, no major improvement in terms of compression performances can be foreseen in this direction. Furthermore, the lack of information about the content of the scene limits the functionalities available to the end-user at the decoder side. For instance, desired features such as the ability to manipulate and personalize the received visual information are not available.

In contrast to the coding techniques of the first generation, the techniques of the second generation introduced by Kunt *et al.* [11, 12] are based on a semantic representation of the scene. They attempt to imitate the functions of the human visual system and rely on psycho-visually meaningful primitives. For this purpose, the image is described in terms of the primitives of the scene, namely the objects which compose the scene. Compression is achieved by taking into account the properties of the human visual system, i.e. by allocating more bits in visually important areas and fewer bits otherwise, rather than by a mere reduction of the spatial and temporal redundancy in the data. As a result of the psycho-visually meaningful representation, these techniques allow for very high compression ratios and visually less annoying artifacts. Furthermore, the information about the content of the scene offers new functionalities at the decoder side [13]. In particular, the end-user can take intelligent actions allowing, among other features, interactivity. The reader is referred to Chap. 1 for a more detailed discussion on the objectives of second generation coding and to Chap. 3 for a description of segmentation-based coding algorithms.

In the framework of second generation coding techniques, motion estimation should be consistent with a representation of the scene in terms of objects. Furthermore, it should aim at reliably estimating the motion of the considered objects rather than providing a mere temporal prediction. This is important in order to assure a high visual quality of the reconstructed sequence. Two more reasons for requiring a motion field representative of the real motion in the scene occur when very low bit rate coding is considered. First, in order to achieve a very high compression, prediction errors resulting from motion compensation might not be transmitted to the decoder. Therefore, it is important not to introduce annoying visual artifacts in the motion compensated frames. Second, it might be desirable to subsample temporally the sequence prior to coding. In this case, the motion information can be used at the decoder side to interpolate the frames discarded at the encoder, provided that the estimated motion field is reliable.

In this context, classical block matching techniques are no longer adequate due to their block-based nature and their inability to reliably estimate the true motion in the scene. To overcome these drawbacks, improved block matching techniques segmenting the scene according to its motion have been proposed, e.g. varying block size [14], locally adaptive multigrid [10, 15], and segmentation of blocks corresponding to failure of the block-based model [15, 16, 17]. Although these improved block matching techniques have lead to higher performances by relaxing the block constraint, the segmentation remains coarse. Moreover, these improved block matching techniques do not comply with a representation of the scene in terms of objects.

These considerations naturally motivate the design of motion estimation techniques which directly take the objects of the scene as support. The design of segmentation-based motion estimation techniques relying on the semantic primitive-based representation of the scene implies to take into account several new considerations.

First, a model has to be defined in order to describe the motion. In segmentation-based algorithms, motion estimation is carried out on the objects of the scene rather than on blocks of pixels. Each of these objects is characterized by a coherent motion. Therefore, in order to compactly represent the motion information, this coherent motion can be modeled by a set of motion parameters. This motion model is referred to as fully parametric [18]. Usually, one object corresponds to a large area of the image when compared to the small block of pixels used in block matching. Consequently, the simple translational model used in block matching is no longer applicable and a more complex motion model has to be introduced. From the perspective of very low bit rate coding, the typically low temporal sampling frequency gives another reason to consider a more complex motion model.

Second, both the moving objects and their motions have to be determined simultaneously. The spatio-temporal segmentation defines the regions which are characterized by a coherent motion and which can therefore be identified as moving objects [19, 20, 21, 22, 23, 24, 25, 26, 27]. The segmentation-based motion estimation estimates the motion parameters relative to each of the moving objects [18, 23, 24, 27, 28, 29]. There is obviously a very strong dependence between these two processes of spatio-temporal segmentation and motion estimation. On the one hand, as the motion estimation depends on the region of support, a good segmentation is needed to precisely estimate the motion. On the other hand, as the moving regions are defined by a coherent motion, an accurate estimate of the motion is required to obtain a good segmentation.

Finally, in order to assure a coherent segmentation through time, tracking techniques have to be introduced [21, 30, 31, 32, 33]. More precisely, these techniques define the trajectories of the detected moving objects along time. Furthermore, the introduction of tracking techniques tends to stabilize the spatio-temporal segmentation and the motion estimation in a cluttered environment. Finally, in the context of video coding, such a knowledge prevents from transmitting the spatio-temporal segmentation for each coded frame.

The above problems of spatio-temporal segmentation, segmentation-based motion estimation and tracking have been widely studied in the field of image sequence analysis and robot vision. However, image sequence coding implies very different goals and only a very few results have been unfortunately reported in this field. More precisely, image sequence coding aims at representing in a compact way the motion information.

For this purpose, a very important consideration specific to coding is the trade-off on the motion estimation accuracy and the amount of overhead information to be sent to the decoder [15, 34, 35]. This involves, among other factors, the control of the segmentation process, the choice of the motion model, and the quantization of the motion parameters. Finally, for conversational and interactive services, coding delay is also an important issue that must be taken into account.

This chapter reviews motion estimation techniques in the context of second generation coding schemes and presents new perspectives. The structure of the chapter is as follows. In Sec. 2, different motion models to represent the motion of an object are derived. Sec. 3 addresses the problem of computing the motion of an object, assuming the object is known. Different techniques are reviewed and discussed. The spatio-temporal segmentation, which is tightly linked to the process of motion estimation, is addressed in Sec. 4. Tracking is the subject of Sec. 5. Finally, Sec. 6 draws the conclusions.

2 MOTION MODELS

In order to describe the motion, all motion estimation techniques rely on the definition of a motion model. According to Anandan *et al.* [18], the following classification of the motion models is introduced: non-parametric, quasi-parametric and fully parametric. Whereas non-parametric models rely on a dense local motion field (e.g. optical flow motion estimation techniques [36, 37]), parametric models represent the motion of a large region by a single set of parameters, referred to as motion parameters. A further distinction is made between quasi-parametric models which represent the motion by a set of parameters as well as a local field (e.g. the depth map or local optical flow) and fully parametric models which completely specifies the motion by a set of parameters. In the following, the terms non-parametric, quasi-parametric and fully parametric motion estimation techniques should be understood as techniques which rely on a non-parametric, a quasi-parametric and a fully parametric motion model respectively.

A fully-parametric motion model is applicable only to those pixels which belong to a same moving entity, namely which undergo a coherent motion. The advantage of the approach is that since all the pixels within the region of support can contribute to the motion estimation, robustness and high accuracy can be expected. Conversely, a non-parametric model is applicable under a wider range of situations, as one motion vector is estimated per each pixel and therefore no hypothesis is made that those pixels belong to a common moving entity (although this model requires an explicit constraint such as smoothness or local uniformity, see Sec. 3). However, due to the local computation, this approach results in limited accuracy. To take advantage of the robustness and accuracy of the fully parametric model and the wide validity of the non-parametric one, a quasi-parametric model has been proposed by Hsu *et al.* in [38]. It combines layers of fully parametric representations as well as local optical flow to cope with local variations.

In the framework of second generation coding techniques, a segmentation-based motion estimation which directly takes the objects of the scene as support is desired. As an object is characterized by a coherent motion, it is natural to apply a parametric model. Furthermore, the amount of overhead motion information to be sent to the decoder is a capital issue. In this context, dense motion fields such as the one resulting from a non-parametric or quasi-parametric model require very high overhead information. Conversely, a fully parametric model ensures a compact representation of the motion information and thus tends towards low overhead information.

Clearly, as the notion of coherent motions is closely related to the specific model considered, the spatio-temporal segmentation depends on the motion model. Even though it is not mandatory, the moving entities defined in this way are likely to correspond to the moving objects in the scene, and will be identified as such in the following.

In the remainder of this section, different parametric motion models are derived by the projection of the $3D$ motion in the scene onto the image plane. Furthermore, the $3D$ motion can be described by a rigid body motion model. Consequently, a representation of the $2D$ motion in the image plane is derived by the projection of this $3D$ motion. For this purpose, a perspective or an orthographic projection is commonly used [39]. Finally, the restriction of the scene model to a patchwork of planar surfaces is assumed to avoid the estimation of the depth map. The latter hypothesis leads to fully parametric motion models.

2.1 Rigid body motion

The relation defining the $2D$ motion in the image plane induced by the $3D$ motion in the scene is derived as follows.

If we assume that a camera is represented by a pinhole, an image is formed by a perspective projection of the real world scene [39]. Let $\vec{R} = (X, Y, Z)^T$ be the Cartesian coordinates system fixed with respect to the camera, and $\vec{r} = (x, y)^T$ be the coordinates in the image plane (i.e. Z is perpendicular to the image plane and defines the optical axis), as illustrated in Fig. 1. Under a perspective projection, $\vec{r}$ is related to $\vec{R}$ by

$$x = \frac{X}{Z} \quad \text{and} \quad y = \frac{Y}{Z}, \tag{6.1}$$

where the focal length has been normalized to 1 without any loss of generality.

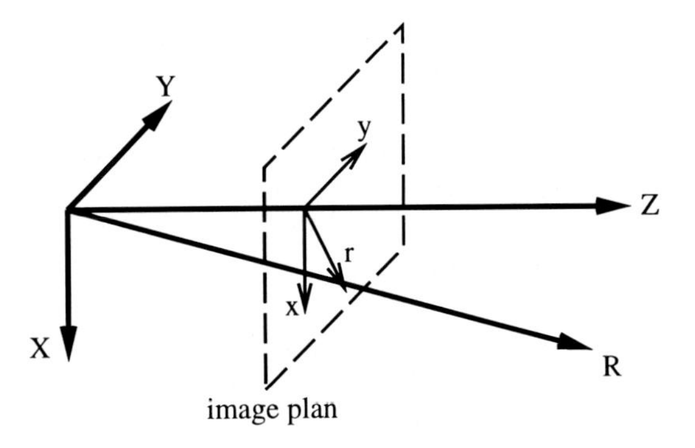

Figure 1 $(X, Y, Z)^T$ are the Cartesian coordinates fixed with the camera and $(x, y)^T$ are the coordinates in the image plane.

The scene is commonly supposed to be composed of rigid bodies. This hypothesis is motivated by the difficulty to cope with a non-rigid body motion, although algorithms to estimate non-rigid body motion have been proposed [40]. Furthermore, it is justified by the fact that a rigid body motion closely approximates a non-rigid body motion when the temporal sampling frequency is sufficiently high and the motion between two frames is sufficiently small. This assumption is therefore made throughout the chapter.

With regard to the $3D$ rigid body motion, two distinct formulations are possible, the first in terms of instantaneous velocities and the second in terms of displacements. In the first case, the motion of a $3D$ rigid body is expressed as [39]

$$\dot{\vec{R}} = \vec{\omega} \times \vec{R} + \vec{T} \ , \tag{6.2}$$

where $\dot{\vec{R}}$ represents the temporal derivative, $\vec{\omega} = (\omega_x, \omega_y, \omega_z)^T$ is the angular velocity and $\vec{T} = (T_x, T_y, T_z)^T$ is the translational motion. Therefore, in this formulation $\vec{\omega}$ and $\vec{T}$ correspond to instantaneous velocities. Introducing Eq. (6.1) in Eq. (6.2) leads to

$$\dot{\vec{r}} = \begin{pmatrix} \dot{x} \\ \dot{y} \end{pmatrix} = \begin{pmatrix} -\omega_x \ xy + \omega_y (x^2 + 1) - \omega_z \ y + \frac{T_x - T_z x}{Z} \\ -\omega_x (y^2 + 1) + \omega_y \ xy - \omega_z \ x + \frac{T_y - T_z y}{Z} \end{pmatrix} . \tag{6.3}$$

In the formulation in terms of displacements, the $3D$ rigid body motion equation is written as

$$\vec{R}' = \Omega \vec{R} + \vec{D} \ , \tag{6.4}$$

where $\vec{R}$ and $\vec{R}'$ are the positions at time t and t' respectively, Ω denotes the rotation matrix and $\vec{D} = (D_x, D_y, D_z)^T$ is the translational displacement.

Therefore, Ω and $\vec{D}$ denote differences in orientation and position over a time interval. Similarly, by introducing Eq. (6.1) in Eq. (6.4), it follows that

$$\vec{r}' = \begin{pmatrix} x' \\ y' \end{pmatrix} = \begin{pmatrix} \frac{x-\Omega_z y+\Omega_y+D_x/Z}{1-\Omega_y\ x+\Omega_x\ y+D_z/Z} \\ \frac{\Omega_z x+y-\Omega_x+D_y/Z}{1-\Omega_y\ x+\Omega_x\ y+D_z/Z} \end{pmatrix} , \tag{6.5}$$

where the rotation matrix Ω has been approximated, assuming small rotation angles, by

$$\Omega = \begin{pmatrix} 1 & -\Omega_z & \Omega_y \\ \Omega_z & 1 & -\Omega_x \\ -\Omega_y & \Omega_x & 1 \end{pmatrix} . \tag{6.6}$$

Equations (6.3) and (6.5) define the $2D$ motion in the image plane induced by the $3D$ motion for the formulation in terms of instantaneous velocities and displacements respectively. According to the nomenclature in [18], Eqs. (6.3) and (6.5) corresponds to quasi-parametric models. Under some realistic assumptions, it can be shown that both formulations in terms of instantaneous velocities and displacements are equivalent (see [19] and Appendix A.1).

2.2 Planar surface motion

The dependency on the depth map Z in Eqs. (6.3) and (6.5) causes two problems. First it is difficult to accurately estimate the depth map. Second, in the context of coding the depth map represents a large amount of overhead information to transmit to the decoder. In order to overcome these problems, the model of the scene is restricted to a patchwork of planar surfaces which can closely approximate $3D$ rigid bodies. This simplification removes the depth map dependency in the motion model and leads to a fully parametric model.

A planar surface is defined as

$$k_x X + k_y Y + k_z Z = 1 , \tag{6.7}$$

or equivalently as

$$k_x x + k_y y + k_z = \frac{1}{Z} . \tag{6.8}$$

For the formulation in terms of instantaneous velocities, introducing Eq. (6.8) in Eq. (6.3) leads to

$$\begin{pmatrix} \dot{x} \\ \dot{y} \end{pmatrix} = \begin{pmatrix} a_1 + a_2 x + a_3 y + a_7 x^2 + a_8 xy \\ a_4 + a_5 x + a_6 y + a_7 xy + a_8 y^2 \end{pmatrix} , \tag{6.9}$$

where $a_1, \ldots, a_8$ are given in Appendix A.2. The 8 parameters model defined by Eq. (6.9) specifies the instantaneous velocity in the image plane due to a moving planar surface under perspective projection.

In a similar way, for the formulation in terms of displacements, the introduction of Eq. (6.8) in Eq. (6.5) gives

$$\begin{pmatrix} x' \\ y' \end{pmatrix} = \begin{pmatrix} \frac{a_1+a_2x+a_3y}{a_7x+a_8y+a_9} \\ \frac{a_4+a_5x+a_6y}{a_7x+a_8y+a_9} \end{pmatrix}, \tag{6.10}$$

where $a_1, \ldots, a_9$ are given in Appendix A.2. Eq. (6.10) is referred to as the perspective model. It defines the displacement in the image plane due to a moving planar surface under perspective projection. Usually, the parameters are normalized by a_9, leading to an 8 parameters model.

When the distance to the scene is much larger than the variation in distance among objects in the scene, perspective projection is closely approximated by orthographic projection [39]. In other words, this simplification is valid when the depth map of the scene is small relative to the distance from the camera. Rather than modeling rays of light passing through the origin (pinhole camera) as in the perspective projection, the orthographic projection supposes rays of light parallel to the optical axis.

Under an orthographic projection, the second order terms in Eq. (6.9) can be neglected as $x = X/Z \ll 1$ and $y = Y/Z \ll 1$. This simplification leads to the affine model

$$\begin{pmatrix} \dot{x} \\ \dot{y} \end{pmatrix} = \begin{pmatrix} a_1 + a_2x + a_3y \\ a_4 + a_5x + a_6y \end{pmatrix}. \tag{6.11}$$

The model defined by Eq. (6.11) allows for the representation of the motion of a planar surface under orthographic projection. For the formulation in terms of displacements, under the same hypothesis of orthographic projection, a similar affine model is obtained from a first order Taylor series expansion of Eq. (6.10). It is useful to note that the affine model can equivalently be expressed in terms of a translation vector, two scaling factors, and two rotation angles [28, 41].

The affine model can be further simplified. For instance, the simple and widely used translational model with 2 parameters results from setting $a_2 = a_3 = a_5 = a_6 = 0$

$$\begin{pmatrix} \dot{x} \\ \dot{y} \end{pmatrix} = \begin{pmatrix} a_1 \\ a_4 \end{pmatrix}. \tag{6.12}$$

According to the nomenclature in [18], Eqs. (6.9) - (6.12) correspond to fully parametric models.

2.3 Choice of an appropriate model

Given the different motion models available, the choice of the most appropriate one remains an open question. Obviously, the answer depends strongly on the type of application. Another difficulty lies in the lack of a measure to assess the performance of a model for a given problem. Finally, the motion model is intimately related to the support of the motion estimation. Whereas many models have been used in the literature, unfortunately little research effort has been devoted to the comparison of the performances obtained when using those different models. Consequently, the choice of the model seems often to be ad-hoc.

Video coding applications, in contrast to image analysis, require sending the motion parameters to the decoder as side information. In this framework, the cost of transmitting a dense motion field (e.g. optical flow or depth map) is usually not affordable. The choice is therefore restricted to fully parametric models (e.g. Eqs. (6.9) - (6.12)). It is obvious that a more complex parametric model can represent more complex motions. Consequently, it allows the motion of a larger region of the image to be precisely described. However, in this case the estimation process tends to be more difficult and in particular more sensitive to noise. Furthermore, estimating more parameters demands a higher computational complexity. Finally, the more complex the motion model is, the higher the amount of overhead information.

The current video coding standards MPEG-1 [1, 2], MPEG-2 [3, 4], H.261 [5] and H.263 [6] rely on a block-based transform coding. In this context, block matching motion estimation techniques [7, 8, 9, 10] have proved to be the most efficient. These techniques partition the image into small blocks and assign a set of motion parameters to each of these blocks. As the region described by one set of parameters is small, its motion can be efficiently represented by a simple model. For this reason, these techniques commonly consider a translational model (Eq. (6.12)), although a generalized block matching is proposed in [42] which supports more complex models. In [43], a block-based differential motion estimation is proposed which relies on a hierarchy of models, namely models of various complexities which can be applied to each block. The difficulty of the approach is in measuring the performance of the respective models in order to define a criterion to choose the optimal one.

In the framework of segmentation-based motion estimation, a motion model is applied to an object in the scene rather than to a small block of pixels. Therefore, the considered region corresponds usually to a larger area of the image. In this context, a simple translational model appears to be inefficient and more complex fully parametric models are needed (e.g. perspective models Eqs. (6.9) and (6.10), and affine model Eq. (6.11)). In particular, the affine model seems to be a good compromise and has been adopted in [22, 25, 27, 29].

3 SEGMENTATION-BASED FULLY PARAMETRIC MOTION ESTIMATION

In this section, the problem of motion estimation for a given segmentation is addressed. More precisely, the estimation of the motion parameters of a fully parametric model on a predefined region is discussed. The following discussion is general and the techniques presented hereafter are applicable to any type of regions (e.g. blocks, objects). The problem of spatio-temporal segmentation will be considered in Sec. 4.

The motion arising in a scene can be decomposed into a global motion due to the camera (e.g. pan or zoom) and a local motion due to the displacement of the objects in the scene. In order to efficiently handle camera motion, a two-stage global/local motion estimation can be carried out as shown in [44, 45, 46]. However, the motion estimation techniques discussed in this section rely only on local motion estimation. The global motion is thus integrated into the local estimates.

In the framework of coding, and in particular conversational and interactive services, coding delay is an important characteristic. For this reason, even though the techniques presented hereafter can be extended to N frames, only the problem of motion estimation between two frames is considered here. Furthermore, in coding applications the motion is commonly measured backwards, more precisely from the frame t towards the frame $t-1$, where t denotes the time index. Namely, for each pixel in the current frame, a correspondence is made in the previous frame. This way, the current frame is predicted from the previous one by

$$\hat{I}(\vec{r},t) = I(T(\vec{r}),t-1) \ , \tag{6.13}$$

and the prediction error or residual is given by

$$I(\vec{r},t) - \hat{I}(\vec{r},t) \ , \tag{6.14}$$

where $I(\vec{r},t)$ denotes the image intensity at location $\vec{r}$ and time t, $\hat{I}(\vec{r},t)$ represents the motion compensated prediction, and $T(\vec{r})$ is the corresponding pixel in the previous frame. Eq. (6.13) represents the principle of motion compensated predictive coding.

As shown in [47], the problem of motion estimation is ill-posed. In order to regularize it, all motion estimation techniques need therefore an additional assumption. This constraint can be implicit or explicit. In the case of a non-parametric motion model, an explicit constraint is required, whereas a parametric motion model implicitly constrains the motion estimation.

Two different approaches have been investigated to estimate the motion parameters of a fully parametric model. The techniques in the first class are composed of two steps: the computation of a dense optical flow using a non-parametric technique, followed by the modeling of the motion vectors by a set of motion parameters [19, 25, 29, 44]. The non-parametric motion estimation algorithm requires the introduction of an explicit constraint (e.g. smoothness constraint or assumption of local uniformity). As the motion parameters are not computed from the luminance signal itself, this approach is referred to as an indirect or regression technique. Its drawback is that its performance depends on the efficiency of the non-parametric motion estimation technique.

The second class of techniques directly estimates the parameters of the motion model as in [18, 23, 26, 27, 28, 46]. The computation of the motion field is implicitly constrained by the motion model itself, and consequently no additional explicit constraint is required. The motion parameters are obtained by minimizing an error norm computed from the luminance signal. As the estimation is carried out on the luminance signal itself, these techniques can be seen as direct. In this framework, differential techniques, which are based on a Taylor series expansion of the luminance signal, are the most widely used [18, 23, 26, 28]. An alternative approach to solve the problem is a matching technique [27, 46]. In contrast to the differential technique, the matching technique does not rely on a model of the luminance. Therefore, it is characterized by its robustness and its resilience to noise.

In the parametric motion estimation techniques above, outliers due to noise or to a badly defined support of the motion estimation which does not correspond to an area characterized by a coherent motion lead to wrong motion parameters. To overcome these problems, robust estimators less sensitive to outliers can be used [48, 49].

All the motion estimation techniques described in this section can be straightforwardly embedded in a multiresolution framework [18]. To that end, a pyramid structure of the input images is used [50], whereas the segmentation labels are simply subsampled. The final motion parameters at one level propagate as initial parameters to the next level where they are iteratively refined. The multiresolution structure leads to a more robust motion estimation and allows for a decreased computational complexity.

Finally, an important remark has to be made on the ultimate goal of motion estimation in the framework of coding. The motion information has to be transmitted to the decoder. This information includes the the segmentation information, as well as the motion parameters resulting from the above motion estimation algorithms. The former is commonly coded by chain code [51] or polygon approximation [52], where as the latter are usually PCM encoded [52, 53]. As far as the motion parameters are concerned, the quantization stage is very important as the quantization step sizes corresponding to each motion parameter determine the accuracy of the motion estimation as well as the amount of overhead to transmit the motion parameters. A uniform scalar quantizer is commonly applied. Despite their low correlations (in contrast to block matching motion estimation techniques where motion parameters are spatially highly correlated), the resulting quantized values can be entropy coded [53]. Against the popular belief that more accurate motion estimation is needed in order to reach higher coding performances, in [15, 34, 35] it is shown that the trade-off on the motion estimation accuracy and the amount of overhead information to be sent to the decoder must be considered. For this purpose, an appropriate motion model and an adequate quantization of the motion parameters have to be selected, while an efficient control of the segmentation process has to be implemented.

The above mentioned techniques that estimate the motion parameters of a fully parametric model on a predefined region are described in more detail hereafter. Their respective advantages and drawbacks are also analyzed. Finally, robust estimators less sensitive to outliers are discussed.

3.1 Constant brightness and the optical flow equation

Most motion estimation techniques assume that image luminance is constant along motion trajectories; in other words, the changes in the image intensity are only due to the motion in the scene

$$I(\vec{r}, t) = I(\vec{r} - \vec{v}(\vec{r}) \cdot \Delta t, t - \Delta t) \ , \tag{6.15}$$

where $I(\vec{r},t)$ denotes the image intensity at location $\vec{r}$ and time t, and $\vec{v}(\vec{r})$ is the motion during the time interval Δt. The first order Taylor series expansion of the right-hand term in Eq. (6.15) leads to the *spatio-temporal constraint equation* or the *optical flow constraint equation*

$$\vec{v}(\vec{r}) \cdot \vec{\nabla} I(\vec{r},t) + \frac{\partial I(\vec{r},t)}{\partial t} = 0 \ , \tag{6.16}$$

where $\vec{\nabla}$ is the spatial gradient.

The matching techniques are based on Eq. (6.15) and directly estimate the displacement $\vec{d}(\vec{r}) = \vec{v}(\vec{r}) \cdot \Delta t$, whereas differential techniques rely on Eq. (6.16).

The main problem of non-parametric motion estimation as defined by Eq. (6.16) is that it is underconstrained. In this case, an explicit constraint is needed to regularize the problem. For instance, Horn and Schunck propose a smoothness constraint [36], and Lucas and Kanade make the assumption that the motion is uniform and translational in a small region [37].

In contrast, when applying a parametric motion model, the motion $\vec{v}(\vec{r})$ becomes a function of the motion parameters $\vec{p}$, i.e. $\vec{v}(\vec{r})$ is replaced by $\vec{v}(\vec{r},\vec{p})$ in Eq. (6.16). When the motion is estimated on a region, this change provides an implicit constraint. Therefore, the problem of motion estimation as defined by Eq. (6.16) no longer requires an explicit constraint.

3.2 Indirect parametric motion estimation

The indirect parametric motion estimation techniques [19, 25, 29, 44] estimate the motion parameters of a fully parametric model in two steps.

A dense optical flow is first estimated using a non-parametric technique. In [19, 25], this initial optical flow is estimated by a differential technique, whereas [29] uses a block matching algorithm (i.e. in fact a trivial parametric technique). The method described in [44] is similar to the one in [29], except that it is rather applied to estimate the camera motion. The motion vectors obtained are then modeled by a set of motion parameters, using a Least Mean Square (LMS) technique [19, 25, 29, 44].

These methods can be seen as indirect, as they compute the motion parameters from a dense motion field rather than from the luminance signal itself. Their drawback is that they depend greatly on the accuracy of the non-parametric

motion parameters and the computational complexity. However, in contrast to the differential and the regression techniques, this quantization is already taken into account in the motion estimation.

Due to the discrete nature of the images, the point to match in the previous frame may not coincide with a pixel location. In this case, the luminance value of this point is evaluated by bilinear interpolation.

In order to reach the absolute minimum in Eq. (6.22), an exhaustive search of the n-dimensional discretized parameters space is required (the equivalent of the full-search in the classical block matching technique). However, depending on the number of parameters, this may require a too high computational complexity. To decrease the computational complexity, a fast non-exhaustive search can be carried out. In [27, 46], a generalization of the 3-step search [8] is used. Furthermore, the algorithm is carried out on a multiresolution structure based on a Gaussian pyramid [50]. The final motion parameters at one level propagate as initial estimates on the next level. A deterministic relaxation scheme is applied during the propagation stage. It compares the motion parameters obtained for neighboring regions and selects the one providing the lowest prediction error. This multiresolution and relaxation scheme allows for the reduction of the computational load, as well as the prevention of local minima due to the non-exhaustive search.

As the matching technique does not rely on a model of the luminance, it is characterized by its robustness. In [46], this technique has proved to outperform differential and regression techniques to estimate the camera motion.

3.4 Robust estimation

In the above parametric motion estimation, the estimation process is spoiled in the presence of samples whose value is far from the prevailing tendency. Those samples are referred to as outliers. They may be the result of impulse noise, a clutter environment, or a badly defined support of the motion estimation which does not correspond to an area characterized by a coherent motion. To overcome these problems, robust estimators less sensitive to outliers can be used [48, 49].

A very meaningful measure to assess the performance of a robust estimator is the breakdown point [49]. It is defined as the smallest fraction of outliers which results in a biased estimation. It indicates therefore the robustness of the estimator.

Estimators can be grouped in three categories: M-estimators, L-estimators and R-estimators [49]. Those three categories are briefly described hereafter.

M-estimators minimizes $\sum_i \rho(r_i)$ where r_i are the residuals (i.e. the difference between the data and the modeled value, see Eq. (6.14)), and ρ is a symmetric positive-definite function with a unique minimum at $r_i = 0$. Least-squares method is an example of such M-estimators where $\rho(r_i) = r_i^2$. Several functions ρ have been proposed to reduce the effect of large residuals, i.e. outliers. One example is the following robust estimator,

$$\rho(r_i) = \frac{\frac{r_i^2}{s^2}}{1 + \frac{r_i^2}{s^2}} , \tag{6.24}$$

where $s = 1.4826\ \mathrm{median}\{\mid r_i \mid\}$. The breakdown point of this estimator is less than $1/(n+1)$ where n is the number of parameters to be estimated.

L-estimators are linear combinations of order statistics. Two examples are the α-trimmed-mean and the median. As the median remains reliable when half of the data samples at the most are contaminated, it yields the maximum breakdown point of 0.5. R-estimators finally are based on rank tests.

4 SPATIO-TEMPORAL SEGMENTATION

The previous section reviews techniques to estimate the motion of a region assuming the segmentation is known. This section addresses the problem of spatio-temporal segmentation to define the support of the motion estimation. In contrast to static segmentation whose goal is to determine regions of homogeneous luminance (see Chap. 3 for a review), the purpose of the spatio-temporal segmentation is to represent the image sequence in terms of regions characterized by a coherent motion. Those regions can thus be identified to the moving objects in the scene. Straightforwardly, the notion of coherent motions depends upon the motion model considered. Hence, the spatio-temporal segmentation is closely linked to the underlying motion model.

Spatio-temporal segmentation is a chicken and egg problem. There is indeed a very strong interaction between the estimation of the moving objects boundaries and their motions. On the one hand, an accurate estimate of the motion is required to obtain a good segmentation. On the other hand, a good segmentation is needed in order to precisely estimate the motion.

To overcome the above dilemma, various methods have been proposed in the literature. In [57], a 3D segmentation based on luminance information is performed. This technique solves the problem of correspondence along the time axis [30]. However, its drawback is that the scene is segmented according to a criterion of uniform luminance instead of coherent motion.

Another approach consists in estimating first the motion in the scene by means of an optical flow. Based on this motion information, moving objects are then defined by the pixels whose motion is consistent with a set of motion model parameters [19, 25]. The initial optical flow estimation, which plays a crucial role, relies on a non-parametric motion model (see Secs. 2 and 3). It results in limited accuracy, especially on motion boundaries.

In order to overcome the drawbacks of the above approaches, spatio-temporal segmentation techniques which simultaneously estimate segmentation and motion have been proposed in [20, 21, 22]. The segmentation is expressed as a relaxation problem based on a Markov Random Field (MRF) modeling and a Bayesian criterion. In this framework, the spatio-temporal segmentation and the motion are simultaneously estimated. Whereas [21] relies on a non-parametric motion model, the techniques in [20, 22] assume a fully-parametric motion model.

Another approach consists in sequentially refining the segmentation and the motion [23, 24, 26, 27]. In [23], a single dominant motion is first estimated. Then, the current image is compared with the warped image, and new regions are defined as the areas corresponding to large prediction errors. The same procedure is applied to each of these new regions recursively. In [24], at each iteration pixels are assigned to regions based on their consistency with the different motions. Motion estimates are then updated using the new regions. In [26], the motion estimation and segmentation is performed on multiple frames and takes into account the information of luminance. Finally, the technique in [27] starts from a static segmentation. Regions corresponding to a failure of the motion estimation are further split by a clustering on the luminance. Conversely, regions characterized by a similar motion are merged by a clustering in the motion parameters space. In the above techniques [23, 24, 26, 27], the support of the motion estimation corresponds to a whole region of the scene. Assuming that the pixels within this region undergo a coherent motion, a parametric motion model can therefore be used (see Secs. 2 and 3). The latter is characterized by its robustness and high accuracy as many pixels contribute to the estimate, although biased estimates may occur when the support of the motion estimation is not well defined and does not correspond to an area characterized by a coherent motion.

As mentioned earlier, coding delay is an important feature for conversational and interactive services. In order to decrease the coding delay, it is desirable to limit the spatio-temporal segmentation to two frames. As a result, a coherent segmentation through time is not granted. In other words, a correspondence should be made among the regions defined at subsequent times. To alleviate this difficulty, tracking techniques [21, 30, 31, 32, 33] can be used, this is the subject of Sec. 5.

In the rest of this section the algorithms introduced in [22, 24, 25, 27], which represent four prevailing trends to tackle the problem of spatio-temporal segmentation are described and discussed in more detail. The problems due to occlusions and disocclusions are also addressed.

4.1 Segmentation of the optical flow

The technique by Wang and Adelson [25] estimates first an optical flow, and then segments the scene iteratively by fitting affine motion models. It is discussed in more detail hereafter.

An optical flow is first estimated (see Sec. 3.2). This local computation is based on a non-parametric motion model. Starting from an arbitrary block-based segmentation as an initial partitioning, the segmentation is iteratively refined by the following algorithm. For each region, affine motion parameters are computed by LMS regression (see Sec. 3.2). In order to group regions which cover the same moving object, a k-means clustering algorithm [58] is applied in the motion parameters space. Hence, this clustering provides a set of motion models representative of the motion in the scene.

Following the clustering, segmentation is refined. More precisely, each pixel location is assigned to one of the motion models. If i designs the i-th motion model, $\hat{\vec{v}}_i(\vec{r})$ the corresponding parameterized motion field, and $\vec{v}(\vec{r})$ the initially estimated optical flow, the pixel location $\vec{r}$ is assigned to the i-th motion model by

$$i(\vec{r}) = \arg\min_i(\vec{v}(\vec{r}) - \hat{\vec{v}}_i(\vec{r}))^2 \ . \tag{6.25}$$

In order to prevent inaccurate optical flow estimates from corrupting the segmentation, pixels where the error between the observed and modeled motions is greater than a threshold are left unassigned.

Before iterating the whole procedure, disjoint regions represented by a single model are split, whereas small regions are eliminated. Convergence is achieved when only a few points are reassigned or a maximum number of iterations is reached.

In order to assure the stability of the segmentation and coherence through time, the current segmentation is used to initialize the segmentation of the next frame.

4.2 Simultaneous estimation of segmentation and motion

The technique by Bouthemy and François [22] simultaneously estimates the spatio-temporal segmentation and the motion. The problem is expressed in a relaxation paradigm based on an MRF modeling and a Bayesian criterion. It is presented in more detail hereafter.

A priori knowledge is introduced to model the intensity and the motion in the image sequence. The MRF model is formulated in terms of local constraints on intensity and motion. Due to the equivalence between MRF and Gibbs distribution [59], the model reduces to energy functions over local neighborhoods. Hence, the probability p that the system is in a particular state is given by

$$p = \frac{1}{Z} e^{-U/T} , \tag{6.26}$$

where Z is a normalizing constant (also known as partition function), T is a constant (corresponding to the temperature in statistical mechanics), and U is an energy function (or objective function). The energy U is expressed as a sum of potentials, defined on so-called cliques, that specify how mutual neighbors contribute to the probability of a particular state.

Adopting a Maximum A Posteriori (MAP) estimation one form of Bayesian estimation leads to the maximization of the a posterior probability of a state given the observed data. Due to the equivalence between MRF and Gibbs distribution, the problem reduces to the minimization of the energy function U. This can be achieved by a stochastic or deterministic relaxation algorithm.

In [22], the energy function U is expressed as the combination of two terms U_1 and U_2. The first term contains the a priori knowledge on the regions to be segmented. Examples of such knowledge are that the regions are homogeneous, have regular boundaries, or are connected. It can be expressed by [22]

$$U_1 = \sum_{c \in C} \mu(1 - \delta_{e_s, e_t}) \ , \tag{6.27}$$

where $c = < s, t >$ denotes the binary clique, C is the set of all binary cliques, μ is a preset parameter ($\mu > 0$), e_s and e_t are the labels at sites s and t, and δ is the Kronecker symbol.

The second term indicates the adequacy between the observed and estimated data. Based on the optical flow constraint equation (Eq. (6.16)) and an affine motion model, it is given by [22]

$$U_2 = \frac{1}{2\sigma^2} \sum_{s \in S} \left[\vec{\nabla} I(s) \cdot \vec{v}(s, \vec{p}_{e_s}) + \frac{\partial I(s)}{\partial t} \right]^2 \ , \tag{6.28}$$

where S denotes the set of sites s where labels are defined (e.g. the original pixel grid [1]), $\vec{v}(s, \vec{p}_{e_s})$ is the parameterized motion field (which therefore depends on the labels e_s used to compute the motion parameters $\vec{p}_{e_s}$), and the error between the parameterized motion field and the true velocity is supposed to follow a zero-mean Gaussian distribution with variance σ^2.

The energy function $U = U_1 + U_2$ is finally minimized by a deterministic relaxation scheme. The labels, which are local, and the motion parameters, which are global, are alternatively and iteratively estimated. A supplementary label is considered in order to determine newly appearing regions. Convergence is achieved when the number of sites whose label has been changed fall below a preset threshold.

To assure a coherent segmentation through time, the labels and the motion field obtained at time t are used to predict the segmentation at time $t + 1$. Therefore, this allows for the linking of the regions through time.

[1] in [22] a site s is rather chosen as a 2×2 block and in this case Eq. (6.28) should be summed at the 4 pixel locations involved in each site s

4.3 Sequential estimation of segmentation and motion based on dominant motion

In this section, the algorithm proposed by Peleg and Rom in [24] which iteratively refines the segmentation and the motion estimates is described in more detail. In this algorithm, the number of different regions is predefined. For the sake of clarity, the simple case of two motions (i.e. one foreground moving object and the background) is considered.

A dominant motion is first estimated by a fully parametric technique. Under the assumption that the moving object is small when compared to the background, the resulting set of motion parameters, denoted by $\vec{p_0}$, corresponds to the motion of the background. The prediction error is then computed using the parameters $\vec{p_0}$. Straightforwardly, the largest region of high error corresponds to the location of the moving object. Hence, a second set of motion parameters $\vec{p_1}$ is estimated for this region. The parameters $\vec{p_1}$ can therefore be identified as the motion of the moving object.

For every pixel location $\vec{r}$ in frame $I(t)$, its location in frame $I(t + \Delta t)$ is computed with both sets of parameters $\vec{p_0}$ and $\vec{p_1}$. The ratio $R(\vec{r})$ is defined as the ratio between the prediction error when using the motion $\vec{p_1}$ and the prediction error when using the motion $\vec{p_0}$

$$R(\vec{r}) = \frac{\mid I(\vec{r}, t) - I(\vec{r} + \vec{v}(\vec{p_1}), t + \Delta t) \mid}{\mid I(\vec{r}, t) - I(\vec{r} + \vec{v}(\vec{p_0}), t + \Delta t) \mid} . \qquad (6.29)$$

To cope with noisy images, the ratio $R(\vec{r})$ is averaged on a 3×3 window. Pixels are then classified in three categories depending on the ratio $R(\vec{r})$: as background when $R(\vec{r}) \gg 1$, as a moving object when $R(\vec{r}) \ll 1$, and as undecided when $R(\vec{r}) \approx 1$. Finally, the parameters $\vec{p_0}$ and $\vec{p_1}$ are refined on the background and on moving object regions respectively, and the whole procedure is iterated.

The algorithm can be generalized to handle several moving objects by using more motion parameters $\vec{p_0}, \vec{p_1}, \ldots, \vec{p_n}$.

4.4 Sequential estimation of segmentation and motion based on motion information and static segmentation

The algorithm introduced by Dufaux *et al.* in [27] sequentially refines the segmentation and the motion estimates by efficiently combining static segmentation and motion information. It is discussed in more detail hereafter.

The motion arising in a scene can be decomposed into a global motion due to the camera (e.g. pan or zoom) and a local motion due to the displacement of the objects in the scene. In order to efficiently handle camera motion, a global motion estimation is first carried out by a frame matching technique [46] to remove the camera motion.

To make the image easier to segment, a spatial pre-filtering is then applied. The purpose of this pre-filter is to produce constant luminance regions delimited by sharp contours. The morphological operator open-close by reconstruction has shown to be well adapted for this task [57]. It indeed produces flat zones while preserving the contour information. Therefore this operator is adopted in [27].

Following the pre-processing stage, the static segmentation is performed by a k-means clustering algorithm [58] on the luminance values. For each of the resulting static regions, motion parameters are computed using a matching technique (see Sec. 3.3). The motion estimation relies on an affine motion model. To decrease the computational complexity and to allow a non-exhaustive search while avoiding local minima, a Gaussian pyramid structure of the input images is built. A deterministic relaxation scheme is applied during the propagation stage to avoid local minima. Due to the use of the static segmentation, the motion estimation is always applied on a region characterized by a coherent motion, therefore allowing a precise and robust estimate. Nevertheless, in order to cope with a badly defined support of the motion estimation due to a failure of the segmentation, the robust estimator defined by Eq. (6.24) less sensitive to outliers is used (see [48, 49] and Sec. 3.4).

Because they are supposed to be the least significant, small or low contrast image features are lost during the segmentation process. However, some of these features might carry a visually important relevance. To partially overcome this problem, regions not well compensated are split further. The decision is taken based on a threshold on the prediction error. The split is then carried out by a static segmentation performed by k-means clustering on each of the selected regions. Furthermore, the new regions are restricted to the areas which correspond to high prediction error. This refinement of the static segmentation allows for the recovery of some of the small or low contrast features having a significantly distinct motion which can therefore be assumed to be visually important.

In the following stage, regions with similar motions are merged by applying a k-medoid clustering algorithm [58] in the motion parameters space. At this stage, the k-medoid is preferred to the k-means clustering for the following reason. The k-means is very sensitive to outliers, whereas in the k-medoid the centroid of each cluster is chosen among the elements of the input data resulting in a more robust clustering and is less sensitivity with respect to outliers. This clustering in the motion parameters space results in regions characterized by coherent motion which can therefore be identified as the moving objects in the scene.

The main advantage of the above spatio-temporal segmentation algorithm is that as the boundaries are computed on the luminance signal, they can be very precisely located. In particular problems related to the presence of occlusions and disocclusions are alleviated. Moreover, the two-stage global/local matching motion estimation algorithm using a robust estimator is characterized by its robustness and its resilience to noise leading to high performances. Therefore, the proposed algorithm is efficient in finding the spatio-temporal meaningful entities existing in the scene.

Figures 2 and 3 show a frame of each test sequences "Foreman" and "Carphone", the pre-filtered image, the initial static segmentation, the final spatio-temporal segmentation obtained with the above algorithm, the resulting motion compensated prediction (using the previous original frame rather than the decoded one) and the corresponding prediction error. The spatio-temporal segmentation has resulted in only 8 and 7 regions for "Foreman" and "Carphone" respectively, whereas the corresponding initial static segmentation has generated 32 and 27 regions. In particular, the moving objects are effectively segmented and the motion boundaries are very precisely located, showing the efficiency of the segmentation method. As the resulting regions are characterized by a coherent motion, a very accurate motion compensated prediction is obtained. Furthermore, the low number of moving objects allows for a compact representation of the motion information leading to high coding performances.

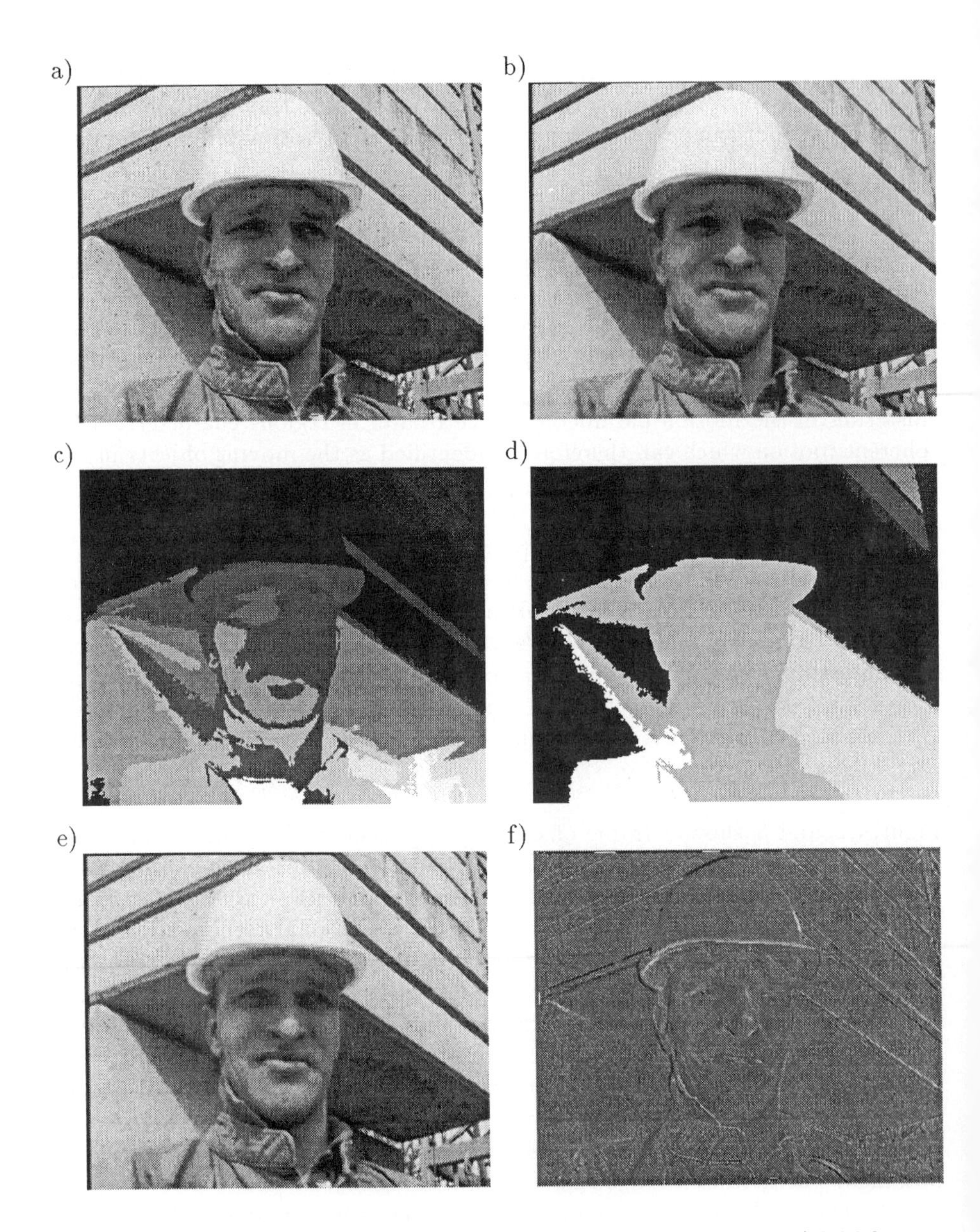

Figure 2 Sequence "Foreman": a) original frame, b) pre-filtered, c) initial static segmentation, d) final spatio-temporal segmentation, e) motion compensated prediction, f) prediction error (rescaled).

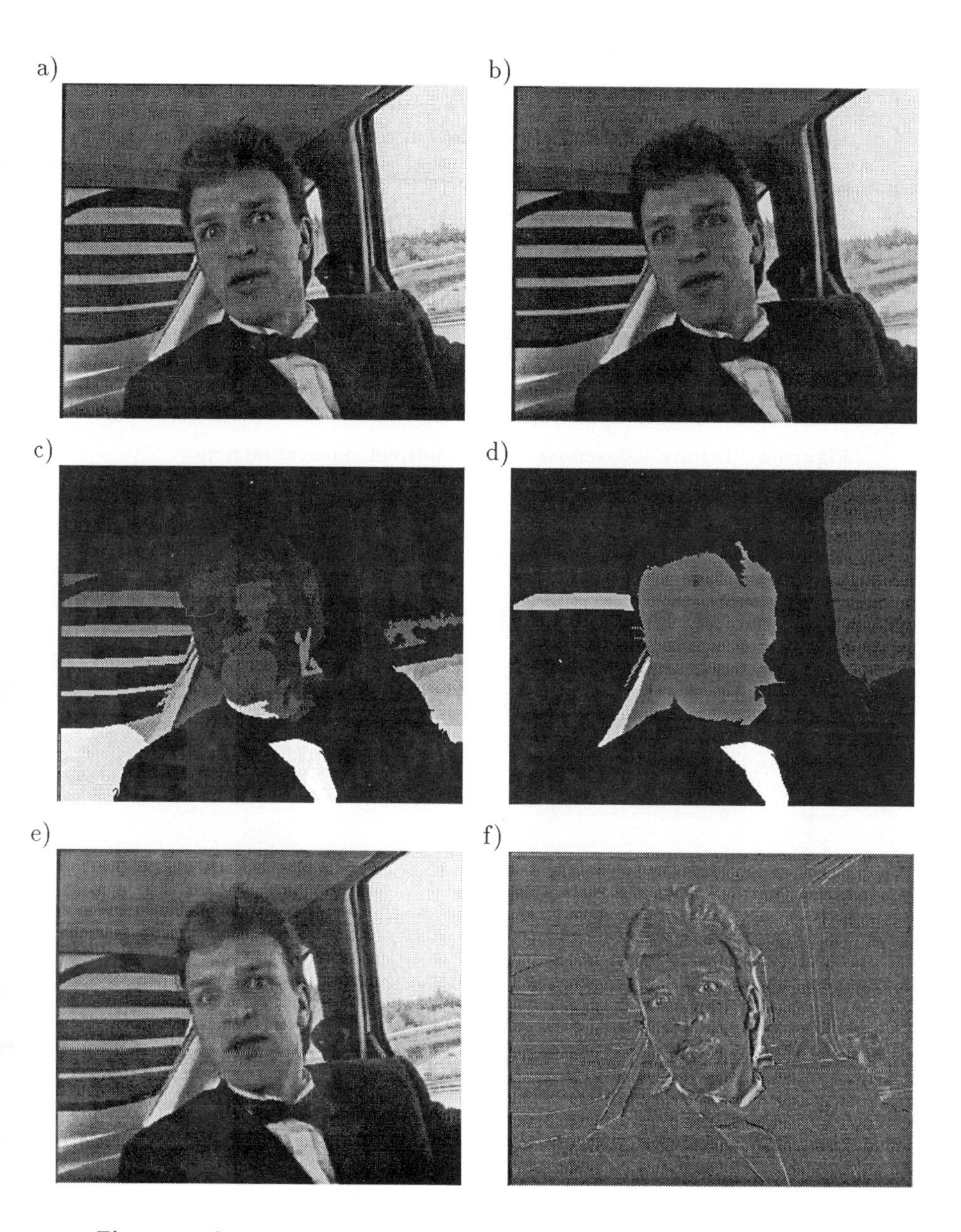

Figure 3 Sequence "Carphone": a) original frame, b) pre-filtered, c) initial static segmentation, d) final spatio-temporal segmentation, e) motion compensated prediction, f) prediction error (rescaled).

4.5 Occlusion and disocclusion

The presence of occlusions and disocclusions raises difficulties during the segmentation process. Figures 4, 5 and 6 illustrate those situations.

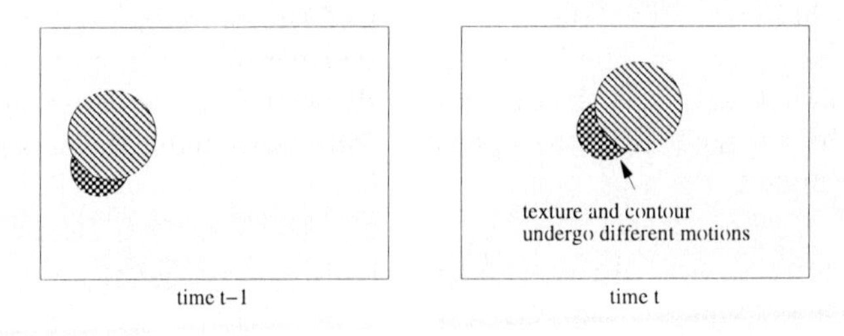

Figure 4 Texture and contour undergo different apparent motions.

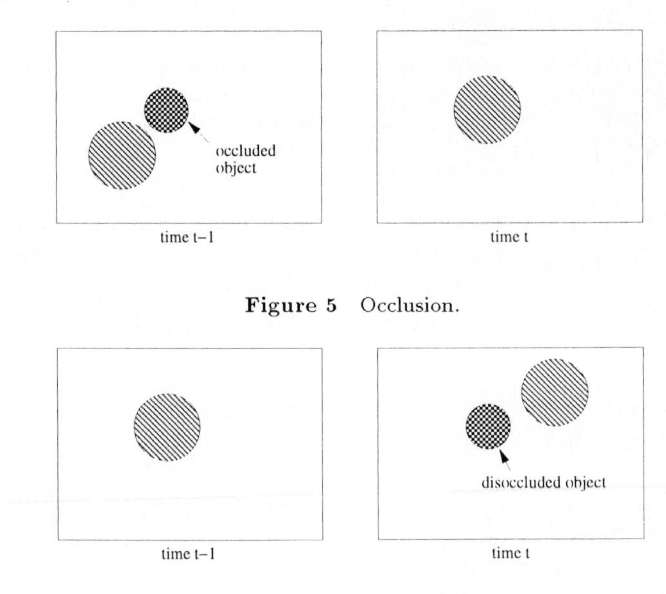

Figure 5 Occlusion.

Figure 6 Disocclusion.

Due to occlusions and disocclusions, the apparent motion of the texture and the contour of an object may be different as illustrated in Fig. 4. More precisely, the motion of the pixels inside a region and the motion of the shape of this region coincide in the absence of occlusion. However, this is no longer the case in the presence of a partial occlusion, as the modification of the shape of the occluded region is defined by the motion of the occluding region. This problem is addressed in Chap. 4 and in [29]. In order to alleviate this problem, the motion corresponding to the contour could be estimated and transmitted along with the motion corresponding to the texture.

Another approach is to use tracking techniques which are presented in Sec. 5. Tracking techniques define the trajectories of the moving objects along time. Therefore, the objects can be tracked in successive frames, even when occlusion occurs [60, 61]. Conversely, in the case of disocclusion, the tracking technique introduced in [62] recognizes the appearing object as being the same as the one that was occluded.

The problem of an area of the background being disoccluded, also referred to as uncovered background, is very pertinent in coding. An efficient solution to overcome this problem is to store the background information in a long term memory [63, 64, 65]. These algorithms are particularly efficient for video-conference or video-phone sequences.

5 TRACKING

The techniques which aim at obtaining a spatio-temporal segmentation must heavily rely on the motion information in order to assures a thorough understanding of dynamic scenes and enables for the derivation of visually meaningful results. However, most techniques only take into account two consecutive frames and as such suffer from many drawbacks. In particular, estimation inaccuracies, noise as well as the lack of decisive information may alter the interpretation of the scene. The coherence of the spatio-temporal segmentation is thus not guaranteed through time.

In order to tackle such problems, a tracking procedure can be applied [21, 30, 31, 32, 33]. It allows the information obtained over past images to be used and as such is a key element to dynamic scene analysis. Although many types of features can undergo tracking, the success of the tracking algorithm depends heavily on the intrinsic meaning of the chosen features. In other words, the tracked features should carry a natural meaning in order to allow a tracking procedure to be successful.

In the context of segmentation-based motion estimation coupled with a spatio-temporal segmentation, one can expect to obtain good estimates of both the objects present in the scene and their respective motion. The use of the latter as features being tracked is thus most natural. In the framework of video coding, the tracking procedure has two main objectives. First of all, a memory of the previous measurements is kept so as to base the current motion estimate not only on the current measurement, but also on the previous ones. The measurements are seen as carrying inherent inaccuracies which must be accounted for. Secondly, a prediction of the objects temporal evolutions can be performed and hereby solve the correspondence problem by linking temporally the successive segmentations. A coherent spatio-temporal segmentation through time is the key to reducing the cost of sending the segmentation information [66].

5.1 Object-Tracking

Arising from the spatio-temporal segmentation approach, the segmented regions are characteristic of the real objects present in the scene and, consequently, their temporal behavior can be expected to be quite predictable and smooth. Their motions are thus very well suited to undergo a tracking procedure [32]. Their evolution through time is referred to as trajectory.

In a formal manner, tracking can be defined as the process of estimating the current dynamic state of the object based on the motion measurements [31]. More precisely, the dynamic state is updated through time based on the successive motion measurements which are seen as inaccurate observations related to the state of the object. Finally, a track is the trajectory of the state and is formed by measurements which are temporally linked. The implementation of a tracking procedure requires the definition of a temporal filtering as well as sometimes an explicit trajectory model. In the latter case, the motion is described by kinematic components which try to approximate the dynamic motion evolution.

The tracking procedure has many attractive points. In a first stage, it allows for the combination of the information present in multiple images. This integration capacity permits the uncertainty on the state estimate to be decreased as the information of many measurements is accumulated through time. The tracking procedure is thus able to explicitly take into account the inaccuracies existing in the spatio-temporal segmentation as well as in the segmentation-based motion estimation. Those inaccuracies arise due to many factors. First of all, distortions are introduced at the level of the image formation (e.g. geometric distortions, blurring, discretization, quantization). Moreover, the spatio-temporal segmentation may be inexact. Furthermore, texture and contour may undergo different motions in appearance (see Sec. 4.5). This gives rise to further inaccuracies in the motion measurements. Another problem is that the motion estimation may fall in a local minimum and thus give an inaccurate estimate of the motion. Finally, in the framework of video coding, the support of the motion estimation is distorted due to the loss of information inherent to the coding process.

In order to exploit the properties and characteristics of the tracking procedure, the trajectories of the objects have to be found. The operation involves solving the correspondence problem [30] through time so as to perform a temporal linkage between successive spatio-temporal segmentations. In other words, the object has to be identified in the successive images of the sequence. Although the correspondence problem can be solved a priori [30], it can be tackled by using the information arising from the tracking procedure. Assuming a smooth continuous motion, the latter permits a prediction of the temporal evolution to be made. Such prediction can be used to roughly define the location of the object through time. An adjustment is then carried out so as to precisely locate the object.

As expressed here above and shown in [60], the tracking procedure allows for the stabilization of the estimation process when partial occlusion problems arise. The tracking approach has the further property that it permits the tackling of the cases when total occlusion occurs [61]. As no measurement is available, the predicted location is assumed to be the real one without performing any adjustment. The object can thus be tracked even if it is totally occluded. When disocclusion takes place, the challenge is to be able to recognize the appearing object as being the same as the one that was occluded. In [62], a solution relying on the statistical concept of validation gate is proposed.

In order to implement the tracking procedure and benefit from all its advantages, a temporal filtering as well as, in some cases, a trajectory model have to be used. The trajectory model may be explicitly defined and permits us to characterize mathematically the temporal evolution of the object motion. The temporal filtering aims at integrating present motion measurements with previous ones while taking into account the inherent inaccuracies.

In Sec. 5.2, a first approach relying on an explicit trajectory model is presented. The temporal evolution is approximated through kinematic components and is used in conjunction with linear estimation techniques. Based on the state space representation and a recursive linear Minimum Mean Square Error estimator (MMSE), the latter approach leads to the Kalman filter algorithm. In Sec. 5.3, an approach relying on the concept of time integration is described. It tracks the object of interest through successive warpings and is able to decrease the inaccuracies in the motion measurements. This technique does not require any trajectory model.

5.2 Kalman Filter Algorithm

Because the segmented regions are closely related to the real objects present in the scene, their respective motion undergoes a temporal evolution which is well approximated by classical kinematic laws. This observation allows for the definition of explicit trajectory models which typically use kinematic components such as position, velocity or acceleration. In order to have a compact representation of the tracking procedure, the state space representation reveals itself to be most appropriate. The kinematic components are embedded in the state vector $\vec{x}$ whose temporal evolution is thus implicitly defined by the chosen trajectory model. Such evolution is described by *the dynamic equation*, also referred to as *Plant equation*. Assuming a discrete temporal sampling and time t_k, the dynamic equation can be written as follows [31]

$$\vec{x}(t_{k+1}) = F(t_k)\vec{x}(t_k) + G(t_k)\vec{u}(t_k) + \vec{v}(t_k) \ . \tag{6.30}$$

The matrix $F(t_k)$ is a known matrix which incorporates the trajectory model. The relationship between the system and the known input vector $\vec{u}(t_k)$ is defined by the known matrix $G(t_k)$. The input vector $\vec{u}(t_k)$ represents the external influences upon the system. It can also represent the unmodeled evolutionary behaviors which are not predicted by the trajectory model and are usually referred to as maneuvers. The process noise $\vec{v}(k)$ assures that the inherent inaccuracies of the trajectory model are taken into account. This noise is supposed to be white, Gaussian and zero-mean with covariance $Q(t_k)$ defined by

$$E[\vec{v}(t_k)\ \vec{v}(t_l)] = \delta_{kl}Q(t_k) \ , \tag{6.31}$$

where $E[\ ]$ and δ indicate respectively the mathematical expectation and the Kronecker delta symbol.

Equation (6.30) describes the state of a dynamic system at time t_k on which measurements are performed. The relationship between $\vec{x}(t_k)$ and measurements $\vec{z}(t_k)$ is given by *the measurement equation*

$$\vec{z}(t_k) = H(t_k)\vec{x}(t_k) + \vec{w}(t_k) \ . \tag{6.32}$$

The relationship between the measurements and the state representation is defined through the known matrix $H(t_k)$. The measurement noise $\vec{w}(t_k)$ stands for the inaccuracies of the measurement procedure and is usually modeled as a zero-mean, white and Gaussian noise with covariance $R(t_k)$ defined by

$$E[\vec{w}(t_k)\ \vec{w}(t_l)] = \delta_{kl} R(t_k)\ . \tag{6.33}$$

In order to obtain an evaluation of the state $\vec{x}(k+1)$, an estimator $\hat{\vec{x}}(t_{k+1})$ of the latter has to be defined. Furthermore, a temporal filtering of the state estimate is obtained by relying not only on the present measurement $\vec{z}(t_{k+1})$ but also on the set of past measurements $Z^k = \{\vec{z}(t_j)\ , j = 1, \ldots, k\}$. The temporal consistency and its robustness to measurement errors is hereby increased. Trying to minimize the variance of the estimate, the MMSE estimator $\hat{\vec{x}}(t_{k+1} \mid k+1)$ and its covariance $P(t_{k+1} \mid k+1)$ are given by

$$\hat{\vec{x}}(t_{k+1} \mid k+1) = E[\vec{x}(t_{k+1}) \mid Z^{k+1}]\ , \tag{6.34}$$

and

$$P(t_{k+1} \mid k+1) = E[(\vec{x}(t_{k+1}) - \hat{\vec{x}}(t_{k+1} \mid k+1))(\vec{x}(k+1) - \hat{\vec{x}}(t_{k+1} \mid k+1))' \mid Z^{k+1}]\ , \tag{6.35}$$

with $()'$ representing the transposition operation.

The notation $\hat{\vec{x}}(t_{k+1} \mid k+1)$ denotes the estimate of the state at time t_{k+1} based on the measurements up to time $k+1$. The same notation convention holds for $P(t_{k+1} \mid k+1)$ and is used in the following.

In order to facilitate the estimation procedure, recursive forms of Eqs. (6.34) and (6.35) are desired. In view of the linearity of the measurement equation, linear estimation techniques can be applied. As expressed here above, the measurement noise $\vec{w}(t_k)$ can be seen as zero-mean, white and Gaussian with covariance $R(t_k)$. Under the further conditions that the process noise $\vec{v}(t_k)$ and the initial state $\vec{x}(t_0 \mid 0)$ are Gaussian, the stochastic processes $\vec{x}(t_k)$ and $\vec{z}(t_k)$ are joint Gauss-Markov processes. Using Eqs. (6.30) and (6.32), the latter property allows us to define a one-step prediction relationship on the one hand and, on the other hand, a recursive relationship to estimate the state. The ensemble of these two relationships forms *the Kalman filter algorithm*. The one-step prediction relationship computes the estimates $\hat{\vec{x}}(t_{k+1} \mid k)$ and $\hat{\vec{z}}(t_{k+1} \mid k)$ by only relying on Z^k. It is written as follows

$$\hat{\vec{x}}(t_{k+1} \mid k) = F(t_k)\hat{\vec{x}}(t_k \mid k) + G(t_k)\vec{u}(t_k)\ , \tag{6.36}$$

$$\hat{\vec{z}}(t_{k+1} \mid k) = H(t_{k+1})\hat{\vec{x}}(t_{k+1} \mid k)\ . \tag{6.37}$$

Furthermore, a prediction of the covariance matrix $P(t_{k+1} \mid k)$ is obtained by:

$$P(t_{k+1} \mid k) = F(t_k)P(t_k \mid k)F'(t_k) + Q(t_k)\ . \tag{6.38}$$

The estimation of the state at time t_{k+1} uses the results of the one-step prediction relationship. It is given by:

$$\hat{\vec{x}}(t_{k+1} \mid k+1) = \hat{\vec{x}}(t_{k+1} \mid k) + W(t_{k+1})\nu(t_{k+1}) \ , \tag{6.39}$$

where

$$\nu(t_{k+1}) = \vec{z}(t_{k+1}) - \hat{\vec{z}}(t_{k+1} \mid k) \ , \tag{6.40}$$

and

$$W(t_{k+1}) = P(t_{k+1} \mid k)H'(t_{k+1})(H(t_{k+1})P(t_{k+1} \mid k)H'(t_{k+1}) + R(t_{k+1}))^{-1} \ . \tag{6.41}$$

The term $\nu(t_{k+1})$ is called innovation. It embodies the new information brought by the measurement $\hat{\vec{z}}(t_{k+1})$ to the state estimate procedure. The innovation can be shown to form an uncorrelated process. As far as $W(t_{k+1})$ is concerned, it is referred to as the filter gain. Its function is to balance the influence of the innovation in the state estimate and thus ensure the robustness of the estimate to flawed or unstable measurements.

Based on the predicted covariance (Eq. (6.38)), the updated covariance matrix is:

$$\begin{aligned} P(t_{k+1} \mid k+1) &= [\mathcal{I} - W(t_{k+1})H(t_{k+1})]P(t_{k+1} \mid k)[\mathcal{I} - W(t_{k+1})H(t_{k+1})]' \\ &\quad + W(t_{k+1})R(t_{k+1})W'(t_{k+1}) \ , \end{aligned} \tag{6.42}$$

where $\mathcal{I}$ stands for the identity matrix.

Up to now, the matrices F, G and H were assumed to vary with time. In cases such as digital video sequences where the time sampling is constant, the latter matrices no longer depend on time. With the further hypothesis that the process and measurement noises are stationary, the discrete-time system described by Eqs. (6.30) and (6.32) is time-invariant. For such a system, the covariance P and filter gain W converge towards constant values and entails that the Kalman filter tends to a steady state [31].

5.3 Time Integration

The definition of an explicit trajectory model may prove to be too complex. The temporal integration technique allows for the implementation of a tracking procedure without requiring any trajectory model [33]. More specifically, the information of the past object motion measurements is kept under the form of a registered image. The latter is obtained by a combination of the registrations of successive images $I(t_k)$ with respect to the object motion. A dynamic internal representation for each object is thus obtained which tracks its respective object. The temporal integration performs also a function of temporal filtering. It is carried out by the combination of past and present registered images which implies a robustness to motion measurements inaccuracies. For a given object, the corresponding sequence of temporally integrated images Av is defined as follows

$$Av(0) = I(t_0) , \tag{6.43}$$

$$Av(t_{k+1}) = m\, I(t_{k+1}) + (1-m)\ \text{register}(Av(t_k), I(t_{k+1})) , \tag{6.44}$$

where $m \in [0,1]$ and register(K, L) signifies the registering of image K towards image L with the measured object motion. The latter is computed between times t_k and t_l.

By computing the object motion between the new frame and the corresponding temporally integrated image, the inaccuracies in the motion measurements are decreased. In the corresponding registered image, the tracked object remains indeed sharp while other objects blur out.

6 CONCLUSIONS

Motion estimation plays a key role in reaching high performances in video coding. In the framework of second generation coding, there is a clear need for the development of a new approach to motion estimation which rely on the semantic representation of the scene in terms of objects.

This chapter reviewed the major points to address in order to develop such a new approach to motion estimation. Namely, those points are: motion models, segmentation-based motion estimation, spatio-temporal segmentation and tracking. Various promising algorithms have been discussed, and simulation results have been presented showing the feasibility of this approach.

APPENDIX A

A.1 EQUIVALENCE BETWEEN VELOCITY AND DISPLACEMENT FORMULATIONS

In order to show the equivalence between the instantaneous velocity and displacement formulations, Eq. (6.5) can be written as

$$\vec{r}' - \vec{r} = \begin{pmatrix} x' - x \\ y' - y \end{pmatrix} = \begin{pmatrix} \frac{-\Omega_x \ xy + \Omega_y (x^2+1) - \Omega_z \ y + (D_x - D_z x)/Z}{1 - \Omega_y \ x + \Omega_x \ y + D_z/Z} \\ \frac{-\Omega_x (y^2+1) + \Omega_y \ xy - \Omega_z \ x + (D_y - D_z y)/Z}{1 - \Omega_y \ x + \Omega_x \ y + D_z/Z} \end{pmatrix} . \tag{A.1}$$

Under the assumptions that the Z component of the translation is small with respect to the distance of the object from the camera ($| D_z/Z | \ll 1$), and that the visual angle as well as the rotation angles are small ($| -\Omega_y \ x + \Omega_x \ y | \ll 1$), a first order Taylor series expansion of Eq. (A.1) leads to a form similar to Eq. (6.3) [19]. Therefore both formulations in terms of instantaneous velocities and displacements are equivalent under these assumptions.

A.2 PLANAR SURFACE UNDER PERSPECTIVE PROJECTION

For a planar surface under perspective projection, the instantaneous velocity in the image plane is given by

$$\begin{pmatrix} \dot{x} \\ \dot{y} \end{pmatrix} = \begin{pmatrix} a_1 + a_2 x + a_3 y + a_7 x^2 + a_8 xy \\ a_4 + a_5 x + a_6 y + a_7 xy + a_8 y^2 \end{pmatrix} , \tag{A.2}$$

where

$$a_1 = \omega_y + T_x k_z , \tag{A.3}$$

$$a_2 = T_x k_x - T_z k_z , \tag{A.4}$$

$$\begin{aligned}
a_3 &= -\omega_z + T_x k_y \ , & \text{(A.5)}\\
a_4 &= -\omega_x + T_y k_z \ , & \text{(A.6)}\\
a_5 &= \omega_z + T_y k_x \ , & \text{(A.7)}\\
a_6 &= T_y k_y - T_z k_z \ , & \text{(A.8)}\\
a_7 &= \omega_y - T_z k_x \ , & \text{(A.9)}\\
a_8 &= -\omega_x - T_z k_y \ . & \text{(A.10)}
\end{aligned}$$

Similarly, for a planar surface under perspective projection, the displacement in the image plane is given by

$$\begin{pmatrix} x' \\ y' \end{pmatrix} = \begin{pmatrix} \frac{a_1+a_2x+a_3y}{a_7x+a_8y+a_9} \\ \frac{a_4+a_5x+a_6y}{a_7x+a_8y+a_9} \end{pmatrix} , \quad \text{(A.11)}$$

where

$$\begin{aligned}
a_1 &= \Omega_y + D_x k_z \ , & \text{(A.12)}\\
a_2 &= 1 + D_x k_x \ , & \text{(A.13)}\\
a_3 &= -\Omega_z + D_x k_y \ , & \text{(A.14)}\\
a_4 &= -\Omega_x - D_y k_z \ , & \text{(A.15)}\\
a_5 &= \Omega_z + D_y k_x \ , & \text{(A.16)}\\
a_6 &= 1 + D_y k_y \ , & \text{(A.17)}\\
a_7 &= -\Omega_y + D_z k_x \ , & \text{(A.18)}\\
a_8 &= \Omega_x + D_z k_y \ , & \text{(A.19)}\\
a_9 &= 1 + D_z k_z \ . & \text{(A.20)}
\end{aligned}$$

Acknowledgements

This work was partly supported by the Television of Tomorrow Program at the MIT Media Laboratory. The authors would like to thank N. Vasconcelos and R. Kermode for providing valuable comments.

REFERENCES

[1] ISO/IEC JTC1 CD 11172. Information technology - Coding of moving pictures and associated audio for digital storage media up to about 1.5

Mbit/s - Part 2: Coding of moving pictures information. Technical report, 1991.

[2] D. LeGall. MPEG: A video compression standard for multimedia. *Commun. of the ACM*, vol. 34, no. 4, pp. 47-58, April 1991.

[3] ISO/IEC DIS 13818. Information technology - Generic coding of moving pictures and associated audio. ITU - T Recommendation H.262. Technical report, March 1994.

[4] D. LeGall. The MPEG video compression algorithm. *Signal Processing: Image Communications*, vol. 4, no. 2, pp. 129-140, April 1992.

[5] CCITT SG XV. Recommendation H.261 -Video codec for audiovisual services at p*64kbit/s. Technical Report COM XV-R37-E, August 1990.

[6] Draft ITU-T. Recommendation H.263 - Video coding for narrow telecommunication channels at $<$ 64kbit/s. Technical report, July 1995.

[7] J.R. Jain and A.K. Jain. Displacement measurement and its application in interframe image coding. *IEEE Trans. Commun.*, vol. COM-29, no. 12, pp. 1799-1808, December 1981.

[8] T. Koga, K. Iinuma, A. Hirano, Y. Iijima, and T. Ishiguro. Motion compensated interframe coding of video conferencing. In *Proc. Nat. Telecommun. Conf.*, pages G5.3.1–G5.3.5, New Orleans, LA, December 1981.

[9] H.G. Musmann, P. Pirsch, and H.J. Grallert. Advances in picture coding. *Proc. IEEE*, vol. 73, pp. 523-548, April 1985.

[10] F. Dufaux and F. Moscheni. Motion estimation techniques for digital TV: a review and a new contribution. *Proc. IEEE*, vol. 83, no. 6, pp. 858-876, June 1995.

[11] M. Kunt, A. Ikonomopoulos, and M. Kocher. Second generation image coding techniques. *Proc. IEEE*, vol. 73, no. 4, pp. 549-575, April 1985.

[12] M. Kunt, M. Benard, and R. Leonardi. Recent results in high compression image coding. *IEEE Trans. Circuits and Syst.*, vol. CAS-34, no. 11, pp. 1306-1336, November 1987.

[13] ISO/IEC JTC1/SC29/WG11. Coding of moving pictures and associated audio information, MPEG-4 proposal package description. Technical report, March 1995.

[14] M.H. Chan, Y.B. Yu, and A.G. Constantinides. Variable size block matching motion compensation with applications to video coding. *IEE Proc.*, vol. 137, no. 4, pp. 205-212, August 1990.

[15] F. Dufaux. *Multigrid Block Matching Motion Estimation for Generic Video Coding.* PhD thesis, Swiss Federal Institute of Technology, Lausanne, Switzerland, 1994.

[16] M.T. Orchard. Predictive motion field segmentation for image sequence coding. *IEEE Trans. Circuits and Systems for Video Technology*, vol. CSVT-3, no. 1, pp. 54-70, February 1993.

[17] F. Dufaux, I. Moccagatta, F. Moscheni, and H. Nicolas. Vector quantization based motion field segmentation under the entropy criterion. *Journal of Visual Communication and Image Representation*, vol. 5, no. 4, pp. 356-369, December 1994.

[18] P. Anandan, J.R. Bergen, K.J. Hanna, and R. Hingorani. Hierarchical model-based motion estimation. In M.I. Sezan and R.L. Lagendijk, editors, *Motion Analysis and Image Sequence Processing*, pages 1–22. Kluwer Academic Publishers, 1993.

[19] G. Adiv. Determining three-dimensional motion and structure from optical flow generated by several moving objects. *IEEE Trans. Pattern Anal. Machine Intell.*, vol. PAMI-7, no. 4, pp. 384-401, July 1985.

[20] D.W. Murray and B.F. Buxton. Scene segmentation from visual motion using global optimization. *IEEE Trans. Pattern Anal. and Machine Intell.*, vol. PAMI-9, no. 2, pp. 220-228, March 1987.

[21] M.J. Black. Combining intensity and motion for incremental segmentation and tracking over long image sequences. In *Europ. Conf. on Computer Vision*, pages 485–493, Santa Margherita, Italy, May 1992.

[22] P. Bouthemy and E. François. Motion segmentation and qualitative dynamic scene analysis from an image sequence. *Int. Journal of Computer Vision*, vol. 10, no. 2, pp. 157-182, 1993.

[23] M. Hoetter and R. Thoma. Image segmentation based on object oriented mapping parameter estimation. *Signal Processing*, vol. 15, no. 3, pp. 315-334, October 1988.

[24] S. Peleg and H. Rom. Motion based segmentation. In *IEEE Proc. Int. Conf. on Pattern Recognition*, pages 109–113, Atlantic City, NJ, June 1990.

[25] J.Y.A. Wang and E.H. Adelson. Representing moving images with layers. *IEEE Trans. Image Proces.*, vol. 3, no. 5, pp. 625-638, September 1994.

[26] P. Schroeter and S. Ayer. Multi-frame based segmentation of moving objects by combining luminance and motion. In *Proc. EUSIPCO 94*, Edinburgh, U.K., September 1994.

[27] F. Dufaux, F. Moscheni, and A. Lippman. Spatio-temporal segmentation based on motion and static segmentation. In *IEEE Proc. ICIP'95*, Washington, DC, October 1995.

[28] S.F. Wu and J. Kittler. A differential method for simultaneous estimation of rotation, change of scale and translation. *Signal Processing: Image Communication*, vol. 2, no. 1, pp. 69-80, May 1990.

[29] M. Pardas, P. Salembier, and B. Gonzalez. Motion and region overlapping motion estimation for segmentation-based video coding. In *IEEE Proc. ICIP'94*, volume II, pages 428–432, Austin, TX, November 1994.

[30] J.K. Aggarwal, L.S. Davis, and W.N. Martin. Correspondence processes in dynamic scene analysis. *Proc. IEEE*, vol. 69, no. 5, pp. 562-572, May 1981.

[31] Y. Bar-Shalom and T.E. Fortmann. *Tracking and Data Association*. Academic Press, 1988.

[32] F. Meyer and P. Bouthemy. Region-based tracking in an image sequence. In *Europ. Conf. on Computer Vision*, pages 476–484, Santa Margherita, Italy, May 1992.

[33] M. Irani, B. Rousso, and S. Peleg. Detecting and tracking multiple moving objects using temporal integration. In *Europ. Conf. on Computer Vision*, pages 282–287, Santa Margherita, Italy, May 1992.

[34] F. Moscheni, F. Dufaux, and H. Nicolas. Entropy criterion for optimal bit allocation between motion and prediction error information. In *SPIE Proc. Visual Communications and Image Processing '93*, volume 2094, pages 235–242, Cambridge, MA, November 1993.

[35] B. Girod. Rate-constrained motion estimation. In *SPIE Proc. Visual Communications and Image Processing '94*, volume 2308, pages 1026–1033, Chicago, IL, September 1994.

[36] B.K.P. Horn and B.G. Schunck. Determining optical flow. *Artif. Intell.*, vol. 17, pp. 185-203, 1981.

[37] B. Lucas and T. Kanade. An iterative image registration technique with an application to stereo vision. In *Proceedings Image Understanding Workshop*, pages 121–130, 1981.

[38] S. Hsu, P. Anandan, and S. Peleg. Accurate computation of optical flow by using layered motion representations. In *IEEE Proc. Int. Conf. on Pattern Recognition*, pages 743–746, October 1994.

[39] B.K.P. Horn. *Robot Vision.* MIT Press, Cambridge, Massachusetts, 1986.

[40] A. Pentland, B. Horowitz, and S. Sclaroff. Non-rigid motion and structure from contour. In *IEEE Proc. Workshop on Visual Motion*, pages 288–293, Princeton, NJ, October 1991.

[41] C.S. Fuh and P. Maragos. Affine models for image matching and motion detection. In *IEEE Proc. ICASSP'91*, volume IV, pages 2409–2412, Toronto, Canada, May 1991.

[42] V. Seferidis and M. Ghanbari. General approach to block-matching motion estimation. *Optical Engineering*, vol. 32, no. 7, pp. 1464-1474, July 1993.

[43] H. Nicolas. *Hiérarchie de modèles de mouvement et méthodes d'estimation associées. Application au codage de séquences d'images.* PhD thesis, University of Rennes I, France, 1992.

[44] Y.T. Tse and R.L. Baker. Global zoom/pan estimation and compensation for video compression. In *IEEE Proc. ICASSP'91*, volume IV, pages 2725–2728, Toronto, Canada, May 1991.

[45] D. Adolph and R. Buschmann. 1.15 Mbit/s coding of video signals including global motion compensation. *Signal Processing: Image Communication*, vol. 3, nos. 2-3, pp. 259-274, June 1991.

[46] F. Moscheni, F. Dufaux, and M. Kunt. A new two-stage global/local motion estimation based on a background/foreground segmentation. In *IEEE Proc. ICASSP'95*, Detroit, MI, May 1995.

[47] T. Poggio, V. Torre, and C. Koch. Computational vision and regularization theory. *Nature*, vol. 317, no. 6035, pp. 314-319, 1985.

[48] P.J. Rousseeuw and A.M. Leroy. *Robust Regression and Outlier Detection.* Wiley, New York, 1987.

[49] P. Meer, D. Mintz, A. Rosenfeld, and D.Y. Kim. Robust regression methods for computer vision: a review. *Int. Journal of Computer Vision*, vol. 6, no. 1, pp. 59-70, 1991.

[50] P. J. Burt and E. H. Adelson. The laplacian pyramid as a compact image code. *IEEE Trans. Commun.*, vol. COM-31, no. 4, pp. 482-540, April 1983.

[51] M. Eden and M. Kocher. On the performance of a contour coding algorithm in the context of image coding Part I: contour segment coding. *Signal Processing*, vol. 8, no. 10, pp. 381-386, 1985.

[52] M. Hoetter. Object-oriented analysis-synthesis coding based on moving two-dimensional objects. *Signal Processing: Image Communication*, vol. 2, no. 4, pp. 409-428, December 1990.

[53] J. Benois, L. Wu, and D. Barba. Joint contour-based and motion-based image sequences segmentation for TV image coding at low bit rate. In *SPIE Proc. Visual Communications and Image Processing '94*, volume 2308, pages 1074–1085, Chicago, IL, September 1994.

[54] H.-H. Nagel. Constraints for the estimation of displacement vector fields from image sequences. In *Proc. Int. Joint Conf. on Artificial Intelligence*, pages 945–951, Karlsruhe, Germany, August 1983.

[55] F. Glazer. Multilevel relaxation in low-level computer vision. In A. Rosenfeld, editor, *Multiresolution Image Processing Analysis*, pages 312–330. Springer-Verlag, 1984.

[56] W. Enkelmann. Investigations of multigrid algorithms for the estimation of optical flow fields in image sequences. *Computer Graphics and Image Processing*, vol. 43, pp. 150-177, August 1988.

[57] P. Salembier and M. Pardas. Hierarchical morphological segmentation for image sequence coding. *IEEE Trans. Image Proces.*, vol. 3, no. 5, pp. 639-651, September 1994.

[58] L. Kaufman and P.J. Rousseeuw. *Finding Groups in Data: an Introduction to Cluster Analysis.* Wiley, New York, 1990.

[59] S. Geman and D. Geman. Stochastic relaxation, Gibbs distributions, and Bayesian restoration of images,. *IEEE Trans. Pattern Anal. and Machine Intell.*, vol. PAMI-6, no. 6, pp. 721-741, November 1984.

[60] M. Irani, B. Rousso, and S. Peleg. Computing occluding and transparent motions. *Int. Journal of Computer Vision*, vol. 12, no. 1, pp. 5-16, 1994.

[61] G.R. Legters and Y.Y. Tzay. A mathematical model for computer image tracking. *IEEE Trans. Pattern Anal. Machine Intell.*, vol. PAMI-4, no. 6, pp. 583-594, November 1982.

[62] F. Meyer. *Suivi de régions et analyse des trajectoires dans une séquence d'images*. PhD thesis, University of Rennes I, France, 1993.

[63] X. Yuan. Hierarchical uncovered background prediction in a low bit-rate video coder. In *Picture Coding Symposium '93*, page 12.1, Lausanne, Switzerland, March 1993.

[64] M. Irani, S. Hsu, and P. Anandan. Mosaic-based video compression. In *SPIE Proc. Digital Video Compression: Algorithms and Technologies*, volume 2419, San Jose, CA, February 1995.

[65] R. Kermode and A. Lippman. Coding for content: enhanced resolution from coding. In *IEEE Proc. ICIP'95*, Washington, DC, October 1995.

[66] V. Garcia-Garduno and C. Labit. On the tracking of regions over time for very low bit rate image sequence coding. In *Picture Coding Symposium '94*, pages 257–260, Sacramento, CA, September 1994.

7

FRACTAL-BASED IMAGE AND VIDEO CODING

Mohammad Gharavi-Alkhansari and Thomas S. Huang

Beckman Institute for Advanced Science and Technology,
University of Illinois at Urbana-Champaign,
Urbana, Illinois 61801,
USA

ABSTRACT

This chapter reviews the theoretical foundations and implementation issues of fractal-based image coding methods. The concepts of fractals, iterated function systems, and local iterated function systems are discussed and different implementations of compression of both still images and image sequences are reviewed.

1 INTRODUCTION

Fractal-based image coding, which is sometimes called *fractal image coding* or *attractor image coding*, is a new method of image compression. In this method, similarities between different scales of the same image are used for compression. The method is rooted in the work of Mandelbrot, who introduced the concept of fractals and the fractal dimension.

This chapter is organized as follows. Section 1 gives a brief review of the concept of fractals and its applications especially in image compression. In Section 2 the principles of iterated function systems (IFS) will be studied. IFS makes the basis of most fractal-based image compression methods. In Section 3, we will see how the this theory has been used for compression of image and video sequences. Finally in Section 4, conclusions will be drawn.

1.1 Fractals and Self-Similarity at Different Scales

In late 70s and early 80s, Mandelbrot showed that many natural and man-made phenomena have the very fundamental characteristic of invariance under change of scale [85, 87, 86]. Mandelbrot coined the name *fractal* for the geometry of these phenomena.

The mathematical definition of fractals suggested by Mandelbrot is that they are sets for which the Hausdorff-Besicovitch dimension D is strictly larger than their topological dimension D_T [86]. However, computing Hausdorff-Besicovitch dimension is often difficult, and in many cases the *fractal dimension* [12] is used instead. Fractal dimension is defined as

$$D = \lim_{d \to 0} \frac{\ln N(d)}{\ln(\frac{1}{d})} \tag{7.1}$$

where $N(d)$ is the minimum number of balls of diameter d which are needed to cover the set[1].

This definition implies that if d is small enough, we can write the approximate power law

$$N(d) = K(1/d)^D \tag{7.2}$$

where K is a constant. This means that as d decreases, $N(d)$ grows with the Dth power of $1/d$, no matter how small d is. D can be considered as a measure of the roughness of a set, where rougher sets have larger Ds [104]. A classical example of a natural fractal set is the coastline of an island, and an example of an artificial fractal set is the Koch curve [86]. In practice, a natural set is considered fractal if its D is stable over a wide range of scales.

Figure 1 shows different steps in the construction of the Koch curve. In this construction we begin with a line segment of length 1 (Figure 1a). Then divide the line into three equal parts and replace the middle part with two line segments of length 1/3, obtaining the graph of Figure 1b. If we apply the operation that generated Figure 1b from Figure 1a on every line segment in Figure 1b, we get Figure 1c. Repeating this process once more results in Figure 1d, and continuing to apply this process infinitely many times, results in the set shown (approximately) in Figure 1e, which is known as the Koch curve.

[1] For more detailed information on the definition of fractals and different dimensions see [42] or [44].

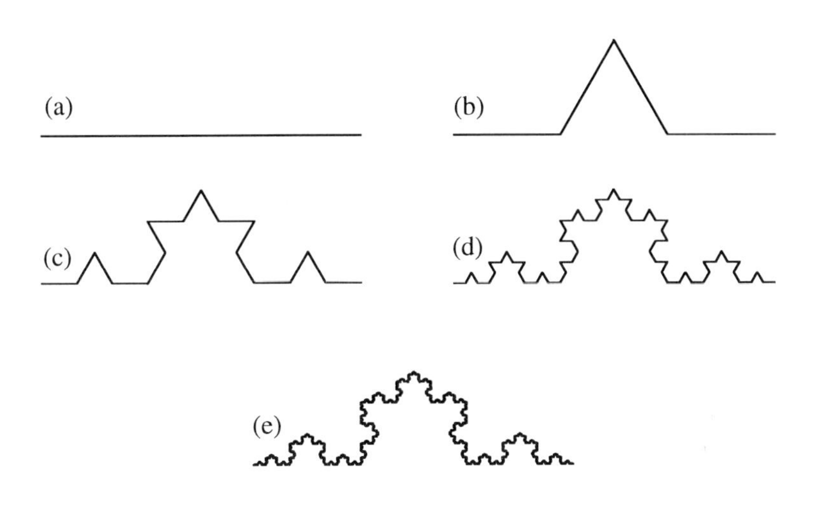

Figure 1 Different stages of construction of the Koch Curve.

It is easy to show that the set shown in Figure 1e has a fractal dimension of

$$D = \frac{\ln 4^n}{\ln 3^n} = \frac{\ln 4}{\ln 3} \approx 1.26 > 1 = D_T.$$

The power law in Equation 7.2 states that dividing d by any factor a, always increases $N(d)$ by a factor a^D, for any small value of d. In the case of integer Ds, this result seems trivial, but when D is not an integer this means that, for example, for a fractal curve, magnifying any part of the curve does not result in a curve that is smoother than the original curve.

This power law means that a fractal set has the same roughness, independent of scale. Therefore one can say that a fractal set is self-similar at different scales in the above general sense. Of course many simple non-fractal geometrical sets, like lines and planes, are also self similar in this sense, but are not fractals as they are not 'rough' sets and always become smooth when magnified enough. Fractals always reveal more and more details under magnification and these details do not diminish by magnification.

Some fractals possess self-similarities in a more restricted sense. The self-similarity of this class of fractals can be either deterministic or statistical. The deterministic self-similarity is when the shape of the set is similar to itself in an deterministic way as the scale changes. A particular kind of fractals of this class are fractals with exact self-similarity which do not change at all under change of scale, e.g. the Koch curve.

The statistical self-similarity is the property of fractals which retain all of their statistical parameters at different scales while a deterministic relationship does not necessarily exist between different scales of the set.

1.2 Applications of Fractals

Fractal geometry in nature is more a rule than an exception. Since the introduction of the concept of fractals by Mandelbrot, the concept has been used in many different branches of science including mathematics, physics, chemistry, geophysics, botany, biology, computer graphics, computer vision, and image processing.

Fractals with statistical self-similarity have been of great interest in the area of computer graphics, where the concept was used to generate complex and strikingly natural-looking graphics of natural scenes using simple rules [86, 98].

In the area of computer vision, the fractal dimension has been used for image modeling, segmentation and shape extraction for natural scenes by Pentland [104, 105] and others [4, 5, 6, 74, 99, 101, 100, 118, 127]. The fractal dimension of different natural objects can be different from each other or from those of man-made objects. On the other hand, under certain conditions, 2-D images taken from some 3-D fractal geometries are also fractals. Pentland [104] used the fractal dimension as a parameter for segmenting images. This method can be used to discriminate between different natural objects in a scene or between man-made objects and natural objects [101, 100].

1.3 Fractals in Image Compression

Fractal geometry has been used for image compression in a few basically different ways.

Fractal curves, specially the Peano curve, were used for scanning images instead of the standard raster scanning [119, 136, 54, 125].

A 'yardstick' method was used for image compression [132, 137, 121] and for shape classification [36].

Fractal dimension has been used as a tool in different aspects of image compression algorithms.

- Fractal dimension was used in a fractal image coder for adjusting error thresholding [60]. Also in [72, 73], fractal dimension was used for image segmentation. After segmentation, fractal dimension was also used as a measure of complexity of the segment to determine how the segments should be coded.
- In the context of image coding, fractal dimension was also used for selecting the optimal scale parameter in an edge detector [38].

Wavelet decomposition has also been used to exploit self-similarities of images at different scales. In this approach, a wavelet decomposition is applied to an image, and the similarity of same-size blocks in different subbands is used to reduce the size of the code needed for representing them [103, 107, 108, 41]. Pentland [103] reports typical compression ratios of 38:1 with 33 dB PSNR for 256×256 pixel images and Rinaldo and Calvagno [107] report compression ratios of about 54:1 (PSNR of 31.4 dB) to 20:1 (PSNR of 35.5 dB) for 512×512 test image Lena.

However, the method that has attracted the most attention, and that will be discussed in more detail in this chapter, is based on the work by Barnsley [17, 9, 10, 21, 26, 12, 23, 27, 24, 19, 11, 18, 28, 13, 20, 14, 15, 29, 22, 25, 16], summarized in [25][2]. His work was based on that of Hutchinson [64], who set up a theory for deterministically self-similar sets, and studied transformations that can generate this kind of sets. These transformations, which Barnsley later named *Iterated Function Systems (IFS)*, were originally used for generating fractals, but because many non-fractal sets can also be deterministically self-similar, these sets can also be generated by IFS. Iterated function systems will be discussed in detail in Section 2.

Barnsley's early work was based on the following assumptions,

- The images of many natural objects can be approximated by members of a class of deterministically self-similar sets.
- These sets can be generated by IFS transformations which have a relatively small number of parameters.

[2]The methods based on Barnsley's work have strong relationship with some of the wavelet methods mentioned earlier. For studies on this, see for example [78, 41].

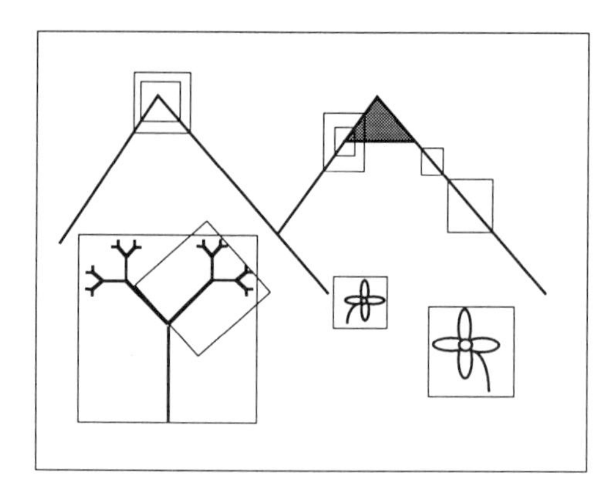

Figure 2 The essence of fractal coding methods is to try to approximate each segment of the image by applying some (contractive) transformation on some bigger segment(s) in the image.

Barnsley observed that even very simple IFSs with very short codes can generate complex sets with infinite details that resemble natural objects. IFS transformations describe the relationship between the whole image and its parts and exploit the similarities that exist between an image and its smaller parts.

Given an IFS, generating the image corresponding to it is quite straightforward and easy. However, the *inverse problem* of finding the IFS which can generate (or closely approximate) a given image has yet to be solved. In other words, the problem was that of how the similarities between the whole image and its parts could be found automatically. Another problem was: what could be done for images for which the smaller parts do not resemble the whole image? To solve the second problem, in 1988, Barnsley generalized the theory of IFS to the theory of *Local Iterated Function Systems* which exploited similarities between parts of the image which were of different sizes. Using this theory, image parts were not required to resemble the whole image; they only needed to be similar to some other bigger parts in the image, as shown in Figure 2. But there was still the first problem: how could these similarities be found automatically? In the early implementations of this theory, these similarities were found by human interaction and hence the images were encoded by interactive computer programs. This resulted in codes for images which were extremely compact in size, but their decoded images had very low quality [12]. This was until the work by Arnaud E. Jacquin (a student of Barnsley) who automated this method for the first time [67, 68, 69, 70]. The code generated by Jacquin's program for an image was not as compact as before, but the compression ratio and the quality of the decoded images looked promising. The work by Jacquin provided a platform for others to continue this line of research. Since then several extensions and generalizations of this method have been found, and many of its properties are better understood, which has resulted in more efficient algorithms. Some of these methods will be discussed in Section 3.

Barnsley and Sloan founded the company 'Iterated Systems, Inc.' in 1987 for the development of products based on the fractal theory, and patented some of the basic algorithms in fractal coding [15, 29]. This company has made different hardware and software products for image and video compression/decompression especially on personal computers. Although many articles have been published on the basics of Barnsley's theory, many of the details of the algorithms used in these products have not been revealed.

1.4 Fractal Techniques in Second Generation Image Coding

Second generation image coding methods take special advantage of the properties of the human visual system and many of them are segmentation-based. In this section we will briefly see how fractals are related into these two features of second generation coding methods.

- *Fractals and the Human Visual System*
  Many researchers have studied the relation between fractals and the human visual system [104, 105, 77].
  - To human eye, many fractal curves and surfaces look very similar to natural curves and surfaces and for this reason they have been extensively used in computer graphics. In model-based coding, this similarity has the potential of being used for coding of natural images by modeling the underlying processes that generates parts of these images.
  - Experiments have shown that the fractal dimension of a curve or set is closely related to human's perception of its roughness [104]. Although fractal dimension alone is not enough for generating a visually good approximation of a set [4, 7], it may be used as one of the parameters for its representation.
  - It is known that human visual system's sensitivity to details in any part of an image is dependent on the amount of activity in the back-

ground of that part of the image. Fractal dimension of image regions has been used as an objective measure of this activity [60].

- *Segmentation Using Fractal Dimension*
  Many researchers have used fractal dimension for image segmentation [104, 72, 112, 6, 120, 79, 81]. Fractal dimension is usually computed locally [122, 90] and is used as the texture feature for segmentation.

- *Edge Detection Using Fractal Dimension*
  In the context of image coding, fractal dimension was also used for selecting the optimal scale parameter in a multiscale edge detector [38]. In this method, edge points were detected by wavelet transform and the dilation parameter is controlled by the fractal dimension.

- *Fractal Coding of Contours*
  One of the first applications of the theory of iterated function systems proposed by Barnsley and Jacquin was in contour coding [23, 67]. A similar method was later used for this purpose by Jacobs et al. [65].

- *Jacquin's Method as a Second Generation Method*
  Fractal coding methods based on Jacquin's method basically use redundancies in an image at different scales, i.e., they use the fact that different parts of the image at different scales are similar. Due to computational complexity limitations, most fractal coding methods find similarities between image 'blocks', after applying limited transformations, even though the most natural choice is finding similarities between 'objects' or 'segments' with more free deformations. The use of blocks instead of segments is more a matter of speed than anything else, especially because fractal coders are usually computationally intensive. In the basic theory, the shape of the domain segments is not restricted in any sense. Simple block splitting methods have been used by many researchers (including Jacquin) for adjusting the size of the blocks to the feature sizes of the image.

  Thomas and Deravi [123, 124] devised a method for merging of blocks using a region growing procedure based on fractal coding. This method results in range 'regions' with rather free shapes that are adapted to the content of the image. Using this method, a region in the image is approximated with another larger region (but with a similar shape) in the same image.

  Franich et al. [52] proposed a method for merging quadtree block splitting method for shape description with the quadtree block splitting method used for fractal coding and used it in an object-based video coding system.

  From this view point, the fractal coding techniques originated by the work of Jacquin, can be well adapted to both first and second generation image coding techniques, although during the recent years most of the advancements of these techniques have been in the direction of combining them with waveform-based coding methods.

2 BASIC THEORY

The essence of most fractal-based image coding methods is to approximate each segment of the image by applying a (contractive) transformation on some bigger segment(s) in the image. One can then reconstruct the image (with some error) by using only the parameters of the transformations [12, 25]. In these methods, most of the information in the image is basically encoded by coding relations among different segments (of different sizes) of the image. The mathematical framework of this theory is presented in the following sections.

2.1 Iterated Function Systems

We begin with a complete metric space (X, d), where $d(.,.)$ denotes the metric[3]. Now, consider a transformation $w: X \mapsto X$, for which there is a constant s such that for all $x, y \in X$,

$$d(w(x), w(y)) \leq s\ d(x, y).$$

If $0 \leq s < 1$, then w is said to be *contractive* (or a *contraction*) with *contractivity factor* s. If w is contractive, then according to the *Contraction Mapping Theorem*,

1. w possess a unique fixed point $x^* \in X$, i.e., $w(x^*) = x^*$.
2. For any $x \in X$, $\lim_{n\to\infty} w^{(n)}(x) = x^*$.

The transformation w defined on X also induces a transformation on subsets of X. This can be done by defining

$$w(B) = \{w(x), \forall x \in B\} \quad \forall B \subseteq X.$$

Let $(H(X), h)$ denote the metric space whose points are non-empty compact subsets of X, and h is the *Hausdorff Distance* [25]. An *Iterated Function System*

[3]For more details on the basic theory brought in this section see [12], [25], or [22].

(IFS) consists of a complete metric space (X, d) and a number of contractive mappings w_i defined on X, i.e. $\{X; w_i, i = 1, \ldots, N\}$. The *fractal transformation* associated with an IFS is the transformation $W : H(X) \mapsto H(X)$ defined by

$$W(B) = \bigcup_{i=1}^{N} w_i(B) \tag{7.3}$$

for all $B \in H(X)$. If the mappings w_i are contractive with contractivity factors s_i, $i = 1, 2, \cdots, N$, then W is also contractive with contractivity factor $s = \max_i s_i$, and W has a unique fixed point $A \in H(X)$ for which

$$A = W(A) = \bigcup_{i=1}^{N} w_i(A),$$

and for all $B \in H(X)$ we have $\lim_{n\to\infty} W^{(n)}(B) = A$. A is called the *attractor* of IFS. w_i's are usually chosen to be affine transformations. For the two-dimensional case, this becomes

$$w_i\left(\begin{bmatrix} x \\ y \end{bmatrix}\right) = \begin{bmatrix} a_i & b_i \\ c_i & d_i \end{bmatrix}\begin{bmatrix} x \\ y \end{bmatrix} + \begin{bmatrix} e_i \\ f_i \end{bmatrix} \tag{7.4}$$

defined on points in R^2. For the three-dimensional case (gray-scale images) this becomes

$$w_i\left(\begin{bmatrix} x \\ y \\ I(x,y) \end{bmatrix}\right) = \begin{bmatrix} a_{1,1,i} & a_{1,2,i} & a_{1,3,i} \\ a_{2,1,i} & a_{2,2,i} & a_{2,3,i} \\ a_{3,1,i} & a_{3,2,i} & a_{3,3,i} \end{bmatrix}\begin{bmatrix} x \\ y \\ I(x,y) \end{bmatrix} + \begin{bmatrix} b_{1,i} \\ b_{2,i} \\ b_{3,i} \end{bmatrix}, \tag{7.5}$$

where $I(x, y)$ denotes the gray-scale value at location (x, y). For an image, the *fractal code* is made up of the parameters of the fractal transformation W, which consists of the number N and the parameters of w_i's. The mappings w_i defined by 7.4 and 7.5 are contractions under suitable constraints on the parameters and therefore the resulting Ws are also contractions.

As an example of an IFS and its attractor in R^2, let us consider an IFS of the form

$$\{R^2; w_1, w_2, w_3\},$$

where

$$w_1\left(\begin{bmatrix} x \\ y \end{bmatrix}\right) = \begin{bmatrix} 0.5 & 0 \\ 0 & 0.5 \end{bmatrix}\begin{bmatrix} x \\ y \end{bmatrix} \tag{7.6}$$

$$w_2\left(\begin{bmatrix} x \\ y \end{bmatrix}\right) = \begin{bmatrix} 0.5 & 0 \\ 0 & 0.5 \end{bmatrix}\begin{bmatrix} x \\ y \end{bmatrix} + \begin{bmatrix} 0.5 \\ 0 \end{bmatrix} \tag{7.7}$$

$$w_3\left(\left[\begin{array}{c} x \\ y \end{array}\right]\right) = \left[\begin{array}{cc} 0.5 & 0 \\ 0 & 0.5 \end{array}\right]\left[\begin{array}{c} x \\ y \end{array}\right] + \left[\begin{array}{c} 0 \\ 0.5 \end{array}\right]. \quad (7.8)$$

The attractor of this IFS can be found by iteratively applying the induced W on any non-empty compact subset of X. Figure 3 shows how a sequence of sets generated by iterative application of W on an arbitrary initial set B converges to the attractor of W and how the attractor is dependent only on W and not the initial set.

In general, A is completely described by W and is independent of B. Therefore, W gives a complete representation of A and the set of parameters that represent W can be considered as a code for A. In the above example, it can be seen that A has a visually complex shape, but W has a very simple mathematical form which can be specified by three affine transformations. Considering the plot of A as a black and white image, the parameters of W make the code for this image.

2.2 The Collage Theorem

Although the theory of generating the attractor of an IFS is well developed, the inverse problem of finding the IFS code for approximating an arbitrary given set, like many other inverse problems in mathematics, has proven to be a rather difficult problem.

Several studies have been made on finding the exact mathematical solution to this inverse problem using tools such as Fourier transform [43], wavelet transform [53, 8], moment method [3, 1, 2, 32, 34, 49, 61, 88, 126, 130, 129, 131, 50, 51], chaotic optimization [89, 88], genetic algorithms [117], combination of the wavelet transform and the moments method [109, 110, 111], fuzzy sets [35] and other methods [94, 37]. However, this problem, in the general case, is not yet solved.

As discussed before, given a W, the decoding process is based on the Contraction Mapping Theorem. The transformation W is applied iteratively on an (arbitrary) initial image until the transformed image does not change significantly. As W is contractive, the convergence of this sequence of images is guaranteed by the Contraction Mapping Theorem.

However, for a given set C, the encoding problem of finding a contractive transformation W such that its attractor A is close to C, is a rather difficult

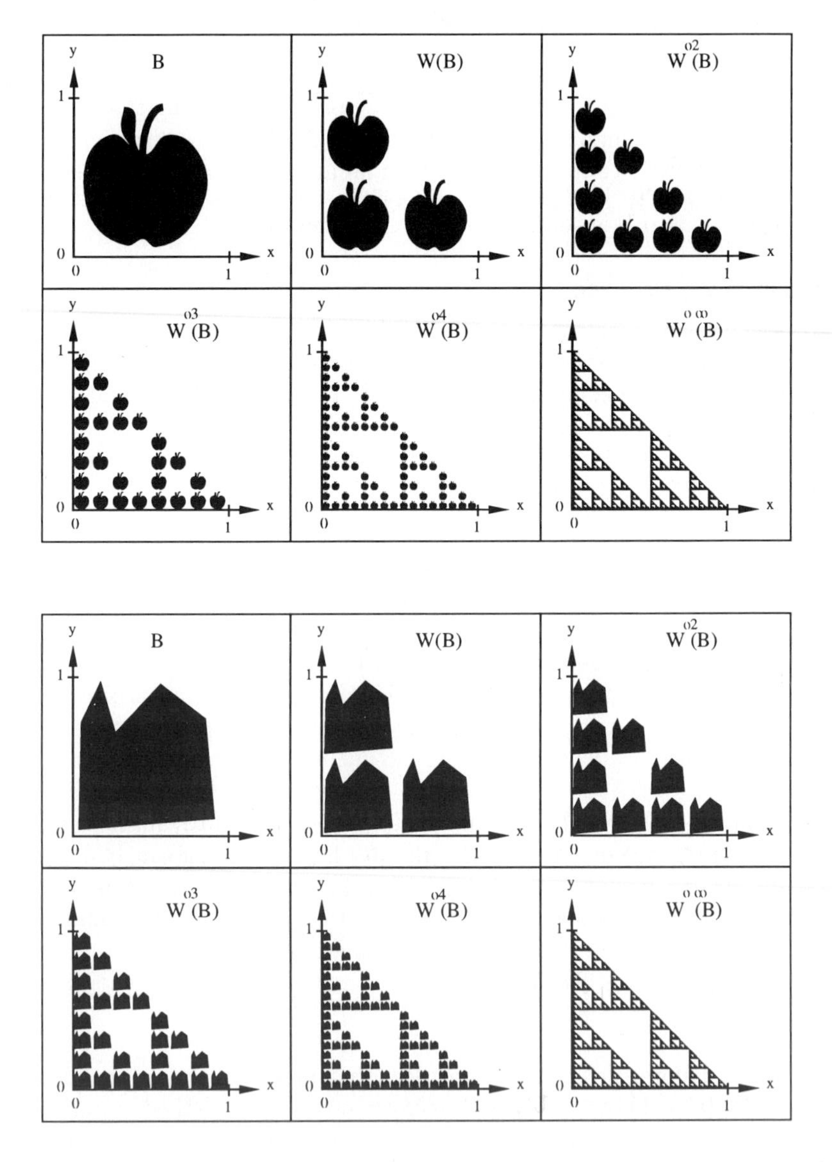

Figure 3 Sequences of sets generated by iterative application of the IFS transformation W defined by 7.3, 7.6, 7.7, and 7.8 on two different arbitrary initial sets (B), converging to the attractor of IFS.

problem. The *Collage Theorem* [21, 12] provides a guideline for solving this problem. It says that for a set C and a contraction W with attractor A,

$$h(C, A) \leq \frac{h(C, W(C))}{1 - s}. \tag{7.9}$$

This means that in order for C and A to be close, it is sufficient that C and $W(C)$ be close, i.e., W may be found in such a way that $W(C)$ be as close to C as possible. $W(C)$ is sometimes called the *collage* of C.

In terms of w_i, we have

$$\left.\begin{array}{l} W(C) \approx C \\ W(C) = \bigcup_{i=1}^{N} w_i(C) \end{array}\right\} \Rightarrow \bigcup_{i=1}^{N} w_j(C) \approx C$$

This can be done by partitioning C into parts C_i,

$$C = \bigcup_{i=1}^{N} C_i$$

such that each part C_i can be closely approximated by applying a contractive affine transformation w_i on the whole C, i.e.,

$$C_i = w_i(C).$$

If we denote $h(C, W(C))$ by ε_E and call it the *encoding error* (or *collage error*), and denote $h(C, A)$ by ε_D and call it the *decoding error*, then according to 7.9,

$$\varepsilon_D \leq \frac{1}{1 - s} \varepsilon_E$$

which gives an upper bound for ε_D in terms of ε_E.

2.3 Local Iterated Function Systems

For most natural images, it is not possible to closely approximate all parts of the image by a small number of transformations applied on the *whole* image. To solve this problem, the theory of Iterated Function Systems was extended to *Local Iterated Function Systems* [22], and its associated fractal transform. In contrast to an Iterated Function System which approximates each part of the

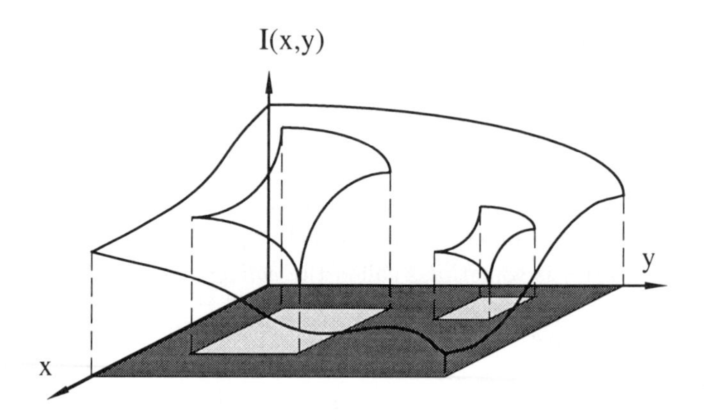

Figure 4 Approximation of a range block by a transformed domain block in a local IFS

image by a transformed version of the whole set, in the Local Iterated Function System each part of the image is approximated by applying a contractive affine transformation on another *part* of the image. In this case, the image C is partitioned into range segments C_i, where $C = \bigcup_{i=1}^{n} C_i$. Then each range segment C_i is approximated by a transformed version of a bigger domain segment D_i, i.e., $C_i \approx w_i(D_i) \Rightarrow C \approx W(C) = \bigcup_{i=1}^{N} w_i(D_i)$ as shown in Figure 4. The decoding process for Local IFS is very similar to that of IFS.

2.4 Resolution Independence

When the above theory is used for image compression, it is implemented in a discrete setting. However, the fractal code generated by encoding a digital image describes relationships (in the form of affine functions) between various segments of the image and is independent of the resolution of the original image. In other words, the fractal code is a *resolution independent* representation of the image and theoretically represents a continuous image approximating the original image. A decoder may decode this code to generate a digital image at any resolution. The resolution of the decoded image may as well be higher than the resolution of the original image. This increase of resolution is sometimes referred to as *fractal zoom*.

The higher resolution obtained is not created by a simplistic technique such as repeating the pixels of the image, but more detail is actually generated in the decoded images. In fact the additional higher resolution information are generated using information from the image at a lower resolution. When an image is reconstructed at the same resolution as the original encoded image, in the decoding process domain blocks of the image are shrunk (lowpass filtering followed by subsampling), which eliminates some of the details of the domain blocks. However, if the image is reconstructed at a higher resolution, in the shrinking of the domain block, the details of the domain block are only shrunk to generate the extra resolution in the range block. In fact, details of the domain blocks are used for missing details of the range block. The details in the domain block are also generated to some extent from details of other domain blocks, used for encoding each part of it. In other words, it is implicitly assumed that if the range block is similar to its corresponding domain block, then the details of the range block (which are *beyond* the resolution of the originally encoded image) are also similar to the details of the domain block (which are *within* the resolution of the encoded image). This assumption is a typical property of self-similarity of fractal sets at different scales, and the resolution independence is a property of the code generated by fractal-based methods.

3 IMPLEMENTATIONS

In view of the theory discussed in the previous section, some of the basic questions to be answered are:

- how to segment the image,
- what transformations to use,
- how to find the parameters of the transformations,
- where to find the matching segments.

These issues will be discussed in this section along with compression results reported for both still images and video sequences.

3.1 Still Images

In 1989 and 1990, Jacquin [67, 68, 69, 70] developed an automatic implementation of the Local IFS method by restricting B_is to squares of two fixed sizes,

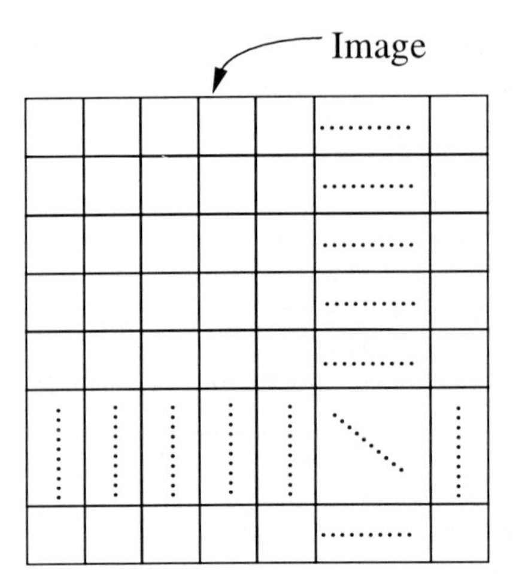

Partition image into range blocks, and for each range block ...

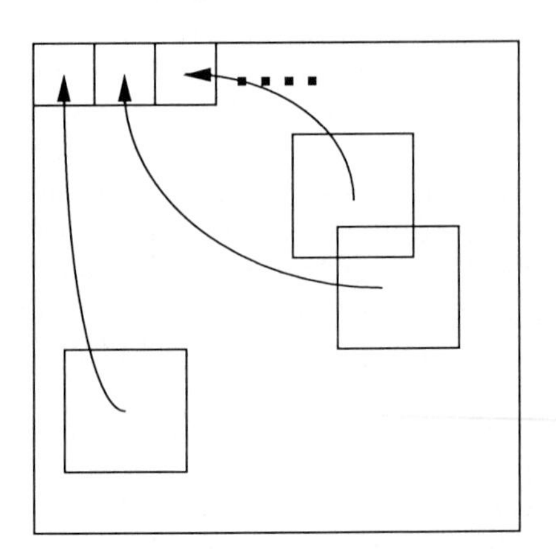

... find a matching block of a bigger size somewhere in the same image.

Figure 5 A demonstration of Jacquin's algorithm

and restricting the affine transformation to the following special case,

$$w_i\left(\begin{bmatrix} x \\ y \\ I(x,y) \end{bmatrix}\right) = \begin{bmatrix} a_{1,1,i} & a_{1,2,i} & 0 \\ a_{2,1,i} & a_{2,2,i} & 0 \\ 0 & 0 & p_{i,0} \end{bmatrix} \begin{bmatrix} x \\ y \\ I(x,y) \end{bmatrix} + \begin{bmatrix} b_i \\ c_i \\ p_{i,1} \end{bmatrix}.$$

where

$$\begin{bmatrix} a_{1,1,i} & a_{1,2,i} \\ a_{2,1,i} & a_{2,2,i} \end{bmatrix} = \begin{bmatrix} \pm a & 0 \\ 0 & \pm a \end{bmatrix} or \begin{bmatrix} 0 & \pm a \\ \pm a & 0 \end{bmatrix}$$

and the origin of the x and y axes is the center of the domain block, $a = 0.5$ and $p_{i,0} < 1$. For each i, the $p_{i,0}$, b_i, c_i are found by search, and $p_{i,1}$ is computed. The essence of Jacquin's method can be summarized as partitioning the image into square range blocks and searching the image for matching domain blocks of twice the size of the range block, as shown in Figure 5. In finding a matching block, we are allowed to apply simple transformations on the domain block, which include shrinking, adding a single value to the gray-scale of the pixels in the block, and scaling by a number less than one. Some shuffling of the pixel locations (*isometric* transformations) are also allowed, which include rotation by multiples of 90 degrees, and/or reflection against vertical or horizontal axes. The encoding process is also enhanced by a two-level hierarchical block splitting method and a range and domain block classification scheme for a faster search.

For the 512 × 512 standard Lena image, PSNRs of 30.1 dB and 31.4 dB were reported at bit rates of 0.57 and 0.6 bits per pixel (bpp) [68, 71].

In 1991, Øien et al. [95] extended this method to

$$w_i\left(\begin{bmatrix} x \\ y \\ I(x,y) \end{bmatrix}\right) = \begin{bmatrix} a_{1,1,i} & a_{1,2,i} & 0 \\ a_{2,1,i} & a_{2,2,i} & 0 \\ d_i & e_i & p_{i,0} \end{bmatrix} \begin{bmatrix} x \\ y \\ I(x,y) \end{bmatrix} + \begin{bmatrix} b_i \\ c_i \\ p_{i,1} \end{bmatrix}.$$

In this case, for each i, values of d_i, e_i, $p_{i,0}$, $p_{i,1}$ are found by least squares methods, and b_i, c_i are again found by search. Using this method, the 512 × 512 Lena image was encoded at a bit rate of 0.5 bpp with a 30.8 dB PSNR.

Monro and Dudbridge [91, 92], in 1992, suggested partitioning an image into small square images, and for each small image, an IFS (and not a Local IFS) was to be found. This is equivalent to a Local IFS with the domain block for each range block being a predetermined block which contains the range block.

Fisher, Jacobs and Boss, studied the effect of using blocks of different shapes, including squares, rectangles, and/or triangles combined with a multi-level hierarchical block splitting method [45, 46, 66]. They also compared the trade offs between compression ratio and signal to noise ratio (SNR) for their method [66]. In two of their experiments, the 512 × 512 Lena image was coded at 0.22 bpp with PSNR of 30.71 dB and at 0.45 bpp with PSNR of 33.40.

In 1993, Gharavi-Alkhansari and Huang [55, 56] extended Jacquin's method and showed that one can use a linear combination of a series of transformed domain blocks instead of only a single domain block.

Thomas and Deravi [123] used blocks of relatively free shapes in Jacquin's algorithm and showed that this could improve the performance of Jacquin's method for simple images. For the 512 × 512 image Lena, they obtained a PSNR of 27.7 dB at 0.30 bpp.

Lepsøy et al. [80] introduced a non-iterative decoding algorithm for fractal-based image compression.

Also Vines and Hayes [128] suggested limiting the search on b_i and c_i by looking for matching domain blocks only in the neighborhood of the corresponding range blocks. Using this method and a multi-level block-splitting scheme, the 512 × 512 Lena image could be compressed at a bit rate of 0.47 bpp with PSNR of 31.5 dB.

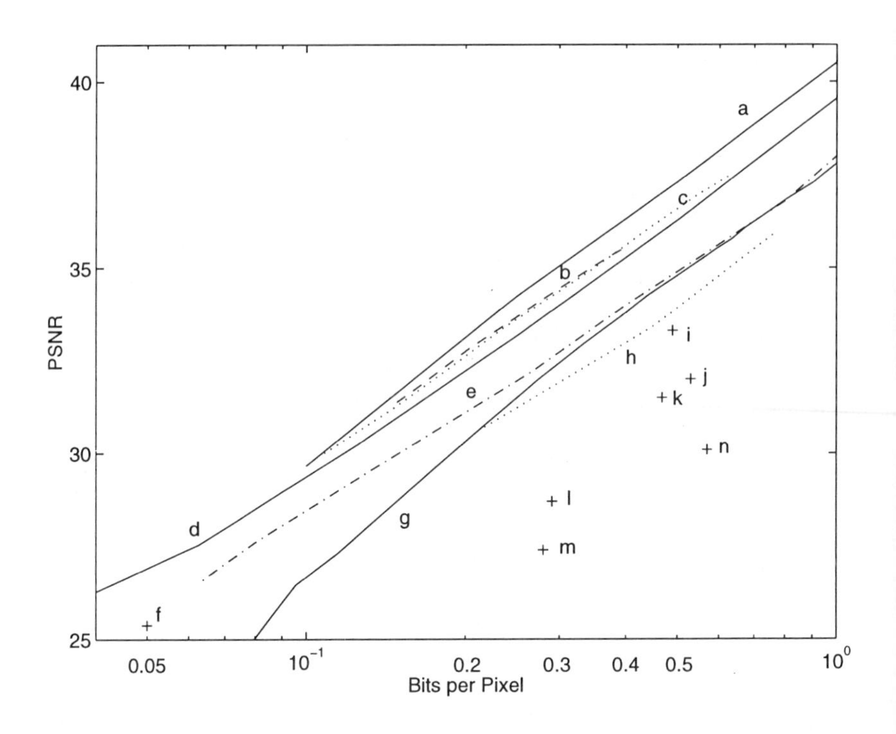

Figure 6 PSNR vs. bit rate for compression of 512 × 512 Lena image with some fractal-based and some non-fractal-based methods. See Table 1 for references.

Up to 1993 most of the attention of the papers published on fractal coding was concentrated on the fractal transform. Since then, more attention has been paid to the entropy coding stages following the fractal transform and on the problem of how the fractal transform parameters could be best modeled for entropy coding. This has resulted in more efficient algorithms.

Barthel et al., in 1994, published results on a fractal-based coding method with a performance of 35 dB PSNR at 0.35 bpp for the 512 × 512 Lena image [30]. In their method, after approximating a range block with a a domain block, any spectrum coefficient of the range block (in DCT domain) that is not well approximated by the domain block, is individually coded using transform coding. These spectral coefficients are then excluded from being approximated by the domain block. Rate-distortion optimality is also used as the criteria for selecting the best possible choice in places where there are several possible alternatives in the encoding process.

Table 1 References for Figure 6

	Year	Researchers	Reference	Method
a	1994	Xiong et al.	[135]	(Non-fractal) Wavelets
b	1994	Rinaldo and Calvagno	[107]	Fractal-Wavelet
c	1994	Barthel et al.	[30]	Fractal-DCT
d	1993	Shapiro	[116]	(Non-fractal) Wavelets
e	1995	Fisher and Menlove	[47]	Fractal
f	1995	Culik and Kari	[40]	Fractal
g	1991	JPEG	[133, 102]	(Non-Fractal) JPEG
h	1992	Fisher et al.	[45, 46, 66]	Fractal
i	1994	Kim and Park	[75]	Fractal
j	1993	Lepsøy et al.	[80]	Fractal
k	1993	Vines and Hayes	[128]	Fractal
l	1994	Lu and Yew	[84]	Fractal
m	1993	Thomas and Deravi	[123, 124]	Fractal
n	1990	Jacquin	[68]	Fractal

Also in 1994, Gharavi-Alkhansari and Huang proposed a generalized image block coding method for unifying the three methods of block transform coding, vector quantization and fractal-based coding methods [57, 58]. In this method every block in the image is approximated by a linear combination of one or more blocks selected from a possibly large dictionary of (not necessarily orthogonal) library blocks. In the case of video coding, block prediction methods like DPCM and adaptive block prediction methods like block motion compensation methods are also special cases of this algorithm. They also proposed that the iterative nature of the fractal image decoders is related to the noncausality of the encoder and using a causal encoder results in a non-iterative decoder that converges in one iteration.

Lin [83] also studied fractal image coding as a generalized predictive coding method and showed how noncausal prediction in fractal coders necessitates an iterative decoding.

In 1994 and 1995, Rinaldo and Calvagno [107, 108] used similarities between blocks in different subbands of image for image coding and reported a performance of 32.78 PSNR at 0.26 bpp [107].

Figure 6 shows the reported performance of some fractal compression methods along with some non-fractal compression methods in terms of PSNR and bit rate for the 512×512 Lena image[4]. Different curves in this plot are assigned letters which are described in Table 1. Curves "g", "d", and "a" are for JPEG, wavelet-based zerotree, and improved wavelet-based zerotree methods which are non-fractal methods and are introduced here only for comparison. The JPEG results brought here are based on the "Independent JPEG Group's free JPEG software" implementation of JPEG. For an implementation based on JPEG standard with a significantly better performance the reader may see [39].

Figure 7 shows the iterative decoding of the 512×512 test image Lena. This image was compressed to 0.43 bits/pixel using a noncausal version of [57, 58, 59]. Beginning with an initial 512×512 black (all zero) image, the decoder generates a sequence of images that converge to the decoded Lena image. Figure 7 shows images resulting from the first five iterations, having PSNRs of (b) 24.22 dB, (c) 27.74 dB, (d) 30.42 dB, (e) 32.21 dB and (f) 33.24 dB. After about 10 iterations, at a PSNR of 34.50 dB the change in the image becomes negligible. The original Lena image is shown in Figure 8 and the final decoded image is shown in Figure 9. Figures 10, and 11 show Lena image compressed at 0.22 bpp (31.2 dB PSNR), and 0.15 bpp (29.2 dB PSNR) using the same method.

3.2 Image Sequences

Fractal-based techniques have also been explored for coding image sequences.

In 1991, Beamount [31] used fractal-based techniques for video compression. He tried two different approaches for this purpose. In one method, he extended Jacquin's method and used three dimensional blocks (or rectangular cubes) of video sequences instead of the 2-D blocks in still images. He reported that although the high compression could be achieved using this method, the quality of the decoded images were not good. Using another method, Beamount applied the 2-D Jacquin's method on individual frames but for each frame (except for the first frame) took the domain blocks from previous frame instead of the same frame. It was reported that 10 frame-per-second 352×288 gray scale video sequences could be coded at a data rate of 80 Kbits/s with "reasonable quality".

[4]The data shown in this plot are brought here only for a rough comparison. PSNR is not always a good measure of image quality. Also in regards to the 512×512, 8 bits per pixel grayscale test image Lena, authors are aware of at least two versions of this image that may have been used by researchers for obtaining these results. Some of the mentioned methods also did not use optimal entropy coders for coding of the fractal transform parameters.

Figure 7 Five iterations of the decoding process

Figure 8 Original 512×512, 8 bits/pixel Lena image

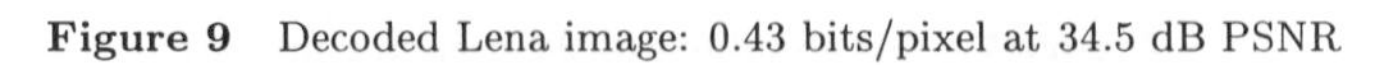

Figure 9 Decoded Lena image: 0.43 bits/pixel at 34.5 dB PSNR

Figure 10 Decoded Lena image: 0.22 bits/pixel at 31.2 dB PSNR

Figure 11 Decoded Lena image: 0.15 bits/pixel at 29.2 dB PSNR

In 1992, Hurd et al. [62], from Iterated Systems, published results on fractal-based video compression claiming compression ratios from 21:1 (average PSNR of 39.2 dB) to 79:1 (average PSNR of 30.8 dB) for a 160×120, 8-bit grayscale video sequence. In their method, they encode the first frame using regular fractal coder. For the following frames, they always use the previous frame as the source of domain blocks. To approximate each range block in one frame they either (1) apply motion compensation and find a matching same-size domain block from previous decoded frame (no contraction) or (2) a single matching larger size domain block (with a contractive transformation applied on it) from the previous decoded frame is found. No residual error is sent for the frames. As the coding of this method is causal, the decoding process is non-iterative. In fact due to the low complexity of the decoding algorithm, this method has a very fast decompression.

In 1993, Hürtgen and Büttgen [63] applied fractal techniques for low bit rate video coding. They applied prediction by frame differencing with no motion compensation. Then for each frame, they applied the fractal transform only to those regions of the frame where prediction failed. For these regions they used a still fractal coding scheme. For range blocks located in these regions, domain blocks from the whole same frame were searched. In contrast to previous methods, for these regions they did not use previous frames. They also used a 3-level block splitting method in their algorithm. The 352×288, 8 1/3 Hz (25 Hz subsampled by 3) Miss America video sequence was reported to be coded at 128 Kbits/sec with an average PSNR of 36–37 dB, and at 64 Kbits/sec with an average PSNR of 34–35 dB, and at 32 Kbits/sec with an average PSNR of 30–32 dB. As the domain blocks for each range block were selected from the same frame, the decoder is iterative in this method.

Also in 1993, Li et al. [82] tried an extension of the still image compression method developed by Monro and Dudbridge [92] to video compression, and showed how compression ratios from 25:1 (average PSNR of 36.2 dB) to 51:1 (average PSNR of 27.2 dB) can be achieved for the 256×256, 15Hz 'Miss America' sequence. In this method the video sequence is partitioned into 3-D blocks. Each block is then partioned into 8 3-D sub-blocks each of which is approximated by a contractive transformation applied on the block that contains it.

In 1994, Lazar and Bruton also extended Jacquin's 2-D algorithm to 3-D, and used 3-D range and domain blocks for image compression. They also used a 3-D block splitting method and the search for selecting domain blocks is done only in the neighborhood of the range block. They reported an average compression ratio of 74.39 at an average PSNR of 32–33 dB , for the 360×280, 8 bit/pixel 30 Hz 'Miss America' video sequence.

Some other researchers have also contribute to the implementations of fractal video coding [106, 76, 134, 93, 48, 52, 33, 96, 97].

As an example of results for fractal video coding, Figure 12 shows three original (a), (c), (e) and decoded (b), (d), (f) frames of the 352×288, 8bpp, 12.5 Hz (25 Hz subsampled by 2) Miss America video sequence. The sequence was coded using a method based on the generalized block coding algorithm described in [57, 58]. In this method, each range block in each frame is approximated by a linear combination of same-size blocks and larger-size blocks taken from the previous frame, and fixed blocks which in this example are DCT basis blocks. The number and type of selected blocks may vary from one block to another and are determined for each range block by an algorithm described in [57, 58]. The test sequence was coded at 80 Kbits/sec with an average PSNR of 36–37 dB.

3.3 Complexity

In terms of complexity, fractal-based image coding is asymmetric, i.e, the complexity of the encoder is typically much higher than that of the decoder. Complexities of these encoders are typically much higher than that of transform coders and vector quantizers. The most time consuming part of the encoding procedure is usually the search for finding the best matching domain blocks. Different techniques have been studied for limiting, structuring or approximating the search procedure [113, 115, 114].

In many implementations of fractal image coders the search is limited to the neighborhood of the range block where finding a good match is more likely. In the extreme case, the search may be totally avoided by using a predetermined domain block at the location of the range block. The search in Jacquin's original method included searching domain blocks that were generated by applying some isometric transformations on the image blocks (e.g. rotation by multiples of 90 degrees or reflection against horizontal or vertical axis). It has been found that it is more likely that best match being the domain block taken from the image rather than among the isometrically transformed versions and therefore in many fractal image and video coders these transformations are not used.

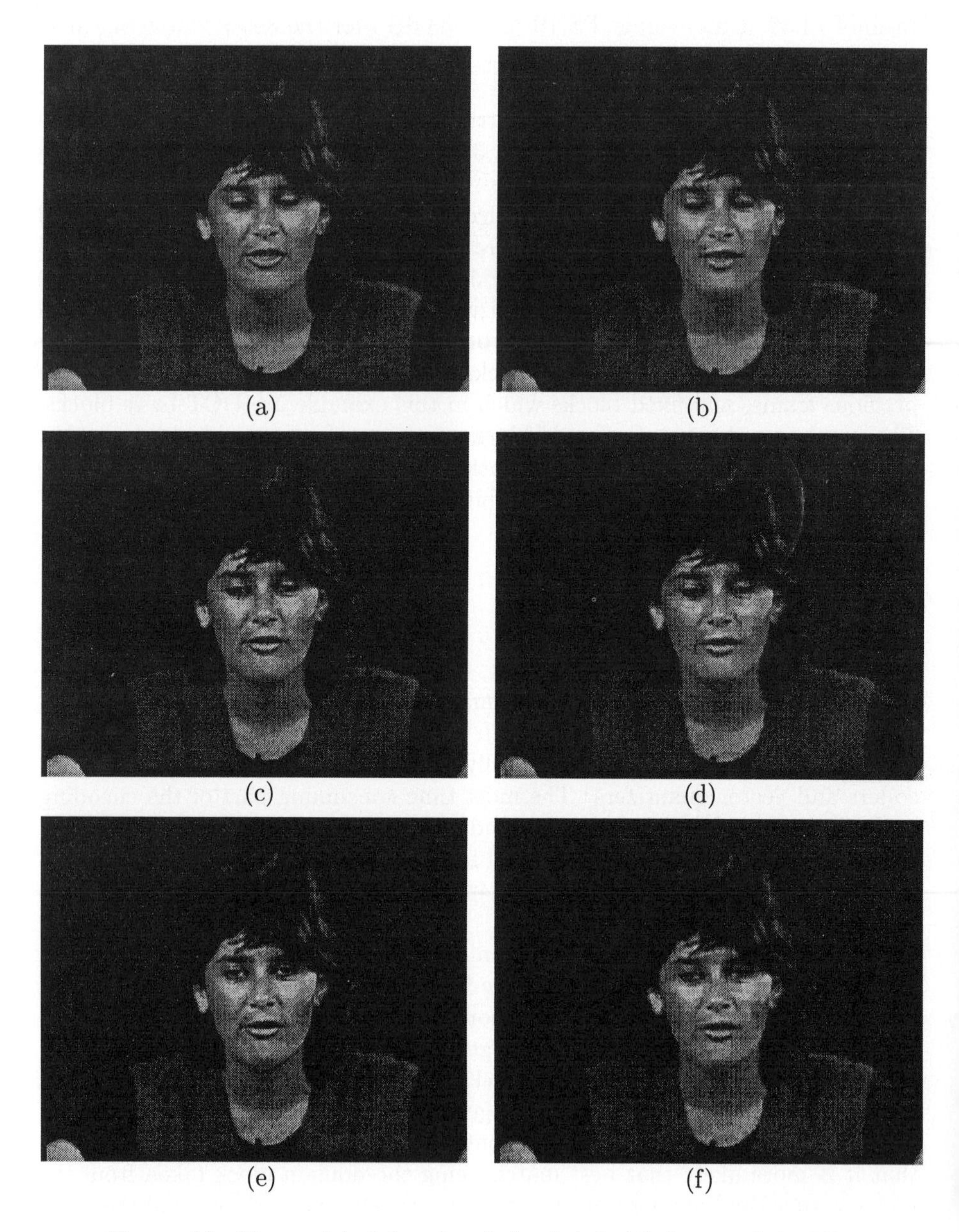

Figure 12 Three original (a,c,e) and decoded (b,d,f) frames of the Miss America sequence.

In some other implementations, the domain and range blocks are classified based on some criteria of structure of the blocks. Then for each range block the block matching search is done only among the domain blocks that are in the same class as of the range block.

Another approach to reduce the search is by doing a coarse to fine search. The search is done first using a coarse measure of similarity of blocks, and then another search with a finer measure of similarity is done among the blocks that have had high similarity using the coarser measure.

On the other hand, complexity of the fractal image decoders is usually much lower than their corresponding encoders and in some cases even less than some transform coding methods. This makes this method more suitable for publishing or broadcasting where the image must be compressed once by a central processor and decompressed many times by smaller receiving processors.

4 CONCLUSIONS

Although the field of fractal-based image coding is relatively young and its methods are different from other image coding methods, the performance of these methods has been comparable to those of the state of the art image compression methods in terms of combination of image quality and compression ratio. However, the complexity of the encoder for fractal-based image coders is typically high. These methods are typically based on the approximation segments of the image with larger segments in the same image.

Acknowledgements

This work is supported in part by Joint Services Electronics Program Grant N00014-90-5-1270 and in part by a grant from Mitsubishi Electric.

REFERENCES

[1] S. Abenda. Inverse problem for one-dimensional fractal measures via iterated function systems and the moment method. *Inverse Problems*,

6(6):885–896, Dec. 1990.

[2] S. Abenda, S. Demko, and G. Turchetti. Local moments and inverse problem for fractal measures. *Inverse Problems*, 8(5):739–750, Oct. 1992.

[3] S. Abenda and G. Turchetti. Inverse problem for fractal sets on the real line via the moment method. *Il Nuovo Cimento*, 104B(2):213–227, Aug. 1989.

[4] A. Ait-Kheddache and S. A. Rajala. Texture classification based on higher-order fractals. In *Proceedings of IEEE ICASSP-88*, pages 1112–1115, Apr. 11–14, 1988.

[5] F. Arduini, C. Dambra, S. Dellepiane, S. B. Serpico, G. Vernazza, and R. Viviani. Fractal dimension estimation by adaptive mask selection. In *Proceedings of IEEE ICASSP-88*, pages 1116–1119. IEEE, 1988.

[6] F. Arduini, S. Fioravanti, and D. D. Giusto. A multifractal-based approach to natural scene analysis. In *Proceedings of IEEE ICASSP-91*, pages 2681–2684. IEEE, 1991.

[7] F. Arduini, S. Fioravanti, and D. D. Giusto. On computing multifractality for texture discrimination. In *Signal Processing VI: Theories and Applications. Proceedings of the Sixth European Signal Processing Conference (EUSIPCO-92)*, pages 1457–1460, Aug. 24–27, 1992.

[8] A. Arneodo, E. Bacry, and J. F. Muzy. Solving the inverse fractal problem from wavelet analysis. *Europhysics Letters*, 25(7):479–484, Mar. 1, 1994.

[9] M. F. Barnsley. Fractal functions and interpolation. *Constructive Approximation*, 2:303–329, 1986.

[10] M. F. Barnsley. Making chaotic dynamical systems to order. In M. F. Barnsley and S. G. Demko, editors, *Chaotic Dynamics and Fractals*, pages 53–68. Academic Press, Inc., New York, 1986. Proceedings of Conference on Chaotic Dynamics, Georgia Tech, March 25-29, 1985.

[11] M. F. Barnsley. Fractal modeling of real world images. In H. O. Peitgen and D. Saupe, editors, *The Science of Fractal Images*, chapter 5. Springer-Verlag, New York, 1988. Based on Lecture Notes for Fractals: Introductions, Basics and Perspectives, in SIGGRAPH'87 (Anaheim, California).

[12] M. F. Barnsley. *Fractals Everywhere.* Academic Press, Inc., New York, 1988.

[13] M. F. Barnsley, editor. *Constructive Approximation*, volume 5. Springer-Verlag, New York, 1989. Special issue on fractal approximation.

[14] M. F. Barnsley. Iterated function systems. In R. L. Devaney, L. Keen, K. T. Alligood, J. A. Yorke, M. F. Barnsley, B. Branner, J. Harrison, and P. J. Holmes, editors, *Chaos and Fractals: The Mathematics Behind the Computer Graphics.* American Mathematical Society, 1989.

[15] M. F. Barnsley. Methods and apparatus for image compression by iterated function systems. United States Patent Number 4,941,193, 1990.

[16] M. F. Barnsley and L. Anson. *The Fractal Transform.* Jones and Bartlett, Apr. 1993.

[17] M. F. Barnsley and S. Demko. Iterated function systems and the global construction of fractals. *Proceedings of the Royal Society of London*, A399:243–275, 1985.

[18] M. F. Barnsley, S. Demko, J. Elton, and J. Geronimo. Invariant measures for Markov processes arising from function iteration with place-dependent probabilities. *Annales de l'Institut Henry Poincare: Probabilites et statiques*, 24(3):367–394, 1988.

[19] M. F. Barnsley and J. Elton. A new class of Markov processes for image encoding. *Advances in Applied Probability*, 20:14–32, 1988.

[20] M. F. Barnsley, J. H. Elton, and D. P. Hardin. Recurrent iterated function systems. *Constructive Approximation*, 5(1):3–31, 1989.

[21] M. F. Barnsley, V. Ervin, D. Hardin, and J. Lancester. Solution of an inverse problem for fractals and other sets. *Proceedings of the National Academy of Sciences USA*, 83:1975–1977, Apr. 1986.

[22] M. F. Barnsley and L. P. Hurd. *Fractal Image Compression.* AK Peters, Ltd., Wellesley, Massachusetts, 1993.

[23] M. F. Barnsley and A. E. Jacquin. Application of recurrent iterated function systems to images. In *Proceedings of the SPIE, Visual Communications and Image Processing*, volume 1001, pages 122–131, 1988.

[24] M. F. Barnsley, A. E. Jacquin, F. Malassenet, L. Reuter, and A. Sloan. Harnessing chaos for image synthesis. In *Computer Graphics Conference Proceedings*, volume 22, pages 131–140. SIGGRAPH, Aug. 1988.

[25] M. F. Barnsley and H. Rising III. *Fractals Everywhere.* Academic Press Professional, Boston, second edition, 1993.

[26] M. F. Barnsley and A. D. Sloan. Chaotic compression. *Computer Graphics World*, pages 107–108, Nov. 1987.

[27] M. F. Barnsley and A. D. Sloan. A better way to compress images. *BYTE*, pages 215–223, Jan. 1988.

[28] M. F. Barnsley and A. D. Sloan. Fractal image compression. In H. K. Ramapriyan, editor, *Proceedings of the Scientific Data Compression Workshop*, pages 351–365, Snowbird, Utah, May 3–5, 1988. NASA Godard Space Flight Center. NASA conference publication 3025.

[29] M. F. Barnsley and A. D. Sloan. Method and apparatus for processing digital data. United States Patent Number 5,065,447, 1991.

[30] K. U. Barthel, J. Schüttemeyer, T. Voyé, and P. Noll. A new image coding technique unifying fractal and transform coding. In *Proceedings of IEEE International Conference on Image Processing*, volume 3, pages 112–116, Austin, Texas, Nov. 13–16, 1994.

[31] J. M. Beaumont. Image data compression using fractal techniques. *British Telecommunications Technical Journal*, 9(4):93–109, Oct. 1991.

[32] D. Bessis and S. Demko. Stable recovery of fractal measures by polynomial sampling. *Physica D*, 47:427–438, 1991.

[33] A. Bogdan. Multiscale (inter/intra-frame) fractal video coding. In *Proceedings of IEEE International Conference on Image Processing*, volume 1, pages 760–764, Austin, Texas, Nov. 13–16, 1994.

[34] C. Cabrelli, U. Molter, and E. R. Vrscay. Recurrent iterated function systems: Invariant measures, a collage theorem and moment relations. In H. . Peitgen, J. M. Henriques, and L. F. Penedo, editors, *Fractals in the Fundamental and Applied Sciences*, pages 71–80. Elsevier Science Publishers B. V. (North-Holland), 1991.

[35] C. A. Cabrelli, B. Forte, U. M. Molter, and E. R. Vrscay. Iterated fuzzy set systems: A new approach to the inverse problem for fractals and other sets. *Journal of Mathematical Analysis and Applications*, 171(1):79–100, Nov. 15, 1992.

[36] C. Chang and S. Chatterjee. Fractal based approach to shape description, reconstruction and classification. In *Proceedings of Twenty-Third Asilomar Conference on Signals, Systems and Computers*, pages 172–176, Oct. 30–Nov. 1, 1989.

[37] J. H. Chen and J. D. Kalbfleisch. Inverse problems in fractal construction: Hellinger distance method. *Journal of the Royal Statistical Society, Series B Methodological*, 56(4):687–700, 1994.

[38] C. K. Cheong, K. Aizawa, T. Saito, and M. Hatori. Structural edge detection based on fractal analysis for image compression. In *IEEE International Symposium on Circuits and Systems*, pages 2461–2464, San Diego, CA, May 10–13, 1992.

[39] M. Crouse and K. Ramchandran. Joint thresholding and quantizer selection for decoder-compatible baseline JPEG. In *Proceedings of IEEE ICASSP-95*, 1995.

[40] K. Culik II and J. Kari. Inference algorithms for WFA and image compression. In Y. Fisher, editor, *Fractal Image Compression: Theory and Application*, pages 243–258. Springer-Verlag, New York, 1995.

[41] G. Davis. Self-quantized wavelet subtrees: A wavelet-based theory for fractal image compression. In *DCC'95: Data Compression Conference*, Snowbird, Utah, Mar. 28–30, 1995.

[42] G. A. Edgar. *Measure, Topology, and Fractal Geometry.* Undergraduate Texts in Mathematics. Springer-Verlag, New York, 1990.

[43] J. H. Elton and Z. Yan. Approximation of measure by Markov processes and homogeneous affine iterated function systems. *Constructive Approximation*, 5:69–87, 1989.

[44] K. J. Falconer. *Fractal Geometry: Mathematical Foundations and Applications.* John Wiley and Sons, Chichester, 1990.

[45] Y. Fisher. A discussion of fractal image compression. In H.-O. Peitgen, H. Jürgens, and D. Saupe, editors, *Chaos and Fractals: New Frontiers of Science*, appendix A, pages 903–919. Springer-Verlag, 1992.

[46] Y. Fisher. Fractal image compression. In P. Prusinkiewicz, editor, *SIGGRAPH '92 Course Notes: From Folk Art to Hyperreality*, pages 1–21, 1992.

[47] Y. Fisher and S. Menlove. Fractal encoding with HV partitions. In Y. Fisher, editor, *Fractal Image Compression: Theory and Application*, pages 119–136. Springer-Verlag, New York, 1995.

[48] Y. Fisher, D. Rogovin, and T. P. Shen. Fractal (self-VQ) encoding of video sequences. In *Proceedings of the SPIE, Visual Communications and Image Processing*, Chicago, Illinois, Sept. 25–28, 1994.

[49] B. Forte and E. R. Vrscay. Solving the inverse problem for measures using iterated function systems. Submitted to Advances in Applied Probability, 1993.

[50] B. Forte and E. R. Vrscay. Solving the inverse problem for function/image approximation using iterated function systems, I. Theoretical basis. In *Proceedings of Fractals in Engineering'94*, pages 143–152, Montreal, Quebec, Canada, June 1–4, 1994. Published in volume 2, number 3 issues of the journal *Fractals* (World Scientific Publishing Corp.).

[51] B. Forte and E. R. Vrscay. Solving the inverse problem for function/image approximation using iterated function systems, II. Algorithm and computations. In *Proceedings of Fractals in Engineering'94*, pages 153–164, Montreal, Quebec, Canada, June 1–4, 1994. Published in volume 2, number 3 issues of the journal *Fractals* (World Scientific Publishing Corp.).

[52] R. E. H. Franich, R. L. Lagendijk, and J. Biemind. Fractal coding in an object-based system. In *Proceedings of IEEE International Conference on Image Processing*, volume 2, pages 405–408, Austin, Texas, Nov. 13–16, 1994.

[53] G. C. Freeland and T. S. Durrani. IFS fractals and the wavelet transform. In *Proceedings of IEEE ICASSP-90*, pages 2345–2348, 1990.

[54] I. Gerner and Y. Y. Zeevi. Generalized scanning and multiresolution image compression. In J. A. Storer and J. H. Reif, editors, *DCC'91: Data Compression Conference*, page 434, Snowbird, Utah, Apr. 8–11, 1991. IEEE Computer Society Press.

[55] M. Gharavi-Alkhansari and T. S. Huang. A fractal-based image block-coding algorithm. In *Proceedings of IEEE ICASSP-93*, volume V, pages 345–348, Minneapolis, Minnesota, Apr. 27–30, 1993.

[56] M. Gharavi-Alkhansari and T. S. Huang. A fractal-based image block-coding algorithm. In *Proceedings of Picture Coding Symposium*, page 1.7, Lausanne, Switzerland, Mar. 17–19, 1993.

[57] M. Gharavi-Alkhansari and T. S. Huang. Fractal-based techniques for a generalized image coding method. In *Proceedings of IEEE International Conference on Image Processing*, volume 3, pages 122–126, Austin, Texas, Nov. 13–16, 1994.

[58] M. Gharavi-Alkhansari and T. S. Huang. Generalized image coding using fractal-based methods. In *Proceedings of the International Picture Coding Symposium*, pages 440–443, Sacramento, California, Sept. 21–23, 1994.

[59] M. Gharavi-Alkhansari and T. S. Huang. Fractal image coding using rate-distortion optimized matching pursuit. Submitted to the 1996 Symposium on Visual Communications and Image Processing, Orlando, Florida., 1996.

[60] B. D. Goel and S. C. Kwatra. A data compression algorithm for color images based on run-length coding and fractal geometry. In *IEEE International Conference on Communications'88*, pages 1253–1256, June 12–15, 1988.

[61] C. R. Handy and G. Mantica. Inverse problems in fractal construction: Moment method solution. *Physica D*, 43:17–36, May 1990.

[62] L. P. Hurd, M. A. Gustavus, and M. F. Barnsley. Fractal video compression. In *Digest of Papers. Thirty-Seventh IEEE Computer Society International Conference (COMPCON)*, pages 41–42, Feb. 24–28, 1992.

[63] B. Hürtgen and P. Büttgen. Fractal approach to low rate video coding. In *Proceedings of the SPIE, Visual Communications and Image Processing*, volume 2094, pages 120–131, Cambridge, Massachusetts, Nov. 8–11, 1993.

[64] J. E. Hutchinson. Fractals and self-similarity. *Indiana University Mathematics Journal*, 30(5):713–747, Sept.–Oct. 1981.

[65] E. W. Jacobs, R. D. Boss, and Y. Fisher. Fractal-based image compression, II. Technical Report 1362, Naval Ocean Systems Center, San Diego, CA, June 1990.

[66] E. W. Jacobs, Y. Fisher, and R. D. Boss. Image compression: A study of the iterated transform method. *Signal Processing*, 29(3):251–263, Dec. 1992.

[67] A. E. Jacquin. *A Fractal Theory of Iterated Markov Operators with Applications to Digital Image Coding.* PhD thesis, Georgia Institute of Technology, Aug. 1989.

[68] A. E. Jacquin. Fractal image coding based on a theory of iterated contractive image transformations. In *Proceedings of the SPIE, Visual Communications and Image Processing*, volume 1360, pages 227–239, Oct. 1–4, 1990.

[69] A. E. Jacquin. A novel fractal block-coding technique for digital images. In *Proceedings of IEEE ICASSP-90*, pages 2225–2228, Apr. 3–6, 1990.

[70] A. E. Jacquin. Image coding based on a fractal theory of iterated contractive image transformations. *IEEE Transactions on Image Processing*, 1(1):18–30, Jan. 1992.

[71] A. E. Jacquin. Fractal image coding: A review. *Proceedings of the IEEE*, 81(10):1451–1465, Oct. 1993.

[72] J. Jang and S. Rajala. Segmentation based image coding using fractals and the human visual system. In *Proceedings of IEEE ICASSP-90*, pages 1957–1960, 1990.

[73] J. Jang and S. A. Rajala. Texture segmentation-based image coder incorporating properties of the human visual system. In *Proceedings of IEEE ICASSP-91*, pages 2753–2756, 1991.

[74] J. M. Keller, R. M. Crownover, and R. Y. Chen. Characteristics of natural scenes related to the fractal dimension. *IEEE Transactions on Patthern Analysis and Machine Intelligence*, PAMI-9(5):621–627, Sept. 1987.

[75] I. K. Kim and R.-H. Park. Image coding based on fractal approximation and vector quantization. In *Proceedings of IEEE International Conference on Image Processing*, volume 3, pages 132–136, Austin, Texas, Nov. 13–16, 1994.

[76] O. Kiselyov and P. Fisher. Self-similarity of the multiresolutional image/video decomposition: Smart expansion as compression of still and moving pictures. In *DCC'94: Data Compression Conference*, page 514. IEEE Computer Society Press, Mar. 29–31, 1994.

[77] D. C. Knill, D. Field, and D. Kersten. Human discrimination of fractal images. *Journal of Optical Society of America. A, Optics and Image Science*, 7(6):1113–1123, June 1990.

[78] H. Krupnik, D. Malah, and E. Karnin. Fractal representation of images via the discrete wavelet transform. In *IEEE 18th Conv. of EE in Israel*, Tel-Aviv, Israel, Mar. 7–8, 1995.

[79] W. S. Kuklinski. Utilization of fractal image models in medical image processing. In *Proceedings of Fractals in Engineering'94*, pages 180–186, Montreal, Quebec, Canada, June 1–4, 1994. Published in volume 2, number 2 and 3 issues of the journal *Fractals* (World Scientific Publishing Corp.).

[80] S. Lepsøy, G. Øien, and T. A. Ramstad. Attractor image compression with a fast non-iterative decoding algorithm. In *Proceedings of IEEE ICASSP-93*, volume V, pages 337–340, Apr. 27–30 1993.

[81] J. Lévy Véhel and P. Mignot. Multifractal segmentation of images. In *Proceedings of Fractals in Engineering'94*, pages 187–193, Montreal, Quebec, Canada, June 1–4, 1994. Published in volume 2, number 2 and 3 issues of the journal *Fractals* (World Scientific Publishing Corp.).

[82] H. Li, M. Novak, and R. Forchheimer. Fractal-based image sequence compression scheme. *Optical Engineering*, 32(7):1588–1595, July 1993.

[83] D. W. Lin. Fractal image coding as generalized predictive coding. In *Proceedings of IEEE International Conference on Image Processing*, volume 3, pages 117–121, Austin, Texas, Nov. 13–16, 1994.

[84] G. Lu and T. L. Yew. Image compression using quadtree partitioned iterated function systems. *Electronics Letters*, 30(1):23–24, Jan. 6, 1994.

[85] B. B. Mandelbrot. *Les Objets Fractals: Forme, Hasard et Dimension.* Flammarion, Paris, 1975. (In French).

[86] B. B. Mandelbrot. *The Fractal Geometry of Nature.* W. H. Freeman and Company, New York, 1982.

[87] B. B. Mandelbrot and R. F. Voss. *Fractals: Form, Chance and Dimension.* Freeman, San Francisco, 1977.

[88] G. Mantica. Techniques for solving inverse fractal problems. In H. . Peitgen, J. M. Henriques, and L. F. Penedo, editors, *Fractals in the Fundamental and Applied Sciences*, pages 255–268. Elsevier Science Publishers B. V. (North-Holland), 1991.

[89] G. Mantica and A. Sloan. Chaotic optimization and the construction of fractals: Solution of an inverse problem. *Complex Systems*, 3:37–62, Feb. 1989.

[90] B. Moghaddam, K. J. Hintz, and C. V. Stewart. A comparison of local fractal dimension estimation methods. *Pattern Recognition and Image Analysis*, 2(1), Mar. 1992.

[91] D. M. Monro and F. Dudbridge. Fractal approximation of image blocks. In *Proceedings of IEEE ICASSP-92*, volume III, pages 485–488, San Francisco, California, Mar. 1992.

[92] D. M. Monro and F. Dudbridge. Fractal block coding of images. *Electronics Letters*, 28(11):1053–1055, May 21, 1992.

[93] D. M. Monro and J. A. Nicholls. Real time fractal video for personal communications. In *Proceedings of Fractals in Engineering'94*, pages 206–209, Montreal, Quebec, Canada, June 1–4, 1994. Published in volume 2, number 2 and 3 issues of the journal *Fractals* (World Scientific Publishing Corp.).

[94] D. J. Nettleton and R. Garigliano. Evolutionary algorithms and a fractal inverse problem. *Biosystems*, 33(3):221–231, 1994.

[95] G. E. Øien, S. Lepsøy, and T. A. Ramstad. An inner product space approach to image coding by contractive transformations. In *Proceedings of IEEE ICASSP-91*, pages 2773–2776, May 14–17, 1991.

[96] B.-B. Paul and M. H. Hayes. Fractal-based compression of motion video sequences. In *Proceedings of IEEE International Conference on Image Processing*, volume 1, pages 755–759, Austin, Texas, Nov. 13–16, 1994.

[97] B.-B. Paul and M. H. Hayes III. Video coding based on iterated function systems. In *Proceedings of IEEE ICASSP-95*, volume 4, pages 2269–2272, Detroit, Michigan, May 9–12, 1995.

[98] H.-O. Peitgen and D. Saupe, editors. *The Science of Fractal Images.* Springer-Verlag, New York, 1988.

[99] S. Peleg, J. Naor, R. Hartley, and D. Avnir. Multiple resolution texture analysis and classification. *IEEE Transactions on Patthern Analysis and Machine Intelligence*, PAMI-6(4):518–523, July 1984.

[100] T. Peli. Multiscale fractal theory and object characterization. *Journal of Optical Society of America A*, 7:1101–1112, 1990.

[101] T. Peli, V. Tom, and B. Lee. Multi-scale fractal and correlation signatures for image screening and natural clutter suppression. In *Proceedings of the SPIE, Visual Communications and Image Processing IV*, volume 1199, pages 402–415, 1989.

[102] W. B. Pennebaker and J. L. Mitchell. *JPEG Still Image Data Compression Standard.* Van Nostrand Reinhold, New York, 1993.

[103] A. Pentland and B. Horowitz. A practical approach to fractal-based image compression. In J. A. Storer and J. H. Reif, editors, *DCC'91: Data Compression Conference*, pages 176–185. IEEE Computer Society Press, Los Alamitos, Apr. 8–11, 1991.

[104] A. P. Pentland. Fractal-based description of natural scenes. *IEEE Transactions on Patthern Analysis and Machine Intelligence*, PAMI-6(6):661–674, Nov. 1984.

[105] A. P. Pentland. Fractal surface models for communications about terrain. In *Proceedings of the SPIE, Visual Communications and Image Processing II*, volume 845, pages 301–306, Oct. 1987.

[106] E. Reusens. Sequence coding based on the fractal theory of iterated transformations systems. In *Proceedings of the SPIE, Visual Communications and Image Processing*, volume 2094, pages 132–140, Cambridge, Massachusetts, Nov. 8–11, 1993.

[107] R. Rinaldo and G. Calvagno. An image coding scheme using block prediction of the pyramid subband decomposition. In *Proceedings of IEEE International Conference on Image Processing*, Austin, Texas, Nov. 13–16, 1994.

[108] R. Rinaldo and G. Calvagno. Image coding by block prediction of multiresolution subimages. *IEEE Transactions on Image Processing*, 4(7):909–920, July 1995.

[109] R. Rinaldo and A. Zakhor. Inverse problem for two-dimensional fractal sets using the wavelet transform and the moment method. In *Proceedings of IEEE ICASSP-92*, volume IV, pages 665–668, San Francisco, California, Mar. 20–23, 1992.

[110] R. Rinaldo and A. Zakhor. Fractal approximation of images. In *DCC'93: Data Compression Conference*, page 451, Mar.–Apr. 1993.

[111] R. Rinaldo and A. Zakhor. Inverse and approximation problem for two-dimensional fractal sets. *IEEE Transactions on Image Processing*, 3(6):802–820, Nov. 1994.

[112] S. K. Rogers et al. Synthetic aperture radar segmentation using wavelets and fractals. In *Proceedings of the IEEE International Conference on Systems Enginering*, pages 21–24, 1991.

[113] D. Saupe. Breaking the time complexity of fractal image compression. Technical Report 53, Institut für Informatik, Universität Freiburg, 1994.

[114] D. Saupe. Accelerating fractal image compression by multi-dimensional nearest neighbor search. In J. A. Storer and M. Cohn, editors, *DCC'95: Data Compression Conference*, Snowbird, Utah, Mar. 28–30, 1995.

[115] D. Saupe and R. Hamzaoui. Complexity reduction methods for fractal image compression. In J. M. Blackledge, editor, *I.M.A. Conf. Proc. on Image Processing; Mathematical Methods and Applications*. Oxford University Press, Sept. 1994.

[116] J. M. Shapiro. Embedded image coding using zerotrees of wavelet coefficients. *IEEE Transactions on Signal Processing*, 41(12):3445–3462, Dec. 1993.

[117] R. Shonkwiler, F. Mendivil, and A. Deliu. Genetic algorithms for the 1-D fractal inverse problem. In *4th International Conference on Genetic Algorithms (ICGA 91)*, pages 495–501, San Diego, CA, July 13–16, 1991.

[118] M. Stein. Fractal image models and object detection. In *Proceedings of the SPIE, Visual Communications and Image Processing II*, volume 845, pages 293–306, Oct. 1987.

[119] R. J. Stevens, A. F. Lehar, and F. H. Preston. Manipulation and presentation of multidimensional image data using Peano scan. *IEEE Transactions on Patthern Analysis and Machine Intelligence*, PAMI-5(5):520–526, Sept. 1983.

[120] C. V. Stewart, B. Moghaddam, K. J. Hintz, and L. M. Novak. Fractaional Brownian motion models for synthetic aperture radar imagery. *Proceedings of the IEEE*, 81(10):1511–1522, Oct. 1993.

[121] M. Temerinac, A. Kozarev, Z. Trpovski, and B. Simsic. An efficient image compression algorithm based on filter bank analysis and fractal geometry. In J. Vandewalle, R. Boite, M. Moonen, and A. Oosterlinck, editors, *Proceedings of Signal Processing VI: Theories and Applications*, page 1373. Elsevier Science Publishers B. V., 1992.

[122] J. Theiler. Estimating fractal dimension. *Journal of Optical Society of America. A, Optics and Image Science*, 7(6):1055–1073, June 1990.

[123] L. Thomas and F. Deravi. Pruning of the transform space in block-based fractal image compression. In *Proceedings of IEEE ICASSP-93*, volume V, pages 341–344, Minneapolis, Minnesota, Apr. 27–30, 1993.

[124] L. Thomas and F. Deravi. Region-based fractal image compression using heuristic search. *IEEE Transactions on Image Processing*, 4(6):832–838, June 1995.

[125] K. S. Thyagarajan and S. Chatterjee. Fractal scanning for images compression. In *Proceedings of Twenty-Fifth Asilomar Conference on Signals, Systems and Computers*, Nov. 4–6, 1991.

[126] D. van Schooneveld. The moment method for invariant measure approximation. Internal report, Department of Mathematics and Computer Science, Delft University of Technology, 1990.

[127] A. M. Vepsalainen and J. Ma. Estimating of fractal and correlation dimension from 2D and 3D images. In *Proceedings of the SPIE, Visual Communications and Image Processing IV*, volume 1199, pages 431–438, 1989.

[128] G. Vines and M. H. Hayes, III. Adaptive IFS image coding with proximity maps. In *Proceedings of IEEE ICASSP-93*, volume V, pages 349–352, Minneapolis, Minnesota, Apr. 27–30, 1993.

[129] E. R. Vrscay. Moment and collage methods for the inverse problem of fractal construction with iterated function systems. In H. . Peitgen, J. M. Henriques, and L. F. Penedo, editors, *Fractals in the Fundamental and Applied Sciences*, pages 443–461. Elsevier Science Publishers B. V. (North-Holland), 1991. June 6–8 1990.

[130] E. R. Vrscay and C. J. Roehrig. Iterated function systems and the inverse problem of fractal construction using moments. In E. Kaltofen and S. M. Watt, editors, *Computers and Mathematics*, pages 250–259. Springer-Verlag, Berlin, 1989.

[131] E. R. Vrscay and D. Weil. Missing moment and perturbative methods for polynomial iterated function systems. *Physica D*, 50:478–492, July 1991.

[132] E. Walach and E. Karnin. A fractal-based approach to image compression. In *Proceedings of IEEE ICASSP-86*, pages 529–532, Tokyo, Japan, Apr. 7–11, 1986.

[133] G. K. Wallace. The JPEG still picture compression standard. *Communications of the ACM*, 34(4):30–44, Apr. 1991.

[134] D. L. Wilson, J. A. Nicholls, and D. M. Monro. Rate buffered fractal video. In *Proceedings of IEEE ICASSP-94*, volume 5, pages 505–508, Adelaide, Australia, Apr. 19–22, 1994.

[135] Z. Xiong, K. Ramchandran, M. T. Orchard, and K. Asai. Wavelet packets-based image coding using joint space-frequency quantization. In *Proceedings of IEEE International Conference on Image Processing*, volume 3, pages 324–328, Austin, Texas, Nov. 13–16, 1994.

[136] K.-M. Yang, L. Wu, and M. Mills. Fractal based image coding scheme using Peano scan. In *Proceedings of IEEE International Symposium on Circuits and Systems*, pages 2301–2304, Espoo, Finland, June 7–9, 1988.

[137] N. Zhang and H. Yan. Hybrid image compression method based on fractal geometry. *Electronics Letters*, 27(5):406–408, Feb. 28, 1991.

8

MODEL-BASED VIDEO CODING

Kiyoharu Aizawa

Department of Electrical Engineering
University of Tokyo
Tokyo, 113, JAPAN

ABSTRACT

Model-based coding is a new approach to image coding, which makes use in some sense of the 3-D properties of the scene. The methods are categorized into three classes: 2-D approach, 3-D approach and hybrid or model-assisted approach. This chapter summarizes these methods and describes 2-D and 3-D model-based coding methods. It also briefly describes some applications of 3-D model-based coding.

1 INTRODUCTION: A DREAM IN 1962

An old text book [74] of information theory published in 1962 contained the following statement;

> is some vocoder-like way of transmitting pictures possible if we confine ourselves to one sort of picture source, for instance, the human face? One can conceive of such a thing. Imagine that we had at the receiver a sort of rubbery model of a human face. Or we might have a description of such a model stored in the memory of a huge electronic computer. First, the transmitter would have to look at the face to be transmitted and "make up" the model at the receiver in shape and tint. The transmitter would also have to note the sources of light and reproduce these in intensity and direction at the receiver. Then, as the person before the transmitter talked, the transmitter would have to follow the movements of his eyes, lips and jaws, and other muscu-

> lar movements and transmit these so that the model at the receiver could do likewise. Such a scheme might be very effective, and it could become an important invention if anyone could specify a useful way of carrying out the operation I have described. Alas, how much easier to say what one would like to do than it is to do it.

This description exactly illustrates the intent of 3-D model-based coding. The description was written in the early age of image coding. Model-based coding is a research area that has recently intensified, but it appears that it has been dreamed of for a long time. Reviews of model-based coding can also be found in [72, 1, 28].

This chapter first reviews the approaches to model-based coding, categorizing them by looking at the point of view of the image source models. A number of proposals have been made to date, and they vary widely in terms of image source models. Works related to these paradigms describe two different scenarios for model-based coding; one is the 3-D model based method illustrated in the description above, and the other is the 2-D model-based method. The 3-D model-based coding is a rather specific approach that utilizes 3-D scene models such as a person's face. The 2-D model based coding is a general approach and it makes use of advanced signal-based models. Two examples are briefly described. Finally, applications and extensions of 3-D model-based approaches are described.

2 MODEL-BASED APPROACHES TO IMAGE CODING

In order to encode image signals efficiently it is necessary to select a suitable image model and use it to represent image signals. In the past, image compression has been largely regarded as a problem of information theory, and image signals have been modeled by using stochastic properties: that is, by exploiting the spatial and temporal correlation in either the spatial domain or the transform domain. For example, waveform coding such as transform coding has been successful in representing image signals as waveforms.

On the other hand, from the image analysis point of view, images can be considered as having structural features such as contours and regions. These image features have been exploited to encode images at very low bit rates, while retaining enough visible structures in the reconstruction to maintain a

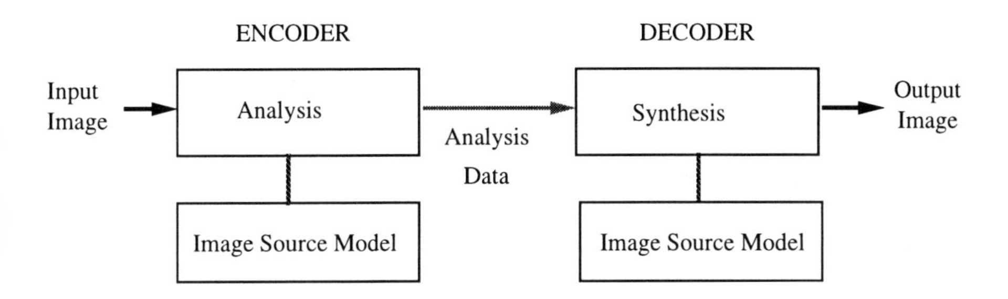

Figure 1 General description of a model-based coding system.

certain level of quality. Notably, in the last few years this structural model-based approach has adopted 3-D structural models of the scene. In this case, the coder and decoder have an object model; the coder analyzes input images and the decoder generates output images using the model (see Figure 1).

Although the details of these model-based approaches vary widely, they can be classified into a few categories on the basis of the models they use. There are two aspects to the image source model used for video coding: segmentation models and motion models. The segmentation model corresponds to entities of the representation of images and the motion model corresponds to movement of the entities. Table 1 shows types of models used in various coding schemes.

For example, traditional coding schemes such as MC-DCT usually divide an image into blocks in the spatial domain and make use of statistical relationship and the motion is usually assumed to be limited to 2-D translation. Thus, traditional methods are very limited in terms of the image source model used. In contrast to traditional coding methods, model-based coding methods make use of variety of image source models, taking account of structural features of the image. There are 2-D approaches and 3-D approaches in model-based image coding schemes. These two approaches are described in the following sections.

2.1 2-D model-based approaches

These coding methods exploit visibly important 2-D features such as edges, contours and regions. Examples are coding methods such as contour-based coding and region-based coding. The former method extracts contours and en-

Table 1 Image coding schemes and their associated image source models

Image Source Models		**Coding Schemes**
Segmentation model	**Motion model**	
Pixel	—	PCM
Statistically dependent pixels Block	2-D translation	MC-DCT etc.
2-D model-based approaches 2-D features : edges, contours, 2-D rigid regions, 2-D flexible regions, deformable triangle blocks, etc.	translation, bilinear transform affine transform, etc.	contour-based coding region-based coding object-based coding 2-D deformable triang. based coding
3-D model-based approaches 3-D global surface model: planes or geometric surfaces parameterized 3-D model	3-D global motion 3-D local deformation	object based coding layered representation 3-D model-based codi
Hybrid and model-assisted approachess combination of the above	combination of the above	2-D/3-D hybrid 3-D and MC-DCT Model-assisted coding

codes shapes and intensities of contours and reconstructs an image from them [45, 7]. The latter method segments images into homogeneous regions and encodes their shapes and intensities [45, 26]. Unlike early work [70, 52] dating back to the early 1980s that encoded only contour information and output binary images, recent methods aim at reconstructing images with natural intensity levels.

In the case of image sequences, moving regions that are detected as changing areas between two successive frames are modeled and coded as arbitrarily-shaped 2D objects translating two dimensionally [32, 33]: both rigid and flexible regions are used for modeling moving 2-D areas. In [76], an image sequence is coded by recursive segmentation relying on homogeneity and region-based motion-compensated texture and contour coding. The motion models are not limited to the simple translation model, rather they exploit affine and bilinear transforms in order to better approximate motion fields of a 3-D moving rigid object, together with linear deformation such as rotation and zooming. The 2-D model-based approach is more advanced than the traditional waveform coding methods in the sense of the image source model, although the model is still signal based. Details of segmentation of video sequences and partition coding are described in Chapters 3 and 4 of this book. Deformable triangle-based motion compensation with affine motion model is described later [61] in this chapter.

2.2 3-D model-based approaches

In the last few years structural model-based approaches have adopted 3-D structural models of scenes. There are two major approaches to 3-D model-based schemes: one approach makes use of surfaces of the object modeled by general geometric models such as planes or smooth surfaces. The other approach utilizes a parameterized model of the object such as parameterized facial models. In the former approach, information such as surface structure and motion information is estimated from image sequences and utilized in image coding. In the latter approach, parameterized models are usually given in advance. At present, these two approaches seem to have different emphases in terms of their applications. In order to distinguish between them we refer to the former approach as *the 3-D feature-based approach* and the latter as *the 3-D model-based approach.*

Several different methods have been proposed in the 3-D feature-based approach. Hötter et al. [31, 59] and Diehl [15] have proposed a method that utilizes a segmented surface model, in which changing regions caused by object motion are detected and modeled by planar or parabolic patches. Ostermann et al. [59, 64], Morikawa et al. [55] and Koch [43] proposed a coding method that utilizes global surface models, in which a smooth surface model of the scene is estimated from an image sequence. Those methods have been applied together with motion compensation and interpolation to improve the performance of conventional waveform coding methods. Differing from coding applications, Wang and Adelson [82, 17] proposed a layered representation of image sequences in which an image sequence is decomposed into a set of overlapping layers which are 3-D planar surfaces taking into account their respective depth. Layered representation is analogous to 'cels' in animation and is suitable for image synthesis applications.

3-D model-based coding, which corresponds to the quotation from the 1962 textbook, utilizes detailed parameterized object models. Obtaining such a detailed model from a general scene is extremely difficult. However, when the object to be coded is restricted to specific classes, special knowledge obtained from a 3-D model of the object can be used in the coding system. For instance, in the case of videophone image transmission, a 3-D model of a face is sufficient to describe scene objects since most of the images consist of a moving head and shoulders. This assumption frees us from the requirement of reconstructing a 3-D model from 2-D images. From this point of view, Aizawa, Harashima [2, 3, 4] and Welsh [84, 85, 83] independently proposed model-based coding schemes, that utilize parameterized 3-D models of a person's face. Prior to their proposals, *semantic coding* had been proposed by Forchheimer [23, 22], but the reconstruction image of his original proposal was too synthetic, although it is in the conceptual same line. Up to now, most of the contributions to 3-D model-based coding have focused on human facial images and the parameterized facial models, which are usually given in advance.

Because 3-D model-based coding uses parameterized models and is more graphics oriented, it is expected to have broader applications than the conventional waveform coding methods. Some interesting applications have been proposed. *Speech-driven image animation*, in which images are synthesized only by speech parameters, and *virtual space teleconferencing* [42, 86] are two particular examples. We will discuss applications of model-based coding in a later section.

Automatic modeling and analysis pose great problems to this approach. It is fair to say that automatic modeling has not yet been reported. Automatic facial motion analysis has been done for restricted conditions such that the model is created in advance and the initial position of the face is known. An experimental real-time coding and decoding system is presented in [38], where it is assumed that the model and the initial position of the face is given, and the facial motion is tracked by using facial feature points which were detected by simple threshold processing. Choi et al. [11, 10] and Li et al. [48] reported a direct estimation of head and facial movements which does not require feature point correspondences. Human motion analysis, including facial motion analysis, is currently a hot topic in the computer vision field.

2.3 Hybrid and model-assisted approaches

3-D model-based coding is at present rather limited and it needs to be combined with other general coding methods such as traditional waveform coding or 2-D model-based methods. There are two different cases for the combination. One case is hybrid coding, in which 3-D model-based coding is combined with signal-based coding methods to improve fidelity and lack of generality of 3-D model-based methods. The other case is model-assisted coding, in which model-based analysis is utilized to control coding parameters of the traditional waveform coding in order to improve the reconstruction of waveform coding

In the former case, in order to improve the fidelity of the reconstructed images and compensate the lack of generality, there have been proposals for 3-D model-based/waveform hybrid coding in which waveform coding is used to compensate errors which occur in the model-based coding process. There are several methods which have been combined with facial model-based coding, including MC/DCT [62, 63], vector quantization [25, 24], and contour coding [53]. In [13], H.261(MC-DCT) is combined with 3-D model based coding, and the simplified geometry of a cylinder is used as the 3-D model. 2-D model-based methods, such as triangular segmentation motion compensation are combined with 3-D model-based coding method to improve the generality [61], and this is described later.

In the latter case, analysis techniques from model-based coding are applied to control dynamic bit assignment and quantization of the traditional waveform coding method to improve the quality of reconstruction of the selected area of the scene. The model-assisted method is employed to enable a system to selectively improve the areas of the scene. In [80, 18], the face area is detected and tracked; the coding parameters of the traditional video coder are controlled in very low bit rate applications.

In the following sections we describe 3-D model-based coding for facial images by K. Aizawa et al. [2] which comes from the area of 3-D model-based techniques. We then discuss the 2-D deformable triangular model-based motion compensation technique of Y. Nakaya et al. [61] which comes from the area of 2-D model-based techniques.

3 3-D MODEL-BASED CODING OF A PERSON'S FACE

Contributions to 3-D model-based coding currently concentrate on facial image analysis and synthesis, since facial images are very important for a broad range of applications including image communication.

This section describes our own contribution [2] to model-based coding, which represents a person's face by utilizing a 3-D facial model, as an example of model-based coding. Related works on the modeling, analysis and synthesis of facial images, performed in the context of model-based coding, are also reviewed.

Our model-based coding system consists of three main components: a 3-D facial model, an encoder and a decoder. This model-based coding system is illustrated in Figure 2. The encoder separates the object from its background, estimates the motion of the person's face, analyzes the facial expressions, and then transmits the necessary analysis parameters. When the new depth information and initially unseen portions of the object are acquired, the encoder updates and corrects the model as required. The decoder synthesizes and generates the output images by using a 3-D facial model and the received analysis parameters.

3.1 Modeling a person's face

Modeling an object is the most important part of model based coding because both analysis and synthesis methods depend strongly on the model they use. For communication purposes, a person's face must be modeled in sufficient detail. Previous studies [69] that have attempted to model human faces with graphics have produced results which lack detail and reality because the imaging methods that relied on computer graphics techniques used only wire frame models and shading techniques to reconstruct the human face.

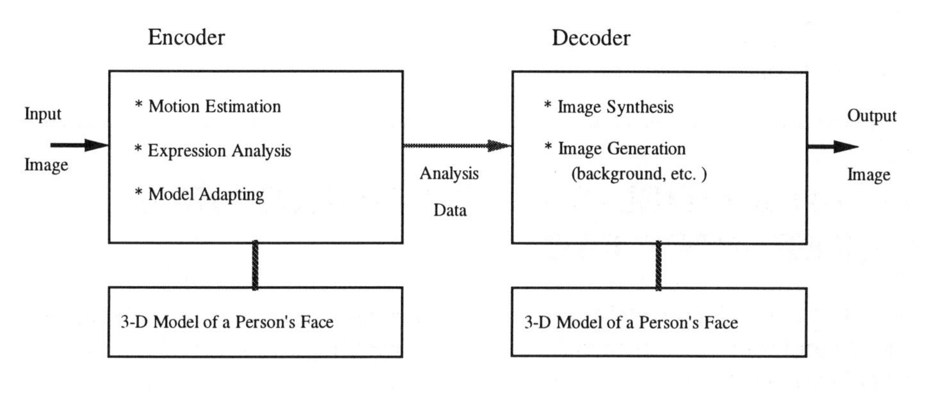

Figure 2 General description of a 3-D model-based coding system for a person's face.

In order to develop an accurate model, we utilize an original facial image and a texture mapping technique. That is, a 3-D wire frame generic face model (Figure 3), which approximately represents a human face and is composed of several hundreds of triangles, is adjusted to fit its feature points positions and outermost contour lines to those of a frontal facial image of the person (Figure 4). This original facial image is then mapped onto the adjusted wire frame model.

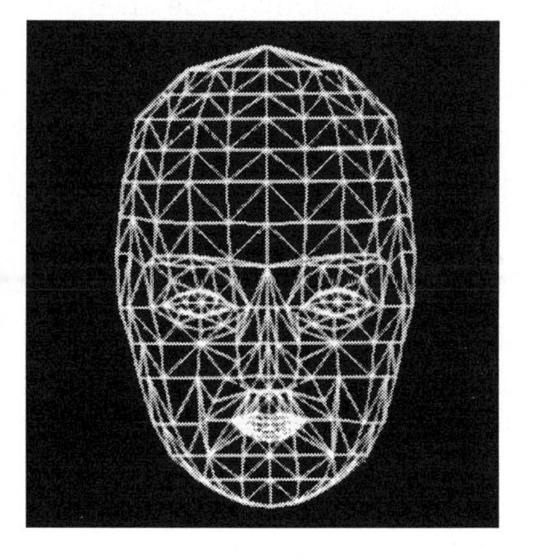

Figure 3 A generic wire frame model

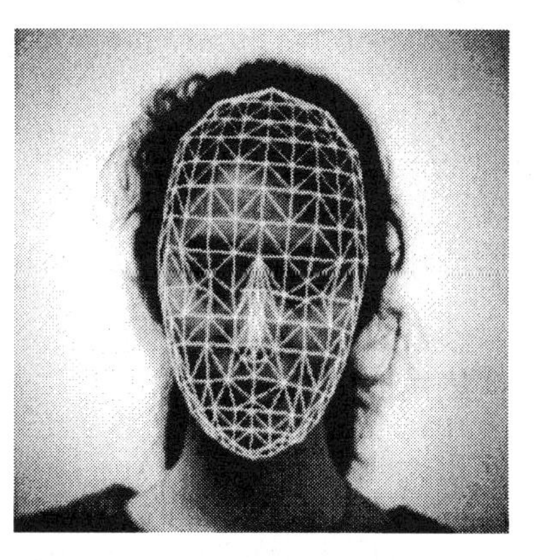

Figure 4 A frontal face image and adjusted wire frame model

When the 3-D facial model is obtained, it can easily be manipulated. For example, Fig. 5 shows a rotated view of the 3-D model. Although the depth is not necessarily a good approximation, the rotated image still appears natural.

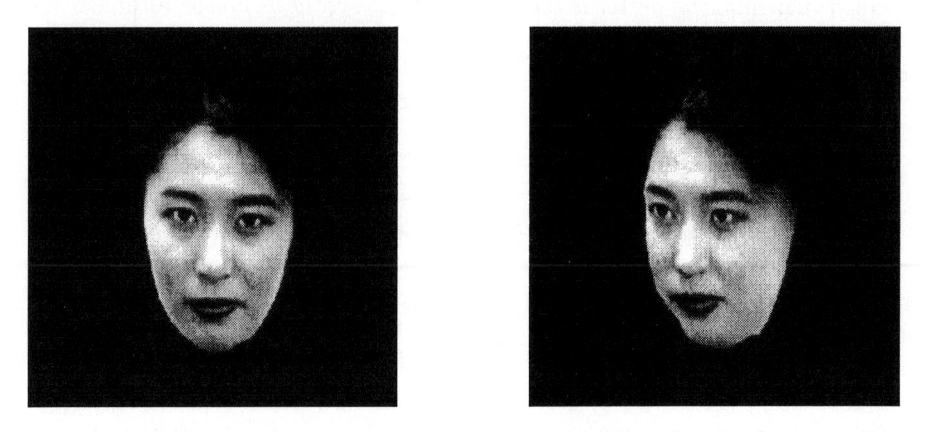

Figure 5 Images synthesized by rotating the 3-D model

Even this modeling process is difficult to automate fully. An interactive system for building a facial model has been developed at the University of Tokyo, and this can be used not only for model-based coding but also by psychologists in experimental studies of facial expressions. This system is also being used and improved at the University of of Illinois at Urbana-Champaign, Carnegie Mellon University and other places.

The use of a 3-D wire frame model and texture of an original image has become a typical approach to modeling faces in model-based coding. However, because the above process uses only a frontal image and it takes no account of the real depth information of the person's face, the 3-D shape of the face can be very different from that of the real face, although the difference is rarely noticeable as long as the face model does not rotate very much from the initial position.

Various approaches have been proposed to improve the geometrical accuracy of the model. A side view of a face is used as well as a frontal view. The 3-D generic wire frame model is adjusted to both frontal and side views of a person [5, 77]. Stereo range data is used for adjusting the 3-D generic facial model [24]. The depth information of the frontal-adjusted model is updated and corrected by using an image sequence [2, 11, 67].

Typically, the wire frame facial models are composed of a surface triangularized divided into 100 - 500 triangular elements. The resolution of triangularization is sometimes not sufficient to describe the details of the face when extreme facial expressions are synthesized on the 3-D face model. The larger the number of triangle elements, the better is the quality of the synthesis image, although the complexity of modeling and analysis will grow. The 3-D range data of a person' face has also been used to model the face [81]. The range data were obtained by using a 3-D scanner (Cyberware) which can acquire both range and color data with sufficiently high density.

3.2 Synthesis of facial movements

Texture mapping of original facial images onto a 3-D wire frame model gives rise to natural looking images. In addition, in order to synthesize naturally animating images, synthesis of facial movements(expressions) plays a very important role in the model-based coding system. Our system has utilized two different methods for synthesizing facial expressions. One technique uses a 'clip-and-paste method' and the other a 'facial structure deformation method'.

The clip-and-paste method is performed by analyzing 3-D head motion parameters, extracting specific expressive regions from an input image, transmitting them, and then placing them in the corresponding region of the synthesized image. The regions that must be clipped are identified on the 3-D wire frame model which moves in unison with the person's head (See Figure 6). Figure 7 shows two frames out of the input image and synthesized image sequences. In this case, the 3-D movement of the head is analyzed by the least mean square method using six white points that are initially plotted on the subject's face. The depth information of these points is first roughly estimated using the 3-D facial model and then successively updated in conjunction with the estimation of the head motion parameters.

The facial structure deformation method simulates facial expressions by deforming the 3-D facial model, which can be done in various ways. The Facial Action Coding System (FACS) [19] has been adopted to describe facial actions. FACS itself was originally used in psychological studies and employs a somewhat qualitative representation. FACS describes a set of minimal basic actions (Action Units or AUs) that a human face can perform. There are 44 possible AUs that are designed to be closely connected with the anatomy of the human face. Any facial actions can be synthesized by parameterizing AUs into deformation rules on the 3-D model. The deformation rules control the location of control nodes of the wire frame facial model. Currently 34 AUs out of 44 are parameterized in our system. Figure 8 shows examples of the

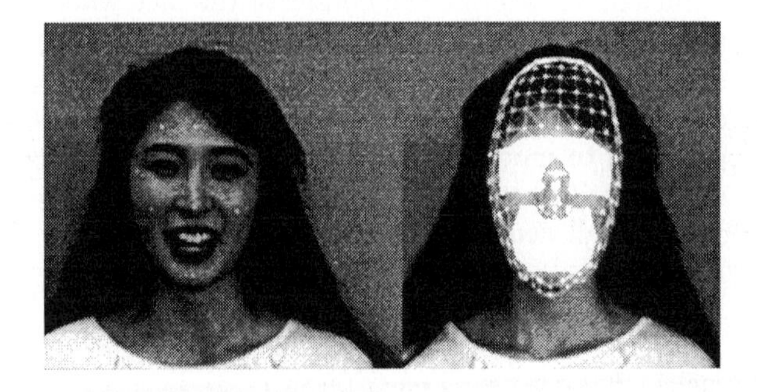

Figure 6 Clip and paste method. Eye and mouth area as clipped regions.

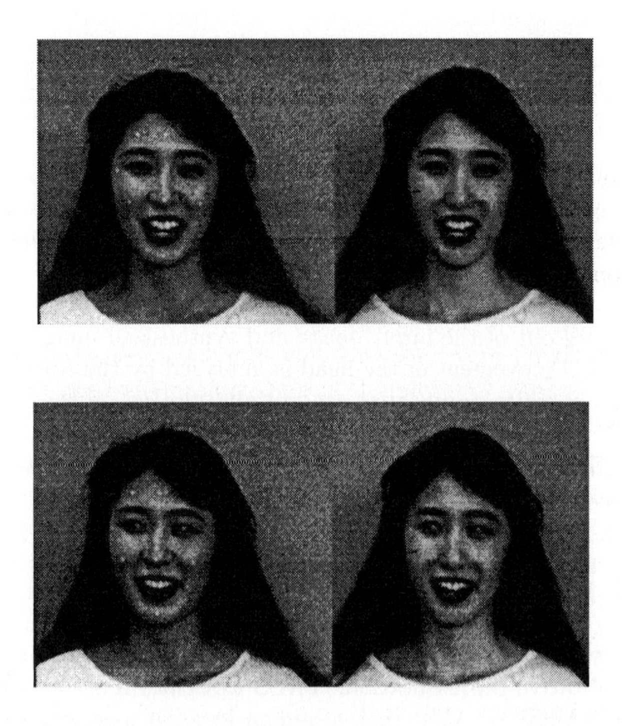

Figure 7 Three frames showing images synthesized by clip-and-paste method. The input images are on the left side and the reconstructed images are on the right side. "White marks" are placed on the input image for motion estimation. The hair and shoulders are modeled by 2-D plane and those of the first frame are translated two-dimensionally.

Table 2 Expected bitrate estimates of 3-D model-based coding scheme

Synthesis method	Information to be transmitted	Expected bitrates
Clip-and-paste method	3-D motion parameters and clipped image signals	1 K - 10 Kbit/s
Facial structure deformation method	3-D motion parameters and facial action unit parameters	< 1 Kbit/s

synthesized images. Four faces in each picture are synthesized by using the same combination of the AU parameters. The estimated bit transmission rates of both clip-and-paste method and structure deformation method is shown in Table 2.

Facial movements can be reproduced in a number of ways when a 3-D wire frame facial model is available. Various parameterizations are listed below:

Texture Level Reproduction:
Reproduce facial movements by updating the texture; for example, clip-and-paste methods [2], template methods [84], model-based/waveform combined approaches [63, 13] etc.

Node Control Level Parameterization:
Synthesize facial movements by controlling the nodes of the wire frame model of the face. Multiple 3-D facial templates are used and interpolated into differing facial expressions [10].

Shape Control Level Parameterization:
Synthesize facial movements by controlling the shape of the wire frame model of the face. For example, shape parameterization [2, 22] by making use of Facial Action Coding System (FACS) [19], heuristic control of shapes of facial components such as eyes and mouth [39] etc.

Muscle Control Level Parameterization:
Parameterize and build muscle models in the wire frame model and synthesize facial actions by controlling the muscles [75, 79, 20, 21, 47].

To date, the majority of computer graphics applications have adopted the shape control method (e.g. [69]). In this work, an attempt has generally been made to implement their own descriptions, though they start with some general description such as FACS. A muscle model approach has been applied [75, 20, 21, 47],

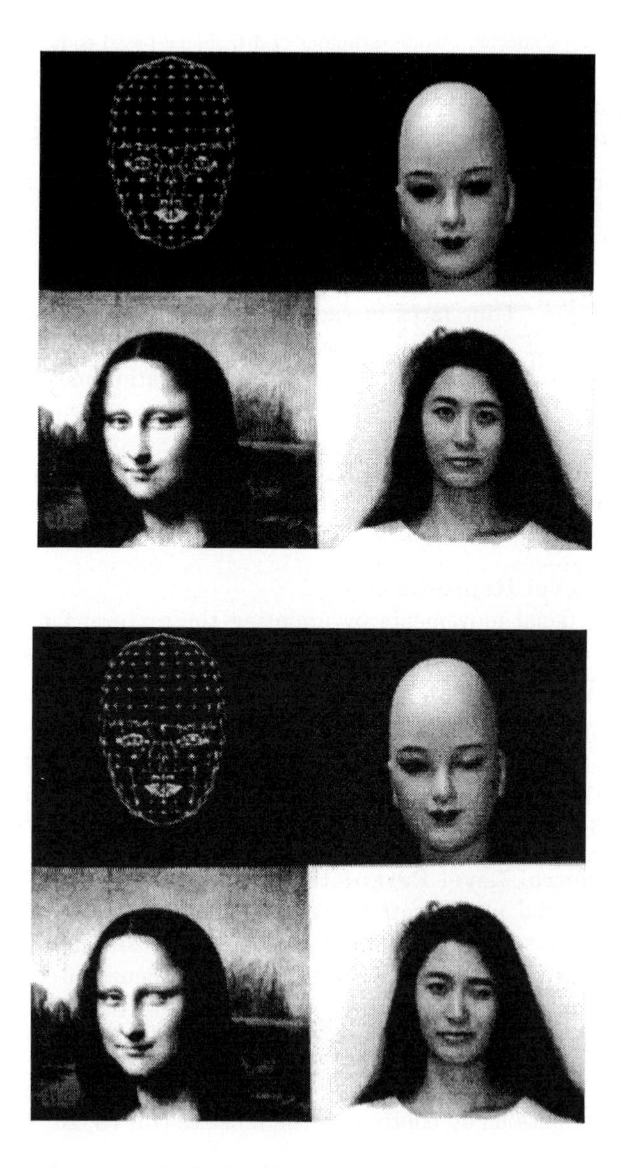

Figure 8 Images synthesized by the structure deformation method. All four facial models are driven by the same combination of action units.

too. Among them, a facial model which has models of three skin layers and muscles has been investigated [79].

In model-based coding proposals, the shape control method and the texture reproduction method have been commonly utilized. For example, one proposal for the texture reproduction approach uses templates for facial components (eyes, mouth), which are stored in advance, and images are updated using these templates [84]. There are a variety of shape control approaches, from heuristic control to more refined control based on facial anatomy such as FACS-based control. Node control level parameterization is not as intuitive as shape control and muscle control parameterization. An interpolation method has been attempted in which differing 3-D facial templates are fused into different facial expressions [10].

3.3 Analysis of facial images

Facial motion analysis based on FACS by Choi et al. [11] will now be described. In this scheme, facial motion is separated into global motion, viz. head motion, and local motion, viz. facial expressions. The head motion is treated as a rigid body motion and the facial expressions are modeled as a linear combination of Action Unit displacement vectors. The head motion and facial expressions are estimated hierarchically by using two consecutive frames; head motion is estimated, head motion is compensated and facial expressions (AUs) are estimated. These steps are iterated to improve precision. Both head motion and facial expression are estimated by direct estimation that does not require a priori knowledge of point correspondences. For example, when analyzing facial expressions, the linear combination of AU parameters is estimated in such a way that the difference between the new frame and the head-motion compensated frame is minimized. It is assumed that the initial location of the head is known.

The results of the analysis experiment which uses the ITU standard image sequence *Claire* is shown in Figure 9. The image synthesized by using the analysis parameters is a fair quality reproduction, even though it only makes use of the texture information of the first frame.

The analysis methods depend strongly on the model chosen and the synthetic output images. Besides, analysis implies several additional problems such as the segmentation of objects, estimation of global motion, and estimation of local motion. In the case of facial images, global motion and local motion correspond

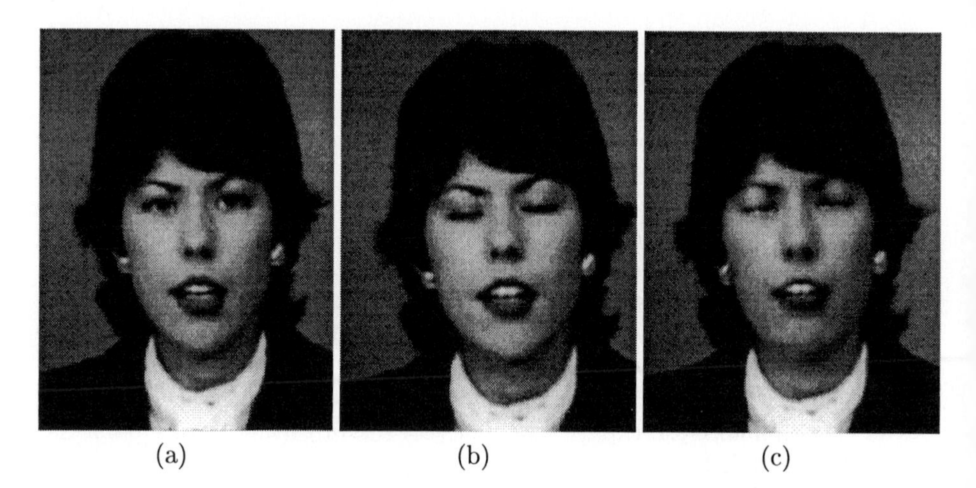

(a) (b) (c)

Figure 9 Analysis of facial motion by using AU constraint. (a) and (b) are the original image frames(#1, #2) and (c) is a image synthesized by using the texture of frame #1 and analysis parameters corresponding to the original image frame#2.

segmentation of objects, estimation of global motion, and estimation of local motion. In the case of facial images, global motion and local motion correspond to head motion and facial expressions, respectively. Motion analyses of facial images have been attempted under some restrictive situations, such as that the model is made in advance and the initial position of the face is known. A tendency of these facial motion analysis methods for model-based coding is that many proposals attempt to use the information of the 3-D model when analyzing the image. Approaches used in analyzing facial motion images are listed below.

Detection of face and facial features

Face region detection by active contour model [84]

Segmentation of moving head-and-shoulder shapes using shape constraints, temporal contour correction and contour smoothness [6].

Head-and-shoulder region segmentation by using stereo cameras [40]

Extraction of facial feature components(eyes, mouth) by using a deformable template [87].

Morphological operations with multi-level thresholding followed by a linear classifier [27].

Detection of elliptical head outline and rectangular eyes-nose-mouth region [18].

Detection and tracking of face and facial features by edge counting method [14].

Head motion (global motion) tracking and facial expression (local motion) analysis
The estimation of head motion by tracking marks plotted on the face [2].
Estimation of global motion of the head using displacement vector field obtained by the block matching technique [44].
Global and local motion analysis with adaptation of the 3-D model taking account of photometric effects [67]. Feature point extraction and tracking by a simple thresholding method [38].
Feature point extraction and tracking by using active contour model [36].
Estimation of facial expressions (Action Units) by using feature point displacements [12].
Direct estimation of head motion and facial expressions (Action Units) using the spatiotemporal gradient [48, 49, 11].
Facial muscle movement estimation based on optical flow [51].
Estimation of facial muscle movement based on physical and anatomical model [79, 21, 20].

4 2-D MODEL-BASED APPROACH USING 2-D DEFORMABLE TRIANGLES FOR ADVANCED MOTION COMPENSATION

3-D model-based coding is expected to have far broader applications than conventional waveform coding techniques. However, 3-D model based coding at present has problems as a complete coding scheme. The first important problem is its lack of generality, that is, the object that can be handled by 3-D model-based coding is restricted and it is not able to deal with unmodeled objects.

In order to cope with this generality problem, one requires more generic model-based approaches based on more general image source models, which can compensate for the lack of generality of 3-D model based schemes. Nakaya has worked on 2-D deformable triangular block-based motion compensation [61] in which an entire image is divided into triangular blocks that deform according to motion in the image. The deformation of each triangle is an affine transform and the computation required for image reproduction is identical to that of 3-D model based coding. This triangle-based motion compensation is quite universal and not limited to facial images. In this section, this work is briefly described and its application to very low bit rate coding such as 16 kbps has been demonstrated. To compensate for the lack of generality, one can combine the generic 2-D triangle-based method and 3-D model-based schemes; the hybrid scheme using both methods is also explained.

4.1 2-D deformable triangle-based motion compensation

As mentioned in Section 2, in the conventional coding methods images are segmented into square blocks and 2-D translational displacement of each block is utilized. The 2-D deformable triangle-based method divides images into triangular blocks and it makes use of an affine motion model.

The affine motion parameters of each triangle are uniquely determined by the displacements of its grid points so that no additional information needs to be transmitted except displacement vectors of triangle nodes. Motion compensation is performed by covering the current frame by triangular patches, estimating the motion of the grid points and synthesizing the prediction image by mapping (warping) the texture of the previous frame onto the corresponding patches of the current frame.

The displacement of grid points is estimated as follows: the initial displacements are first computed by applying the usual block matching at grid points, and the initial displacements are then adjusted successively so that the prediction error after warping is minimized. The advantages of the triangle-based motion compensation are: (1) it can deal with linear deformation such as zooming and rotation; (2) it well approximates the motion field of a 3-D moving object; (3) it effectively alleviates blocking artifacts.

Figure 10 shows how the triangle motion prediction works. In the triangular segmentation of the previous frame and the segmentation of the current frame the motions are shown for "a zooming circle" and "a rotating square" sequences. The prediction images of these examples are compared to that of the conventional block matching as shown in Figure 11. It shows that the triangle-based method alleviates many blocking artifacts compared to the block-matching method.

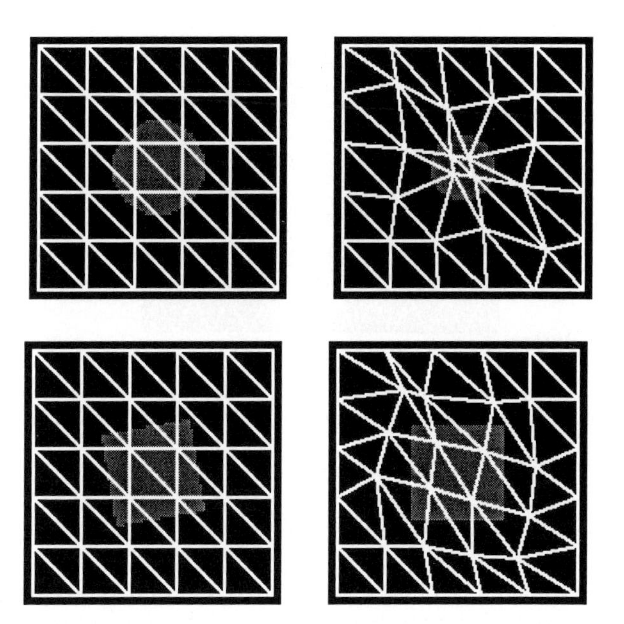

Figure 10 Motion estimation of the grid points. From left to right, current frame and previous frame of zooming circle sequence, and current frame and previous frame of rotating square sequence. Motion is estimated backward so that the current frame is divided into regular triangles and the previous frame into deformed triangles caused by motion.

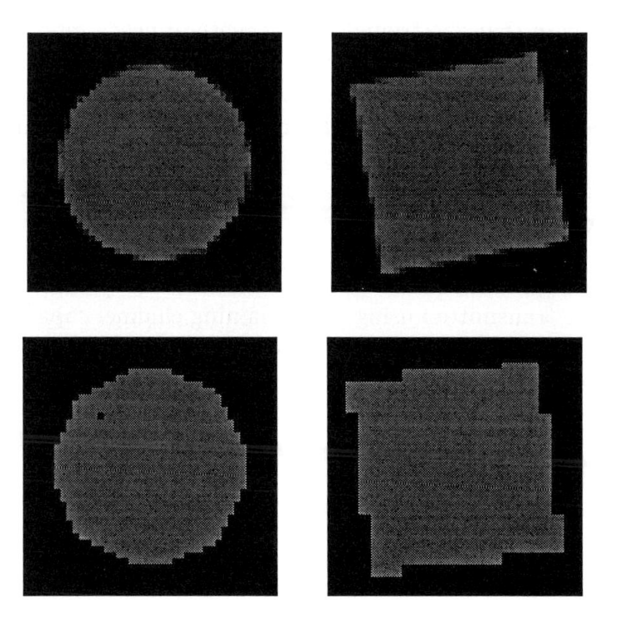

Figure 11 Prediction images. (Top) Two images are predicted by the triangle-based method. (Bottom) Two by the conventional block matching method.

Figure 12 Segmentation used for 16 kbps coding by the triangle-based MC. The number of motion vectors is 95 and that of triangular patches is 160.

4.2 Coding at Very Low Bit Rates

The prediction performance of the triangle-based method is compared with the conventional block matching method in terms of the number of motion vectors in the image. One of the segmentations used in this experiments is shown in Figure 12. When the number of motion vectors is small, each segmentation is larger. The simulation is performed using the monochrome version of the *Miss America* sequence of 352 × 288 pels (CIF) and the frame rate was 10 Hz. An original image was used for prediction. Averaged SNR is shown in Figure 13. The results indicate that the triangle-based method needs much fewer motion vectors than the conventional block matching to achieve the same performance.

Color Sequence of *Miss America* (CIF format) at a 10 Hz frame rate was used in the experiments coded at 16 kbps. Coding schemes in comparison are triangle based MC-DCT (Triangle), block-matching MC-DCT (Block). The segmentation used by triangle-based MC is identical to that shown in Figure 12. The block matching MC-DCT used 99 blocks of 32×32 pixels. In these two methods, the motion information was transmitted initially and error information was subsequently transmitted using the remaining channel capacity. The number of triangles and blocks for the triangle-based method and block matching methods are optimized so that the best coding performance is obtained. RM8 [89] is also down-scaled and tested at 16 kbps in Figure 14.

The coding performance of these methods is shown in Figure 14. It shows that the triangle-based method performs the best and the quality is adequately maintained, while block-matching MC-DCT and RM8 are degraded. RM8 collapses around the 100th frame when the motion is large because of the inadequate buffer control. Images coded by those methods are compared in Figure 15.

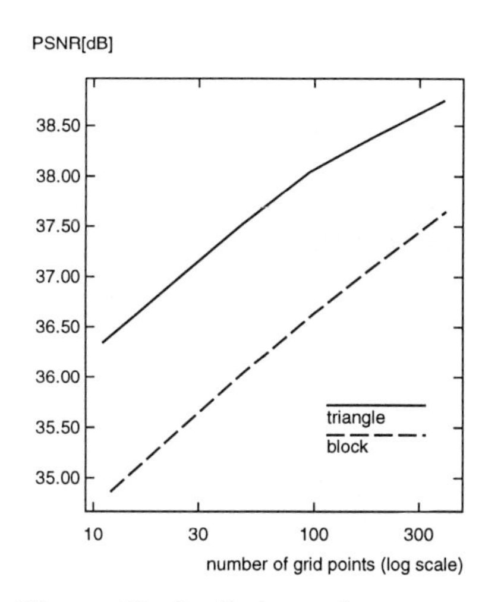

Figure 13 Prediction performance. Averaged SNR versus the number of motion vectors for "Miss America".

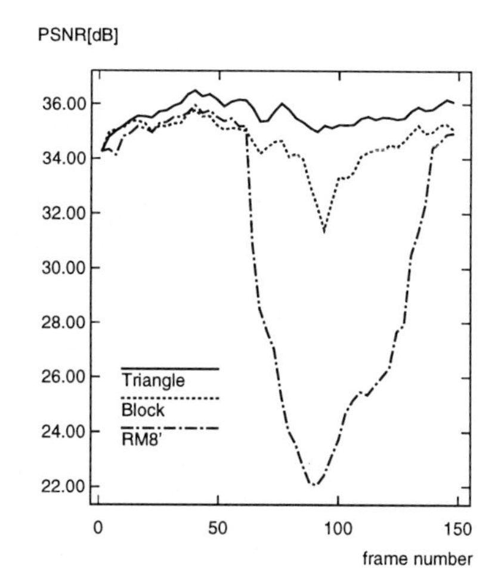

Figure 14 PSNR of the decoded images of the Miss America sequence. "Triangle" means the triangle based MC-DCT. "Block" and "RM8" are block based MC-DCT; however, buffer control mechanism are different between Block and RM8 such that Block method initially sends motion information but RM8 does not.

5 HYBRID OF 3-D AND 2-D MODEL-BASED CODING SCHEMES

Hybrid coding of 3-D model-based method and 2-D triangle-based method has been investigated. By the combination of these two methods, the lack of generality of 3-D model-based coding can be compensated by the 2-D triangle based method. As shown in Figure 16, in the hybrid scheme the face area is coded by 3-D model-based coding and the other area is coded by 2-D triangle-based motion compensation.

In this particular experiment, the error signal is not coded to reduce bit rates. 3-D model is assumed to be made beforehand and positioned at the right place

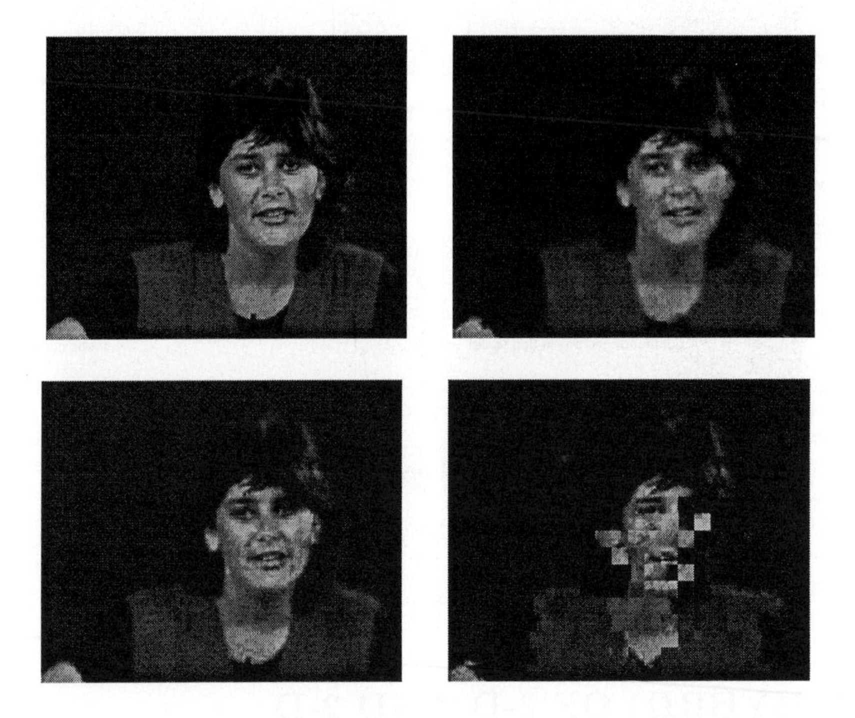

Figure 15 Decoded images. From upper left to lower right, original, the conventional block-matching MC-DCT, the triangle based MC-DCT, scale-down RM8

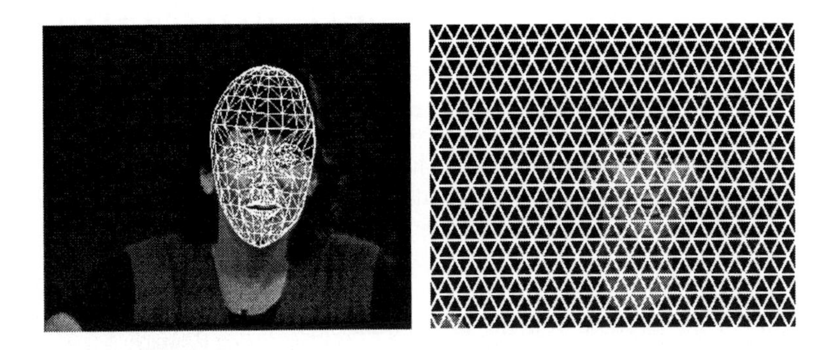

Figure 16 3-D/2-D hybrid coding; Picture format is CIF.

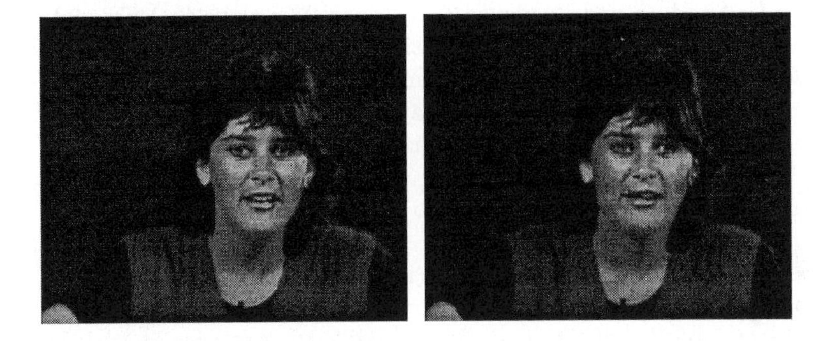

Figure 17 Decoded image by the hybrid method. Left is original frame and right is the decoded images by using the hybrid method (5 kbps).

at the beginning of the sequence. Feature points such as corners of eyes and mouth are automatically detected and tracked by using a simple thresholding processing. The shapes of facial components are deformed by the movements of the feature points. The advantage of the hybrid scheme is that the face area is coded by very low bit rates by using the 3-D model and more information can be assigned to other area so that the coded picture quality is improved.

The coded pictures are shown in Fig. 17. The transmission bitrate is 5 kbps: 1 kbps is assigned to 3-D model-based coding and 4 kbps is assigned to motion information for triangle-based motion compensation. The picture is coded at a rather low rate, but the impression on the coded sequence is adequate.

6 APPLICATIONS OF 3-D MODEL-BASED CODING

Because 3-D model-based coding makes use of 3-D properties of the objects, it has a much broader application area than conventional waveform coding methods. There already exist some applications and extensions of 3-D model based coding. The following lists some of the related work and ideas.

(1) Virtual space teleconferencing [42, 86, 29]

The idea of virtual space teleconferencing is a kind of advanced model-based coding, in which various 3-D CG databases are used together with the 3-D model-based coding scheme. The communication partners are coded by 3-D model-based coding and displayed together with various 3-D CG data. This will provide an advanced communication interface with realistic sensations. They have developed a real time system in which, to avoid complexity of image processing, they use a camera fixed to a helmet and a number of marks plotted on the face for detecting facial expressions.

(2) Structured video and virtual studio [34, 56, 50, 35, 30, 68, 82]

Because model-based coding describes images in highly structured way, it can potentially provide new creative and communicative environments. By making use of 3-D properties of the scene, it is possible to create new scenes from pre-existing material. (Figure 8 is a simple example.) Video modeling will provide a way to handle and edit video materials and compose new scenes employing common computer graphics technology. It will also be able to provide video indexing for video database applications.

The virtual studio is an application of computer graphics and image analysis techniques to program production for broadcasting. Computer generated studio settings are integrated with clipped images of persons, and scenes are generated taking camera manipulation into account which is detected by either mechanical sensors or analysis of an image sequence.

(3) Speech/text-driven facial animation system for an advanced man-machine interface [57, 58, 37, 85, 16]

One of the applications outside image communications is enhancements to current pre-recorded message systems and voice activated databases. By using text data or voice to drive a 3-D facial model, a *talking head* substantially improves the interface between the human user and the message system. In the same context, the synthesized talking head can be utilized in the human-computer interfaces. In comparison to the image communication applications that require great complexity to analyze video, this kind of applications requires much less computation. A prototype real-time working system has been developed.

(4) Real-time implementation of model-based coding system [38]

A prototype real-time system has been developed. The image motion analysis used in the system is rather simple. A threshold operation is applied to the input image, and feature points which are considered to represent basic shapes of facial components (eyes, mouth) are detected. It is assumed that the initial modeling is done beforehand and initial position of the face is known. The experimental system which analyzes and encodes motion images in real-time works at 9600bits/s [38].

(5) Lip synchronization using speech-assisted video processing [8, 9]

Frame skipping at low rate image transmission makes mouth motion jerky and losses lip synchronization. Speech assisted interpolation has thus been proposed. A mouth shape model is utilized and driven by information extracted from speech.

REFERENCES

[1] K. Aizawa and T.S. Huang, Model-based image coding: Advanced image coding technique for very low bit-rate applications, Proc. IEEE Vol. 83, No. 2, pp. 259-271, Feb. 1995.

[2] K. Aizawa, H. Harashima, and T. Saito, Model-Based Analysis Synthesis Image Coding for a Person's Face, Image Communication, Vol. 1, no. 2, pp. 139-152, 1989.

[3] K. Aizawa, H. Harashima, and T. Saito, Model-Based Synthetic Image Coding System, Picture Coding Symposium '87, Stockholm, 1987.

[4] K. Aizawa, T. Saito and H. Harashima, Construction of a 3-dimensional personal face model for knowledge-based image data compression, National Conference Record of IEICEJ, Musashino, Japan, Sep. 3-6 1986, 542, p. I-221 (in Japanese).

[5] T. Akimoto and Y. Suenaga, 3-D Facial Model Creation Using Generic Model and Front and Side Views of Face, IEICE Trans. Inf. & Sys. Vol. 75-D, No. 2, pp. 191-197 (1992).

[6] M. Buck, Segmentation of moving head-and-shoulder shapes, Picture Coding Symposium 90, 9.2, Boston, Mar. 1990.

[7] S. Carlsson, Sketch based coding of gray level images, Signal Processing, 15, pp. 57-83, 1988.

[8] T. Chen, H. P. Graf and K. Wang, Lip synchronization using Speech-assisted video processing, IEEE Signal Processing Letters, Vol. 2, No. 4, pp. 57-59, April 1995.

[9] H. H. Chen, W. Chou, B. G. Haskell and T. Chen, Speech recognition for acoustic-assisted video coding and animation, SPIE VCIP 95, pp. 274-283, May, 1995.

[10] C.S. Choi, T. Okazaki, H. Harahsima and T. Takebe, A System of Analyzing and Synthesizing Facial Images, IEEE International Symposium of Circuits and Systems, pp. 2665-2668, Singapore, 1991.

[11] C.S. Choi and T. Takebe, Analysis and Synthesis of Facial Image Sequences in Model-Based Image Coding, IEEE Trans. Video Technologies, Vol. 4, No. 3, pp. 257-275, June 1994.

[12] C.S. Choi, K. Aizawa, H. Harashima, and T. Takebe, Analysis and Synthesis of Facial Expressions in Model-Based Image Coding, Picture Coding Symposium '90, Boston, 1990.

[13] M.F. Chowdhury, A.F. Clark, A.C. Downton, E. Morimatsu and D.E. Pearson, A Switched Model-Based Coder for Video Signals, IEEE Trans. Circuit and Systems on Video Technology, Vol. 4, No. 3 pp. 216-227, June 1994.

[14] L.C. Desilva, K. Aizawa and M. Hatori, Detection and tracking of Facial Features by using Edge Pixel Counting and Deformable Circular Template Matching, IEICE Trans. on Information Systems, Vol. E78-D No. 9, Sept. 1995.

[15] N. Diehl, Object-Oriented Motion Estimation and Segmentation in Image Sequences, Signal Processing: Image Communication, Vol. 3, pp. 23-56, 1991.

[16] H. Dohi and M. Ishizuka, Realtime synthesis of a realistic anthropomorphous agent toward advanced human-computer interaction, Human Computer Interaction: Software and Hardware Interfaces, pp. 152-157, Elsevier, 1993.

[17] E. H. Adelson, J. Y. A. Wang and S. A. Niyogi, Mid-level vision: New directions in vision and video, IEEE ICIP 94 Vol. II pp. 21-25, Nov. 1994.

[18] A. Eleftheriadis and A. Jacquin, Model-assisted coding of video teleconferencing sequences at low bit rates, IEEE ISCAS 94, Vol. 3 pp. 177-180, May 1994.

[19] P. Ekman and W. V. Friesen, Facial action coding system, Consulting psychologists press 1977.

[20] I. A. Essa, T. Darrel and A. Pentland, Tracking facial motion, IEEE Workshop on Motion of Nonrigid and Articulated Objects pp. 36-42, Nov. 1994.

[21] J. Fischl, B. Miller and J. Robinson, Parameter Tracking in a Muscle-Based Analysis/Synthesis Coding System, Picture Coding Symposium '93, Lausanne, Switherland, 1993.

[22] R. Forchheimer and T. Kronander, Image Coding-From Waveforms to Animation, IEEE Trans. on ASSP, Vol. 37, no. 12, pp. 2008-2023, December, 1989.

[23] R. Forchheimer, O. Fahlander, and T. Kronander, Low Bit-Rate Coding Through Animation, Picture Coding Symposium'83, Davis, CA, pp. 113-114, March, 1983.

[24] T. Fukuhara, K. Asai, and T. Murakami, Model-Based Image Coding Using Stereoscopic Images and Hierarchical Structuring of New 3-D Wire-Frame Model, Picture Coding Symposium '91, Tokyo, 1991.

[25] T. Fukuhara, K. Asai, and T. Murakami, Hierarchical Division of 3-D Wire-Frame Model and Vector Quantization in a Model-Based Coding of Facial Images, Picture Coding Symposium '90, Boston, 1990.

[26] M. Gilge, T. Englehardt and R. Mehlan, Coding of arbitrarily shaped segments based on a generalized orthogonal transform, Image Communication, 1,2, pp. 153-180, 1989.

[27] H.P. Graf, T. Chen, E. Petajan, E. Cosatto, Locating faces and facial parts, Int. Workshop on Automatic Face- and Gesture-Recognition, pp. 41-46, Zurich, 1995.

[28] H. Harashima, K. Aizawa and T. Saito, Model-based analysis synthesis coding of videophone images - conception and basic study of intelligent image coding, IEICE Trans., Vol. E72-5, pp. 452-459, 1989.

[29] H. Harashima and F. Kishino Intelligent image coding and communications with realistic sensation -recent trends-, IEICE Vol. E74, No. 6, pp. 1582-1592, June, 1991.

[30] M. Hayashi et al., Virtual Studio Virtual Reality Application to Video Production, ITEJ Tech. Rep. Vol.1 6, No. 80, pp. 49-54, Dec. 1992.

[31] M. Hötter and J. Ostermann, Analysis Synthesis Coding Based on Planar Rigid Moving Objects, Int. Workshop on 64 kbps Coding of Moving Video, Hannover, 1988.

[32] M. Hötter, Object-Oriented Analysis-Synthesis Coding Based on Moving Two-Dimensional Objects, Image Communication, Vol. 2, no. 4, pp. 409-428, Dec. 1990.

[33] M. Hötter, Optimization and efficiency of an object-oriented analysis-synthesis Coder, IEEE Trans. CSVT, Vol. 4, No. 2, April, 1994.

[34] H. Holtzman, 3-D video modeling, CHI Conference on Human Factors in Computing Systems, CA, May, 1992.

[35] R. Hsu and H. Harashima, Detecting Scene Changes and Activities in Video Databases, ICASSP 94, V 33-36, Adelaide, 1994.

[36] T.S. Huang, S. Reddy, and K. Aizawa, Human Facial Motion Modeling, Analysis and Synthesis for Video Compression, Proc. Visual Communication and Image Processing SPIE, pp. 234-241, Boston, 1991.

[37] M. Kaneko, A. Koike and Y. Hatori, Synthesis of moving facial images with mouth shape controlled by text information, IEIECJ Tech. Rep. IE89-4, 1989 (in Japanese).

[38] M. Kaneko, A. Koike, and Y. Hatori, Real-Time Analysis and Synthesis of Moving Facial Images Applied to Model-Based Image Coding , Picture Coding Symposium '91, Tokyo, 1991.

[39] M. Kaneko, A. Koike, and Y. Hatori, Coding of Facial Image Sequence Based on a 3-D Model of the Head and Motion Detection, Journal of Visual Communication and Image Representation, Vol. 2, No. 1, pp. 39-54, March, 1991.

[40] F. Kappei and G. Heipel, 3-D model based image coding, Picture Coding Symposium 88, Torino, 1988.

[41] T. Kimoto and Y. Yasuda, Hierarchical Representation of the Motion of a Walker and Motion Reconstruction for Model-Based Image Coding, Optical Engineering, Vol. 20, no. 7, pp. 888-903, 1991.

[42] F. Kishino and K. Yamashita Communication with realistic sensation applied to a teleconferencing system, IEICE Tech. Rep. IE89-35, 1989 (in Japanese).

[43] R. Koch, Dynamic 3-D Scene Analysis through Synthesis Feedback Control, IEEE Trans. PAMI, Vol. 15 No. 6, pp. 556-568, 1993.

[44] A. Koike, M. Kaneko and Y. Hatori, Model-based image coding with 3-D motion estimation and shape change detection, Picture Coding Symposium 90, Boston, 1990.

[45] M. Kunt, A. Ikonomopoulos and M. Kocher, Second Generation Image Coding Techniques, Proc. IEEE, 73,4, pp. 795-812, 1985.

[46] A. Lanitis, C.J. Taylor, T.F. Cootes and T. Ahmed, Automatic interpretation of human faces and hand gestures using flexible models, Int. Workshop on Automatic Face-and Gesture-Recognition, pp. 98-103, Zurich, 1995.

[47] F. Lavagetto and S. Curinga Object-oriented scene modelling for interpersonal video communication at very low bit-rate, Image Communication, Vol. 6, No. 5, pp. 379-396, Oct. 1994.

[48] H. Li, P. Rovivainen, R. Forchheimer, 3-D Motion Estimation in Model-Based Facial Image Coding, IEEE Trans. PAMI, Vol. 15, No. 6, pp. 545-555, 1993.

[49] H. Li, R. Forchheimer, Two view facial motion estimation, IEEE Trans. CSVT, Vol. 4, No. 3, pp. 276-287, June, 1994.

[50] H.D. Lin and D.G. Messerschmitt, Video composition methods and their semantics, ICASSP 91, pp. 2833-2836, Toronto, 1991

[51] K. Mase, An Application of Optical Flow - Extraction of Facial Expressions, Proc. of MVA90, Tokyo, 1990.

[52] A. Matsunaga and Y. Yasuda, Video transmission over low bit rate channel, Picture Coding Symposium 83, 1983.

[53] T. Minami, I. So, T. Mizuno, and O. Nakamura, Knowledge-Based Coding of Facial Images, Picture Coding Symposium 90, Boston, 1990.

[54] B. Moghaddam and A. Pentland, Maximum likelihood detection of faces and hands Int. Workshop on Automatic Face- and Gesture-Recognition, pp. 122-128, Zurich, 1995.

[55] H. Morikawa and H. Harashima, 3-D Structure Extraction Coding of Image Sequences, Journal of Visual Communication and Image Representation, Vol. 2, No. 4, pp. 332-344, 1991.

[56] H. Morikawa and H. Harashima, Incremental Segmentation of Moving Pictures -An Analysis by Synthesis Approach-, IEICE Trans. Information and Systems, Vol. E76-D, No. 4, April, 1993.

[57] S. Morishima, K. Aizawa and H. Harashima, An Intelligent Facial Image Coding Driven by Speech and Phoneme, ICASSP 89, Glasgow, 1989.

[58] S. Morishima, A human-machine interface using media conversion and model-based coding schemes, CG International'92, Springer-Verlarg.

[59] H. G. Musmann, M. Hötter and J. Ostermann, Object-Oriented Analysis-Synthesis Coding of Moving Images, Image Communication, Vol. 1, no. 2, pp. 117-138, Oct., 1989.

[60] H. G. Musmann, A Unified View on Video Compression: A Model-Based Perspective, Workshop on Very Low Bitrate Video Compression, Urbana, IL, May 1993.

[61] Y. Nakaya, H. Harashima, Motion compensation based on spatial transformations, IEEE Trans. Video Technologies, Vol. 4, No. 3, pp. 339-356, June 1994.

[62] Y. Nakaya and H. Harashima, Model-based/Waveform Hybrid Coding for Low-Rate Transmission of Facial Images, IEICE Trans. Commun., Vol. E75-B, No. 5, pp. 377-384, 1992.

[63] Y. Nakaya, K. Aizawa, and H. Harashima, Texture Updating Methods in Model-Based Coding of Facial Images, Picture Coding Symposium 90, Boston, 1990.

[64] J. Ostermann, Modelling of 3-D moving objects for an analysis-synthesis coder, Proc. SPIE, Sensing and Reconstruction of Three-Dimensional Objects and Scenes, vol. 1260, pp. 240-249, 1990.

[65] J. Ostermann, An Analysis-Synthesis Coder Based on Moving Flexible 3-D-Objects Picture Coding Symposium 93, Lausanne, Switzerland, 1993.

[66] G. Bozdaği, M. Tekalp, L. Onural, An improvement to MBASIC algorithm for 3-D motion and depth estimation, IEEE Trans. Image Processing, Vol. 3, No. 5, pp. 711-716, Sep. 1994.

[67] G. Bozdaği, M. Tekalp, L. Onural, 3-D Motion Estimation and Wire-Frame Model Adaptation Including Photometric Effects for Model-Based Coding of Facial Image Sequences, IEEE Trans. Circuit and Systems on Video Technology, 1994 Vol. 4, No. 3, pp. 246-256, June 1994.

[68] J.-I. Park, N. Yagi, K. Enami, K. Aizawa and M. Hatori, Estimation of Camera Parameters from Image Sequence for Model-Based Video Coding, IEEE Trans. Circuit and Systems on Video Technology, pp. 288-296, June, 1994.

[69] F.I. Parke, Parameterized Models for Facial Animation, IEEE Computer Graphics and Applications, Vol. 12, pp. 61-68, Nov. 1982.

[70] D.E. Pearson and J. A. Robinson, Visual communication at very low bit Rates, Proc. IEEE, 73, 4, pp. 795-812, 1985.

[71] D. Pearson, Model-Based Image Coding, Proc. GLOBECOM 89, Dallas, 16.1, pp. 554-558, Nov., 1989.

[72] D. Pearson, Developments in Model-Based Video Coding, Proc. IEEE, Vol.83, No. 6, pp. 892-906, June 1995.

[73] D. E. Pearson, Texture mapping in model-based image coding, Image Communication, 2,4, pp. 377-396, 1990.

[74] J. R. Pierce, Symbols, Signals and Noise, Chapter 7, Efficient Encoding, Hutchinson of London, 1962.

[75] S. M. Platt and N.I. Badler, Animating Facial Expressions, Computer Graphics, Vol. 13, pp. 245-242, Aug. 1981.

[76] P. Salembier, L. Torres, F. Meyer, C. Gu, Region-Based Video Coding Using Mathematical Morphology, Proceedings of the IEEE, Vol. 83, No. 6, pp. 843-857, June 1995.

[77] L. Tang, M. Pouyat, K. Aizawa and T.S. Huang, Accuracy of Modelling Human Face Using a Generic Model, Picture Coding Symposium 93, Lausanne, Switherland, 1993.

[78] M. Tanimoto and S. Nakashima, Basic experiment of 2D-3D conversion for a new 3D visual communication, Picture Coding Symposium 90, Boston, 1990.

[79] D. Terzopoulous, K. Waters, Analysis and Synthesis of Facial Image Sequence Using Physical and Anatomical Models, IEEE Trans. PAMI, Vol. 15, No. 6, pp. 569-579, 1993.

[80] H. Ueno, K. Dachiku, K. Ozeki, F. Sugiyama, A study on facial region detection in the standard video coding method, Int. Workshop. on 64 kbit/s coding of moving video, 5-2, Sept. 1990.

[81] K. Waters and D. Terzopoulos, Modeling and Animating Faces Using Scanned Data , The Journal of Visualization and Computer Animation, Vol. 2, pp. 129-131, 1991.

[82] J. Y. A. Wang and E. H. Adelson, Representing Moving Images with Layers, IEEE Trans. Image Processing, Vo. 3, No. 5, pp. 625-638, Sep. 1994.

[83] W. J. Welsh, Model-Based Coding of Moving Images at Very Low Bit Rate, Picture Coding Symposium 87, Stockholm, 1987.

[84] W. J. Welsh, Model-Based Coding of Videophone Images, Electronics & Communication Engineering Journal, Vol. 3, no. 1, pp. 29-36, Feb, 1991.

[85] W. J. Welsh, A. D. Simons, A. D. Hutchinson, R. A. Searby Synthetic face generation for enhancing a user interface, Proc. Image Com 90, pp. 177-182, France, 1990.

[86] G. Xu, H. Agawa, Y. Nagashima, F. Kishino and Y. Kobayashi: Three-Dimensional Face Modeling for Virtual Space Teleconferencing Systems, IEICE Trans.,Vol. E 73, No.10, 1990.

[87] A. L. Yullie, P. W. Hallinan and D. S. Cohen Feature extraction from faces using deformable templates International Journal of Computer Vision, Vol. 8, No.2, pp. 99-111, 1992.

[88] J. Y. Zheng, Y. Nagashima, and F. Kishino, 3-D Modeling From Continuous Aspect Views, Picture Coding Symposium 91, Tokyo, 1991.

[89] CCITT SG XV WP XV/1, Description of reference model 8 (RM8) Doc. 525, July, 1988.

9

A PERCEPTUALLY MOTIVATED THREE-COMPONENT IMAGE MODEL

Nariman Farvardin* and Xiaonong Ran**

**Electrical Engineering Department and Institute for Systems Research, University of Maryland, College Park, Maryland 20742*

***Systems Technology, Corporate Technology Group, National Semiconductor Corporation, 2900 Semiconductor Drive, Santa Clara, CA 95052*

ABSTRACT

Some psychovisual properties of the human visual system (HVS) are discussed and interpreted in a mathematical framework. The formation of perception on monocular images is described by minimization problems based on the special properties of human binocular vision. The edge information, which is found to be of primary importance in visual perception, forms the constraint in the minimization problems. The smooth areas of an image influence human perception together with the edge information. After the concept of edge strength is introduced, it is demonstrated that strong edges are of higher perceptual importance than weaker edges (textures). The notion of a stressed image is introduced and used in the extraction of strong edges; the stressed image is further decomposed into the primary component of strong edges and the smooth variation component. The image is, therefore, decomposed into primary, smooth and texture components. Coding schemes are developed for the three components; the primary component is encoded in intensity and geometric information, and the smooth and texture components are encoded using waveform coding techniques, leading to a hybrid of waveform coding and second generation coding techniques. The above hybrid system is of both high subjective and objective performance, especially at very low bit rates, and is further perceptually tuned for smooth and texture components based on the contrast-sensitivity of the HVS. This perceptually tuned hybrid system can be applied directly to components of color images and to the intra-coded frames in motion video sequences. The above image model has been generalized to a complex-valued 1-D case for an efficient representation of planar curves. Likewise, it can be generalized to a 3-D case for video coding and processing. We are also pursuing an approach for video coding based on the digital image warping techniques, in which the primary components of the three-component image models provide perceptually meaningful features for the specification of warping.

1 INTRODUCTION

The growing interest and need to store and/or transmit digital imagery and video and the practical limitations on the storage and transmission capacity, have led to a significant amount of research activity in image compression (also called image coding) over the past two decades. In any image compression system the goal is to reproduce a good replica of the original image with as small a number of bits as possible. The major thrust in image compression can be divided into the following two categories: (i) *waveform coding* techniques and (ii) *second-generation coding* techniques including *feature-based* techniques.

Waveform coding techniques revolve around information-theoretic principles in which upon selecting a distortion criterion (in most cases squared-error) and certain probabilistic assumptions on the image data, the goal becomes that of minimizing the average distortion between the original image and its reconstruction under an average bit rate constraint. Waveform coding techniques, which by and large constitute the bulk of the research in image coding, have led to a plethora of different coding techniques. Among these, the most noteworthy are: (i) various adaptive forms of two-dimensional (2-D) discrete cosine transform (DCT) coding techniques, (ii) 2-D subband and wavelet image coding techniques and (iii) various forms of vector quantization (VQ) of images. Several versions of transform-based methods have been proposed in the past two decades, [1] - [4]. The work in DCT coding has led to an international still image compression standard known by the acronym JPEG (Joint Photographic Experts Group) [5], and several international video compression standards, H.261 [6], MPEG-1 [7], H.262 (or MPEG-2) [8] and H.263 [9]. Subband image coding was first introduced by Woods and O'Neil [10] in 1986 and since then has been the subject of much research [11] - [14]. More recently, similar multi-resolution methods using wavelet transforms have received much attention in the context of image coding [15]. Since the pioneering work of Linde, Buzo and Gray [16], VQ has been widely studied for image coding both in spatial domain and in frequency domain in conjunction with transform or subband coding. A survey of VQ methods in image coding can be found in [17]. Additional details on these waveform coding techniques and various combinations thereof can be found in [18], [19]. Generally, in designing waveform coding techniques, the human visual system (HVS) and its sensitivity to the coding error do not play a central role. However, there have been some efforts to incorporate specific properties of the HVS in the coder design, [20] - [24].

Generally speaking, waveform coding techniques provide good-to-excellent quality results at compression ratios of 20:1 - 10:1 (bit rates of 0.4 - 0.8 bits/pixel, assuming 8 bits/pixel for the original). With such compression ratios, the average distortion is typically small and hence the corresponding reconstructed image is of high perceptual quality. At lower bit rates however, waveform coding produces specific types of artifacts such as blockiness, ringing and blurring around the edges, etc. In contrast with waveform coding, second generation image coding methods are closely tied to the HVS and its separate sensitivities to the strong edge and texture information. Instead of considering the image as a waveform, these methods attempt to describe the image by a collection of physically significant entities such as regions or contours; this leads to a more compact representation and hence significantly higher compression ratios (e.g., 100:1), albeit at the cost of a different type of distortion [25], [26]. The sketch-based image coding of [27], [28] is motivated by similar ideas: Upon extracting the contours, their geometric and intensity information are encoded resulting in what is called the sketch image; the difference between the original and the sketch image is called texture and is encoded using waveform coding. This approach can be thought of as an extension of an earlier work by Yan and Sakrison [29]. In the sketch-based scheme [27], the contours are extracted using the Laplacian-Gaussian operator (LGO) [30] followed by a gradient operator. This procedure has two drawbacks: (i) It results in location errors of the estimated edges which, in turn, lead to providing incorrect intensity variations on the two sides of the edges and (ii) it makes the unrealistic assumption that all edges are one pixel wide. These problems can lead to large errors in constructing the sketch image [31].

In this chapter, an approach is adopted based on an amalgamation of second generation and waveform coding techniques. Specifically, we have formalized some previous psychovisual studies [32], [33] to characterize the features of the image signal that are responsible for the formation of perception in the HVS. Some observations on the binocular nature of the human vision in [32] are used to formulate the formation of perception as a minimization of the intensity variation energy; this has led to the notion of "strong edges" which apparently plays a significant role in perception. Additionally, we have used the psychovisual observations in [33] to mathematically formulate the interaction between the strong edges and areas of smooth intensity variation. To characterize the strong edges, we have introduced the concept of the "stressed image" and defined the strong edges as the high curvature energy pixels of the stressed image. The stressed image is generated by a space-variant low-pass filtering of the original image. The above formalism has led to a three-component image model consisting of (i) the strong edge, (ii) smooth intensity variation and (iii) texture components.

Coding schemes are developed for the three components. The geometric and intensity information of the strong edges is encoded almost losslessly. The texture and smooth components are encoded using waveform coding methods. Both entropy-coded adaptive DCT and entropy-coded subband coding are considered. Within this framework, we have demonstrated the advantages of the proposed approach in terms of objective and subjective performance improvements. In addition, we have developed a modification of the DCT-based scheme in which the well-known contrast sensitivity of the HVS [34] - [36], is used for the perceptual tuning of the parameters of the coder. It is shown that the combination of this perceptual tuning and the 3-component based approach further improves the subjective performance of the coder, especially at low bit rates.

This chapter is organized as follows. In Section 2, some psychovisual properties of the human perception are presented and discussed. This is followed by the characterization of the strong edge information in Section 3 and development of an algorithm for the extraction of strong edges along with the description of the three-component image model in Section 4. The encoding of the primary component is described in Section 5. Some examples of the three-component decomposition are also provided in this section. This is followed by the description of two encoding schemes for the smooth and texture components in Section 6. Section 7 includes the description of the overall encoding schemes as well as simulation results. Section 8 is devoted to the contrast sensitivity of the HVS and how it is used for the perceptual tuning of the encoding system. Section 9 is devoted to a discussion on video coding using the three-component image models. Section 10 finally concludes the chapter.

2 PSYCHOVISUAL ASPECTS OF THE HUMAN VISUAL SYSTEM

In this section, some observations are made on psychovisual aspects of the HVS. The objective is to extract and to discriminate different properties of image signals that are of significance to human perception. Some interpretations in the context of image coding are provided. In the first two observations, the *binocular* vision is used in a unique way, viewing two unrelated images, to arrive at conclusions about how humans perceive *monocular* images.

2.1 Edge Information of Image Signals

As described in [32], in natural binocular vision, when two views are presented with two forms that are different (in the sense that they do not admit of being combined into the image of a single object), the images of both forms will generally be seen at the same time superposed on one another in the field of view. Usually, in some locations of the field of view, one image dominates the other, and vice versa in other parts of the field.

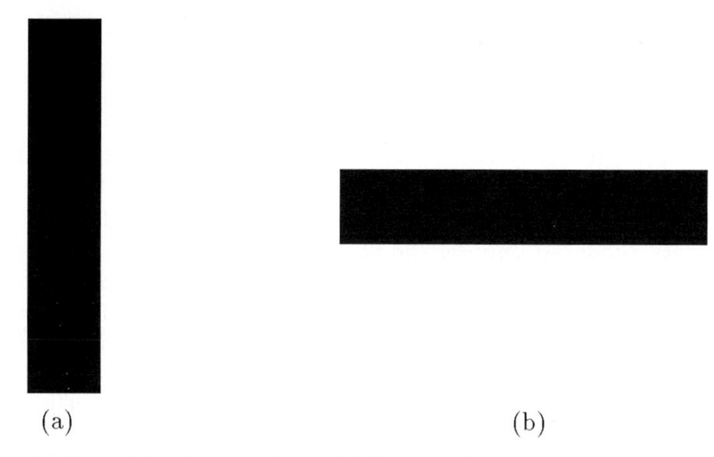

Figure 1 (a) Vertical bar for the left eye; (b) Horizontal bar for the right eye.

When broad black-and-white figures are displayed to both views, the general rule is that the dominating image along an edge and in its vicinity will be the one that owns the edge. As an example, consider Fig. 1 where one vertical bar and one horizontal bar are presented. When the vertical bar is seen by the left eye and the horizontal bar by the right eye, without devoting exclusive

Figure 2 Binocular perception of the images of the vertical and horizontal bars.

attention to any of the two,[1] the total perceptual effect will be an image similar to Fig. 2.

As shown in Fig. 2, the perception will be a cross that is black over the center square while the background appears white. The four arms of the cross are perfectly black at their ends and almost entirely white near the center square, with transitions in between.

Based on this phenomenon, it may be concluded that the collection of the individual pixel intensity values, without any interaction between them (the most primitive property of the image), is not what produces the visual perception; otherwise, the image in the field of view would be formed at each pixel with an intensity produced only by the two corresponding pixel intensities of the two images according to a certain law, and thus the perception would be a uniform combination of the two pictures. Obviously, the image shown in Fig. 2 cannot be constructed by a *uniform* combination of the two images in Fig. 1.

We conjecture that here it is the edge information (to be defined next) of the image signal that arouses our visual perception. For the images in Fig. 1, the edge information is described by (i) the *locations* of variations of intensity values and (ii) the related *intensity values* at these locations. For example, in Fig. 1 (a), the locations of the intensity variations are along the border of the vertical black bar, and the related intensity values are those intensity values immediately inside and outside the border. The above actually describes

[1] A technique for viewing two pictures independently with the two eyes is described in [32].

what is generally referred to as an *edge*. This edge basically has two related intensity values, zero corresponding to black and 255 corresponding to white; these two intensity values represent the intensity variation across the edge. We call the contour formed by the lower intensity value, the *lower edge brim* and the contour formed by the higher intensity value, the *upper edge brim*. In Fig. 1 (a), the rectangular contour of the lower edge brim is just inside the border, while the one for the upper edge brim is just outside. Since the locations and intensity values of the lower and upper edge brims completely specify the corresponding edge, an edge can be defined through its brims that are mathematically characterized later in this chapter.

Let us now go back to our conjecture that the edge information is responsible for the formation of the visual perception. Remarkably, this conjecture is verified in this case by actually producing Fig. 2 *only* from the knowledge of lower and upper edge brims of the two images in Fig. 1, simply by minimizing the variation in intensity values as described below.

Let a generic digital image of size $M \times M$ be denoted by an array of real numbers $\{x_{i,j},\ i,j = 0,1,\ldots,M-1\}$, where $x_{i,j}$ is the intensity value of the pixel on the ith row and jth column. Note that the image can be defined as a set $\mathcal{X}$ of triples: $\mathcal{X} \equiv \{(i,j,x_{i,j}),\ i,j = 0,1,\ldots,M-1\} \subset \mathrm{I\!R}^3$. Let the image shown in Fig. 2 be denoted by $\mathcal{X}_c$. We define the *variation energy* of intensity values of an image $\mathcal{X}$ as follows,

$$V_{\mathcal{X}} = \sum_{i=0}^{M-2} \sum_{j=0}^{M-2} [(x_{i,j} - x_{i,j+1})^2 + (x_{i,j} - x_{i+1,j})^2]. \tag{9.1}$$

The lower and upper edge brims of an image $\mathcal{X}$ are a subset of $\mathcal{X}$. Let the subsets of lower and upper edge brims of the images (a) and (b) in Fig. 1 be denoted by $\mathcal{B}_a$ and $\mathcal{B}_b$, respectively. The information in $\mathcal{B}_a$ and $\mathcal{B}_b$ can be combined by forming a set $\mathcal{B}$ from $\mathcal{B}_a \cup \mathcal{B}_b$ in the following way. Let us first define two projection functions, $f_1 : \mathrm{I\!R}^3 \to \mathrm{I\!R}^2$ where $f_1((i,j,x)) = (i,j)$, and f_2: $\mathrm{I\!R}^3 \to \mathrm{I\!R}$ where $f_2((i,j,x)) = x$. Denote by $f_k(\mathcal{A})$ the image of a set $\mathcal{A}$ under f_k, $k = 1,2$. For a set $\mathcal{A} \subset \mathrm{I\!R}^3$, we also define $\mathcal{A}^{(i,j)} = \{s : s \in \mathcal{A} \text{ and } f_1(s) = (i,j)\}$. Then the set $\mathcal{B}$ is defined as,

$$\mathcal{B} = \{(i,j,\overline{x}_{i,j}) : (i,j) \in f_1(\mathcal{B}_a \cup \mathcal{B}_b)\}, \tag{9.2}$$

where $\overline{x}_{i,j}$ is the average value of the elements of the set $f_2\left((\mathcal{B}_a \cup \mathcal{B}_b)^{(i,j)}\right)$. In other words, $\mathcal{B}$ is a "linear" combination of $\mathcal{B}_a$ and $\mathcal{B}_b$ in the above sense.

Now the objective is to generate the image $\mathcal{X}_c$ solely from the information in $\mathcal{B}$, or to find the image of minimum information while containing the information in $\mathcal{B}$. We call this concept the *minimum information principle.*[2] We quantify the information content of an image $\mathcal{X}$ by its variation energy $V_{\mathcal{X}}$, and define $\mathcal{X}_c = \{(i, j, x^c_{i,j})\}$ to be the solution of the minimization problem,

$$\min_{\{x_{i,j}\}} V_{\mathcal{X}}, \quad \text{subject to } \mathcal{X} \cap \mathcal{B} = \mathcal{B}, \tag{9.3}$$

where $\mathcal{X} = \{(i, j, x_{i,j})\}$. For example, an image with uniform intensity values has zero variation energy, and thus contains the smallest amount of information by the above definition. Because of the quadratic nature of the objective function $V_{\mathcal{X}}$, it may be written in a matrix notation, $\mathbf{x}^T L_V \mathbf{x} + \mathbf{x}^T H \mathbf{x}_{\mathcal{B}} + \mathbf{x}_{\mathcal{B}}^T D \mathbf{x}_{\mathcal{B}}$, where $\mathbf{x}$ is a vector in $\mathbb{R}^{M^2 - |\mathcal{B}|}$ containing the elements $x_{i,j}$, $(i, j) \notin f_1(\mathcal{B})$ in a certain order, $\mathbf{x}_{\mathcal{B}}$ is a vector in $\mathbb{R}^{|\mathcal{B}|}$ with the elements $x_{i,j}$, $(i, j) \in f_1(\mathcal{B})$, the matrix L_V is a non-negative definite (positive definite when $\mathcal{B} \neq \emptyset$) matrix, and the superscript ^{T} indicates vector transpose. When $\mathcal{B} \neq \emptyset$, the unique solution of problem (9.3) is given by the vector $\mathbf{x}_c \equiv -L_V^{-1} H \mathbf{x}_{\mathcal{B}}/2$. For the given example, the resulting image $\mathcal{X}_c$ is shown in Fig. 2; this image, obtained from the above minimization problem, is very similar to the corresponding illustration shown in Fig. 73 of [32].

The above verification of our conjecture as a result of reproducing the image of the perception through solving the minimization problem (9.3) indicates that the necessary and sufficient information to be transmitted and/or stored for the images in Fig. 1 is the edge information, if the final receiver of the images is the HVS. The mechanism governed by the minimization problem (9.3) may be taken as the one that forms our perception.

2.2 Strong Edges and Textures

We now address some issues on the relative perceptual importance of different types of edges. Shown in Fig. 3 are two images, (a) and (b), separately presented to the two eyes. Image (a), examined by the left eye, is a black cross, while image (b) for the right eye is a network of slanted black lines over a white background. Without any special attention to either image, the usual binocular perception would be similar to the image shown in Fig. 4. That is, the image of the cross prevails along its edges; only at some distance from these edges does the network pattern become visible [32].

---

[2] The term "minimum information principle" is motivated by a concept, referred to as "*no news is good news*," introduced in [37].

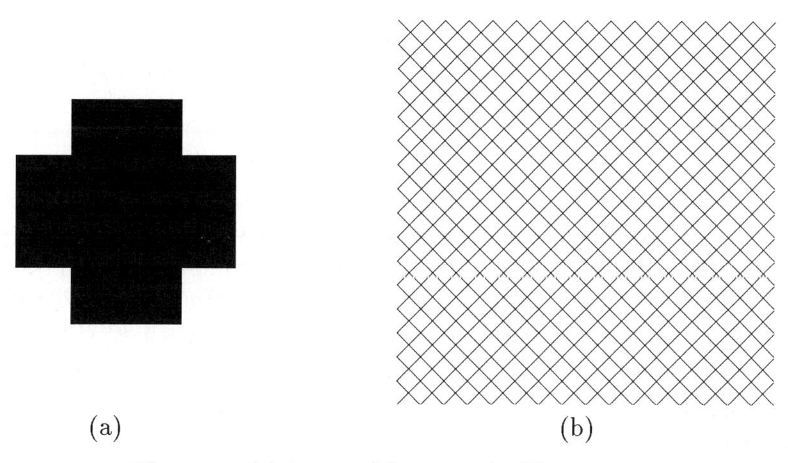

Figure 3 (a) A cross; (b) a network of lines.

Clearly, the two images in Fig. 3 do not have edge information of the same importance, for otherwise in the vicinity of the edges of the cross the network pattern should not disappear. Thus, in this case, the edge information cannot be combined "linearly" as in (9.2). We speculate that an edge has a *strength* associated with it. Edges of relatively high (low) strength have stronger (weaker) influence on human perception, and may be called *stronger (weaker) edges.* Weaker edges commonly belong to the category of *textures* in the usual sense, and thus are referred to as textures. As compared with textures, we may simply refer to stronger edges as *strong edges.*

Figure 4 Binocular perception of the images of the cross and the network of lines.

Apparently, edges with larger intensity variations and shorter widths (distance between the corresponding edge brims) are relatively more significant to human perception, and are generally called sharp edges. However, edges do not individually influence the perception; rather, they locally interact with one another. For example, the edges in the network of Fig. 3 (b) actually have the same intensity variation and width as those of the cross, but they have less perceptual significance when compared with the edges of the cross. Based on this observation, we conjecture that every edge has an original strength that is proportional to (i) the intensity variation between its two edge brims and (ii) its width. Neighboring edges interact with each other in an inhibitive way. The more closely two edges stand, the more severely their strengths are weakened. The resulting strength of an edge after it interacts with the neighboring edges will be called the *strength* of the edge. Thus stronger edges are those edges of higher intensity variation and shorter width, while weaker edges are those of lower intensity variation and longer width. Generally, stronger edges are relatively isolated while weaker edges are relatively crowded with other edges.

We now use this qualitative description of edge strength to explain the phenomenon of Fig. 4. Let us denote the subsets of lower and upper edge brims of images (a) and (b) in Fig. 3 by $\mathcal{B}_a$ and $\mathcal{B}_b$, respectively, and combine the information in $\mathcal{B}_a$ and $\mathcal{B}_b$ based on the following arguments. In the neighborhood of weak edges, strong edges dominate; this domination diminishes gradually as a function of the distance from the stronger edge. After a certain distance from the stronger edge, weaker edges show up and eventually dominate at distances far enough from the stronger edge.

Our experiments indicate that the phenomenon for Figs. 3 and 4 can be explained as follows. In the close neighborhood of the edges represented by $\mathcal{B}_a$, the image for the perception is formed by solving a problem similar to (9.3) with $\mathcal{B}_a$ as the fixed-point set; we call this image $\mathcal{X}_A$. For distances sufficiently far from $\mathcal{B}_a$, where the influence of the strong edge diminishes, the image for the perception is formed by solving a problem like (9.3) with $\mathcal{B}_b$ as the fixed-point set; we call this image $\mathcal{X}_B$. In between these two extreme cases, the perception is formed according to $\lambda\mathcal{X}_A + (1-\lambda)\mathcal{X}_B$ where λ decreases quadratically from 1 to 0 as a function of the distance from the strong edge. The results indicate that this linear combination with the chosen weighting factor results in an image (shown in Fig. 4) which agrees with human perception as well as the description given in [9, pp. 496-497]. Since stronger edges are of higher importance to human perception, edges should be treated in accordance with their strengths to achieve high efficiency in an image information transmission system, especially at low bit rates.

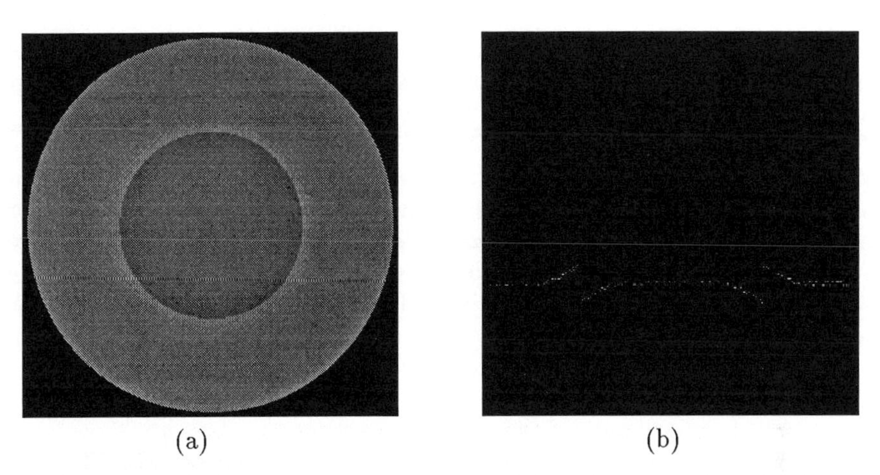

(a) (b)

Figure 5 (a) Two concentric disk image; (b) intensity values along the diameter of the disks.

2.3 Smooth Areas of Image Signals

We now investigate the influence of smooth areas of an image on human perception. Smooth areas are characterized by slow changes in the intensity value. Let us consider Fig. 5 (a) (also shown in [33], [37]) where two concentric disks are shown with the intensity values along the diameter illustrated in Fig. 5 (b). While the perception of the image in Fig. 5 (a) varies from person to person, generally, one feels that it is a small dark disk in the middle of a larger and brighter one. However, as shown in Fig. 5 (b), in the regions near the center of the disks and the outer border of the bigger disk, the intensity values are, in fact, identical.

To describe this phenomenon quantitatively, let us denote the set of lower and upper edge brims by $\mathcal{B}$ representing the two circular edges of the inner and outer disks in Fig. 5 (a). We denote the image in Fig. 5 (a) by $\mathcal{X}$ and the image of human perception by $\mathcal{X}_p = \{(i, j, x^p_{i,j})\}$. Then we conjecture that $\mathcal{X}_p$ is the solution to the following minimization problem:

$$\min_{\{y_{i,j}\}} \lambda \sum_{i=0}^{M-1} \sum_{j=0}^{M-1} (y_{i,j} - x_{i,j})^2 + V_{\mathcal{Y}}, \tag{9.4}$$

subject to $\mathcal{Y} \cap \mathcal{B} = \mathcal{B}$, where $\lambda \geq 0$ is a weighting factor on the squared errors between $\mathcal{X}$ and $\mathcal{Y}$ in the objective function, $\mathcal{Y} = \{(i, j, y_{i,j})\}$, and $\mathcal{B} \neq \emptyset$. Notice that when $\lambda = 0$, $\mathcal{X}_p$ would be an image of two disks with constant intensity values for the inner disk and a linear transition of intensity values on the outer disk, and may well be a description of the human perception. When λ gets larger, $\mathcal{X}_p$ would be closer to $\mathcal{X}$ in the Euclidean sense, and would have larger variation energy than the case with $\lambda = 0$.

We can summarize the psychovisual results described in this section by stating that basically the edge information in an image is responsible for our perception; stronger edges are of higher importance to our perception and the smooth areas influence our perception together with the edge information. In the next two sections, a three-component image model is introduced based on the above observations.

3 CHARACTERIZATION OF STRONG EDGES

In what follows we provide for strong edges a precise characterization that leads to the basic idea for extracting them.

3.1 Description of Strong Edges

First we provide a mathematical description of edges, strong or weak, and then develop a scheme for the discrimination of strong edges. Let us define the second-order directional variations $D^r_{i,j}$ and $D^c_{i,j}$ for $\mathcal{X}$ at pixel (i,j),

$$D^r_{i,j} = x_{i,j-1} - 2x_{i,j} + x_{i,j+1}, \; i = 0, 1, \ldots, M-1, \; j = 1, \ldots, M-2, \quad (9.5)$$

$$D^c_{i,j} = x_{i-1,j} - 2x_{i,j} + x_{i+1,j}, \; i = 1, \ldots, M-2, \; j = 0, 1, \ldots, M-1, \quad (9.6)$$

and use them to describe the edge brim points. More specifically, we define the *pixel row-curvature energy* $C^r_{i,j}$ and the *pixel column-curvature energy* $C^c_{i,j}$ by

$$C^r_{i,j} \equiv \begin{cases} (D^r_{i,j})^2 & i = 0, 1, \ldots, M-1 \text{ and } j = 1, \ldots, M-2, \\ 0 & \text{otherwise,} \end{cases} \quad (9.7)$$

and,

$$C^c_{i,j} \equiv \begin{cases} (D^c_{i,j})^2 & i = 1, \ldots, M-2 \text{ and } j = 0, 1, \ldots, M-1, \\ 0 & \text{otherwise,} \end{cases} \quad (9.8)$$

respectively, and will refer to them collectively as *pixel curvature energies.* We then define the set $\mathcal{B}_T(\mathcal{X})$ of edge brim points of $\mathcal{X}$ as follows,

$$\mathcal{B}_T(\mathcal{X}) \equiv \left\{(i,j,x_{i,j}) : C^r_{i,j} > T \text{ or } C^c_{i,j} > T\right\}, \tag{9.9}$$

where $T \geq 0$. A comparison of the above description of edges with traditional ways [30], [27] and [28][3] can be found in [31].

The definition in (9.9) gives a set of edge brim points regardless of whether or not they correspond to strong edges. To circumvent this problem, let us introduce the concept of a *stressed image* $\mathcal{X}^s = \{(i,j,x^s_{i,j})\}$, associated with the image $\mathcal{X}$, and first state here the properties that the stressed image $\mathcal{X}^s$ is required to possess, and then, in the next subsection, develop a scheme to generate the stressed image. At strong edges, the stressed image $\mathcal{X}^s$ is required to closely approximate the original image $\mathcal{X}$, i.e., the squared-errors, $(x_{i,j} - x^s_{i,j})^2$, to be small at these pixels; the pixel curvature energies of $\mathcal{X}^s$ of these pixels have no additional constraint. In other areas such as smooth and texture areas, the pixel curvature energies of $\mathcal{X}^s$ is required to be small while with a loose constraint on the squared-error $(x_{i,j} - x^s_{i,j})^2$. Thus $\mathcal{B}_T(\mathcal{X}^s)$, as compared with $\mathcal{B}_T(\mathcal{X})$, would only contain the edge brim points corresponding to strong edges because large pixel curvature energies occur only at the strong edges. Note that this stressed image is generally smooth, except at strong edges where it usually assumes large pixel curvature energies, and thus is "stressed". In the edge extraction schemes in [30], [27] and [28] the textures are suppressed by uniformly smoothing the image. While here the image is still smoothed to suppress the textures, the smoothing is performed *non-uniformly.* The term "stressed image" is used to indicate this non-uniformity of smoothing.

3.2 Generating the Stressed Image

In this subsection, we consider the generation of a stressed image $\mathcal{X}^s$ from an original image $\mathcal{X}$. Since the main property of the stressed image $\mathcal{X}^s$ is described in terms of (i) the squared-errors $(x_{i,j} - x^s_{i,j})^2$ and (ii) the pixel curvature energies of $\mathcal{X}^s$, let us consider the following quantity at pixel (i,j),

$$E_{i,j}(\lambda^1_{i,j}, \lambda^2_{i,j}, \lambda^3_{i,j}) = \lambda^1_{i,j}(x_{i,j} - x^s_{i,j})^2 + \lambda^2_{i,j}C^r_{i,j} + \lambda^3_{i,j}C^c_{i,j}, \tag{9.10}$$

[3]The actual schemes introduced in [30], [27] and [28] are combinations of techniques of Gaussian filtering and gradient operator; they work well for edges that are one-pixel wide, but not for edges of multi-pixel width.

where parameters $\lambda^1_{i,j}, \lambda^2_{i,j}, \lambda^3_{i,j}$ are three non-negative real numbers. We define the summation of these $E_{i,j}$ as

$$E(\mathcal{X}^s, \mathcal{X}, \Lambda) \equiv \sum_{i=0}^{M-1} \sum_{j=0}^{M-1} E_{i,j}(\lambda^1_{i,j}, \lambda^2_{i,j}, \lambda^3_{i,j}), \tag{9.11}$$

where Λ represents the collection of the parameters $\lambda^1_{i,j}, \lambda^2_{i,j}$ and $\lambda^3_{i,j}$. Now we consider the following minimization problem for a given parameter set Λ

$$\min_{\{y_{i,j}\}} E(\mathcal{Y}, \mathcal{X}, \Lambda), \tag{9.12}$$

where $\mathcal{Y} \equiv \{(i, j, y_{i,j})\}$. Note that the objective function $E(\mathcal{Y}, \mathcal{X}, \Lambda)$ is a convex function of $\mathcal{Y}$, since each $E_{i,j}(\lambda^1_{i,j}, \lambda^2_{i,j}, \lambda^3_{i,j})$ is a convex function of $\{y_{i,j}\}$ [38]. Therefore, $\{y^*_{i,j}\}$ is a solution of this minimization problem if and only if it is a solution of the following system of linear equations,

$$\nabla_{\{y_{i,j}\}} E(\mathcal{Y}, \mathcal{X}, \Lambda) \equiv 0, \tag{9.13}$$

where $\nabla_{\{y_{i,j}\}} E(\mathcal{Y}, \mathcal{X}, \Lambda)$ denotes the gradient of $E(\mathcal{Y}, \mathcal{X}, \Lambda)$ with respect to $\{y_{i,j}\}$. The objective function in (9.12) can be written in matrix notation as:

$$[y_{i,j}]^T L[y_{i,j}] - 2[\lambda^1_{i,j} x_{i,j}]^T [y_{i,j}] + [\lambda^1_{i,j} x_{i,j}]^T [x_{i,j}], \tag{9.14}$$

where vectors $[y_{i,j}] \in \mathbb{R}^{M^2}$ and $[\lambda^1_{i,j} x_{i,j}] \in \mathbb{R}^{M^2}$ contain elements $y_{i,j}$ and $\lambda^1_{i,j} x_{i,j}$, respectively, in a certain order (one-to-one function) r: $\{0, 1, \ldots, M - 1\} \times \{0, 1, \ldots, M-1\} \rightarrow \{0, 1, \ldots, M^2-1\}$, i.e., the $(r(i_0, j_0))$th element in the vectors $[y_{i,j}]$ and $[\lambda^1_{i,j} x_{i,j}]$ are y_{i_0,j_0} and $\lambda^1_{i_0,j_0} x_{i_0,j_0}$, respectively. The matrix $L = [\ell_{u,v}]$ is a non-negative definite matrix, where $\ell_{u,v}$ is the coefficient of the term $y_{r^{-1}(u)} y_{r^{-1}(u)}$ when $u = v$, and half of the coefficient of the term $y_{r^{-1}(u)} y_{r^{-1}(v)}$ when $u \neq v$, in $E(\mathcal{Y}, \mathcal{X}, \Lambda)$. With this matrix notation, (9.13) simplifies to the following:

$$L[y_{i,j}] = [\lambda^1_{i,j} x_{i,j}]. \tag{9.15}$$

The existence and uniqueness of the solution for the minimization problem (9.12) when $\lambda^1_{i,j} > 0$ for all (i, j) is guaranteed by the fact that the matrix L is positive definite if $\lambda^1_{i,j} > 0$ for all (i, j) [31]. From now on, let us assume that the parameter set Λ is such that all $\lambda^1_{i,j}$ are greater than zero, and therefore talk about the solution of problem (9.12), that is $L^{-1}[\lambda^1_{i,j} x_{i,j}]$. Problem (9.12) is solved iteratively by the Gauss-Seidel iteration[4] using the Multi-Grid technique

[4] The convergence is guaranteed by the above fact [39, p. 125], [40, p. 109], [41].

to speed up the convergence. The Gauss-Seidel iteration and the Multi-Grid technique are well known [39]. The details of the implementation of these two techniques are included in [41].

Now let us consider the influence of the choices of $\lambda^1_{i,j}, \lambda^2_{i,j}$ and $\lambda^3_{i,j}$ on the solution, $\mathcal{Y}^* = \{(i,j,y^*_{i,j})\}$, in problem (9.12). A close examination of (9.11) indicates that if $\lambda^1_{i,j}$ is large compared to $\lambda^2_{i,j}$ and $\lambda^3_{i,j}$, $(x_{i,j} - y^*_{i,j})^2$ would be small, and the constraint on the pixel curvature energies of $\mathcal{Y}^*$ would be loose; on the other hand, if $\lambda^2_{i,j}$ and $\lambda^3_{i,j}$ are large as compared to $\lambda^1_{i,j}$, the pixel curvature energies of $\mathcal{Y}^*$ would be small, and the constraint imposed on the squared-error $(x_{i,j} - y^*_{i,j})^2$ would be relaxed. To make this observation more precise, we may take the relationship between the original image $\mathcal{X}$ and the solution $\mathcal{Y}^*$ governed by the minimization problem (9.12) as a filtering operation with input $\mathcal{X}$ and output $\mathcal{Y}^*$ [31]. This filter is space-varying low-pass with cutoff frequencies at (i,j) controlled by the parameters $\lambda^1_{i,j}$, $\lambda^2_{i,j}$ and $\lambda^3_{i,j}$ in such a way that larger values of $\lambda^1_{i,j}$ give rise to higher cutoff frequencies in both directions, while larger values of $\lambda^2_{i,j}$ and $\lambda^3_{i,j}$ lead to lower cutoff frequencies in row-direction and column-direction, respectively [31], [41]. Note that it is not the absolute values of $\lambda^1_{i,j}$, $\lambda^2_{i,j}$, and $\lambda^3_{i,j}$ but rather the ratios $\lambda^2_{i,j}/\lambda^1_{i,j}$ and $\lambda^3_{i,j}/\lambda^1_{i,j}$ that influence the solution $\mathcal{Y}^*$.

The minimization problem (9.12) has an interesting interpretation in that the solution $\mathcal{Y}^*$ can be thought of as the stable configuration of a mechanical structure, a part of that is depicted in Fig. 6. In this structure, at each pixel location (i,j) we have a vertical spring with both ends fixed on a floor (height $= 0$) and a ceiling (height $= 255$), and a cylinder fixed on it at height $x_{i,j}$. This cylinder is constrained to move only in the vertical direction. For each row and column there is a flexible bar of the shape shown in Fig. 6 with a slot in the middle. The cylinders associated with each (i,j) are fitted inside the corresponding column bars, and the column bars, in turn, are fitted inside row bars. Shown in Fig. 6 is the structure at pixel $(M-1,0)$. We now suppose that this mechanical structure assumes a configuration described by the heights $y_{i,j}$ of the cylinders. Then the potential energy of this configuration $\{y_{i,j}\}$ is approximated by $E(\mathcal{Y},\mathcal{X},\Lambda)$ defined in (9.11), where $\lambda^1_{i,j}(x_{i,j} - y_{i,j})^2$, $\lambda^2_{i,j}C^r_{i,j}$ and $\lambda^3_{i,j}C^c_{i,j}$ approximate the potential energies of the spring, the row bar and the column bar at pixel (i,j), respectively. The parameters $\lambda^1_{i,j}, \lambda^2_{i,j}$ and $\lambda^3_{i,j}$ control the rigidness of the spring, the row bar, and the column bar at pixel (i,j) [42].

To visualize the formation of the stable configuration for this mechanical structure, we note that if all the bars are taken out, the cylinders on the vertical springs for each pixel (i,j) will assume their stable positions at $x_{i,j}$, i.e., this

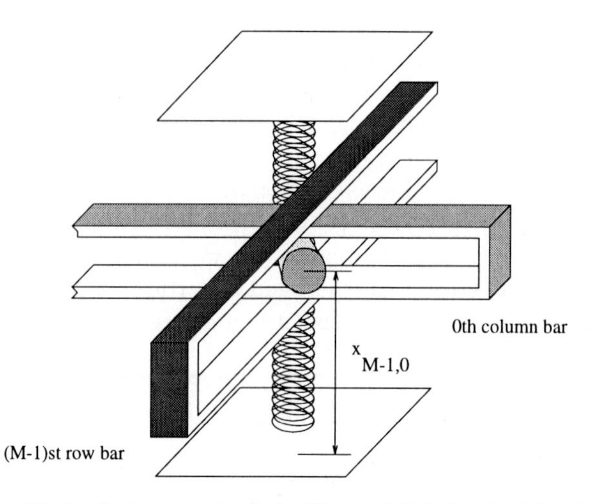

Figure 6 Mechanical structure of the Energy Minimization Model at pixel $(M-1, 0)$.

configuration represents the original image $\mathcal{X}$. After sliding in all the row and column bars whose rigidness (or flexibility) are determined by the parameters, $\{\lambda^2_{i,j}\}$ and $\{\lambda^3_{i,j}\}$, the configuration of the structure will change, and will reach the final stable configuration $\{y^*_{i,j}\}$ that has the minimum potential energy, i.e., $\mathcal{Y}^*$ is the solution of problem (9.12). Due to this analogy to the mechanical system, we refer to (9.12) as an *Energy Minimization Model* (EMM) problem.

To obtain the stressed image $\mathcal{X}^s$ from the original image $\mathcal{X}$, it suffices to solve the EMM problem with a proper parameter set Λ, namely, small ratios $\lambda^2_{i,j}/\lambda^1_{i,j}$ and $\lambda^3_{i,j}/\lambda^1_{i,j}$ at locations of strong edges and larger ratios at other places. However, the parameter set Λ cannot be specified a priori, since this requires the knowledge of the locations of strong edges – the very purpose of generating the stressed image.

We can now explain qualitatively an approach for solving this problem with the help of the above mechanical structure. Assume that in the process of forming the stressed image $\mathcal{X}^s$, the bars and springs first have uniform rigidness, and the stable configuration reached by the structure is a uniformly smoothed version, denoted by $\mathcal{Y}^1 \equiv \{(i, j, y^1_{i,j})\}$, of the original image $\mathcal{X}$. Then allow the bars to have more flexibility at locations corresponding to large pixel curvature energies (where the bars are most severely bent), and let the structure stabilize to a new configuration, denoted by $\mathcal{Y}^2 \equiv \{(i, j, y^2_{i,j})\}$. Again we measure the pixel curvature energies of this new configuration $\mathcal{Y}^2$, and allow a higher flexibility of the bars at places of large pixel curvature energies. The process is continued by repeating the above procedure. Notice that the textures are suppressed in the first configuration $\mathcal{Y}^1$, and will continue to be suppressed in later configurations, while the approximation of strong edges (corresponding to large pixel curvature energies) will become gradually better in later configurations as we change the flexibilities of the bars to better accommodate the strong edges.

To express the above idea mathematically, let us start with a uniform parameter set Λ, i.e., parameter $\lambda^k_{i,j} = \lambda^k$ for all (i,j), $k = 1,2,3$, and solve problem (9.12) to get $\mathcal{Y}^1$. Then let us update the parameter set Λ by changing the ratios, $\lambda^2_{i,j}/\lambda^1_{i,j}$ and $\lambda^3_{i,j}/\lambda^1_{i,j}$, according to $C^r_{i,j}$ and $C^c_{i,j}$ of $\mathcal{Y}^1$. The updating strategy should be such that large values of $C^r_{i,j}$ and $C^c_{i,j}$ give rise to small ratios, $\lambda^2_{i,j}/\lambda^1_{i,j}$ and $\lambda^3_{i,j}/\lambda^1_{i,j}$, respectively. In the algorithm proposed here, $\lambda^2_{i,j}/\lambda^1_{i,j}$ and $\lambda^3_{i,j}/\lambda^1_{i,j}$ are set to be inversely proportional to $C^r_{i,j}$ and $C^c_{i,j}$, i.e.,

$$\lambda^2_{i,j}/\lambda^1_{i,j} = \beta/C^r_{i,j} \text{ and } \lambda^3_{i,j}/\lambda^1_{i,j} = \beta/C^c_{i,j} \tag{9.16}$$

where β is a constant. We then solve problem (9.12) again with the new parameter set to obtain $\mathcal{Y}^2$ and repeat the above iteration until the relative variation of the objective function in (9.12) for the two consecutive iterations is smaller than a given threshold. The final solution $\mathcal{Y}^K$, where K is the total number of iterations, will be called a stressed image (previously denoted by $\mathcal{X}^s$) of the image $\mathcal{X}$.

4 EXTRACTION OF STRONG EDGE INFORMATION AND THE THREE-COMPONENT IMAGE MODEL

As indicated before, the strong edge information can be extracted by identifying pixels with large curvature energy in the stressed image $\mathcal{X}^s$, and is represented by $\mathcal{B}_T(\mathcal{X}^s)$ in (9.9). While generating $\mathcal{B}_T(\mathcal{X}^s)$ based on (9.9) is straightforward, the choice of the threshold T is crucial and its fine-tuning computationally expensive. To circumvent these difficulties, a further selection of the pixels in $\mathcal{B}_T(\mathcal{X}^s)$ is made by identifying pixels of large *local maximum* curvature energy. Specifically, the definition of edge brim points in (9.9) is modified as follows:

$$\begin{aligned}
(i,j,x^s_{i,j}) \quad & \in \mathcal{B}_T(\mathcal{X}^s) \\
\text{if} \quad & D^r_{i,j-1}D^r_{i,j} \le 0,\ D^r_{i,j}D^r_{i,j+1} \le 0, C^r_{i,j} > T; \\
\text{or if} \quad & D^r_{i,j-1}D^r_{i,j} > 0,\ D^r_{i,j}D^r_{i,j+1} \le 0, C^r_{i,j} > \max\{C^r_{i,j-1}, T\};
\end{aligned}$$

$$
\begin{array}{ll}
\text{or if} & D^r_{i,j-1}D^r_{i,j} \le 0,\ D^r_{i,j}D^r_{i,j+1} > 0, C^r_{i,j} > \max\{C^r_{i,j+1}, T\}; \\
\text{or if} & D^r_{i,j-1}D^r_{i,j} > 0,\ D^r_{i,j}D^r_{i,j+1} > 0, C^r_{i,j} > \max\{C^r_{i,j-1}, C^r_{i,j+1}, T\}; \\
\text{or if} & D^c_{i-1,j}D^c_{i,j} \le 0,\ D^c_{i,j}D^c_{i+1,j} \le 0, C^c_{i,j} > T; \\
\text{or if} & D^c_{i-1,j}D^c_{i,j} > 0,\ D^c_{i,j}D^c_{i+1,j} \le 0, C^c_{i,j} > \max\{C^c_{i-1,j}, T\}; \\
\text{or if} & D^c_{i-1,j}D^c_{i,j} \le 0,\ D^c_{i,j}D^c_{i+1,j} > 0, C^c_{i,j} > \max\{C^c_{i+1,j}, T\}; \\
\text{or if} & D^c_{i-1,j}D^c_{i,j} > 0,\ D^c_{i,j}D^c_{i+1,j} > 0, C^c_{i,j} > \max\{C^c_{i-1,j}, C^c_{i+1,j}, T\}.
\end{array}
\tag{9.17}
$$

The modified set $\mathcal{B}_T(\mathcal{X}^s)$ contains pixels not only of high curvature energy, but also of curvature energies larger than those of the neighboring pixels with the same sign of the second-order variations.

4.1 Generation of Edge Brim Contours

To complete the process of extracting strong edges, let us now consider the generation of edge brim contours from $\mathcal{B}_T(\mathcal{X}^s)$ described in (9.17). Define a contour as a sequence, $\overline{b} \equiv \{(i^k, j^k, x^s_{i^k,j^k}),\ k = 0, 1, \ldots, m-1\}$, such that

$$
|i^{k-1} - i^k| \le 1,\ |j^{k-1} - j^k| \le 1, \tag{9.18}
$$

and $(i^{k-1}, j^{k-1}) \ne (i^k, j^k)$, for $k = 1, \ldots, m-1$, and the *maximum-variation* of intensity values

$$
\sigma_{\overline{b}} \equiv \max_{k=0,1,\ldots,m-1} |x^s_{i^k,j^k} - \overline{x}| \le T_c, \tag{9.19}
$$

where $T_c \ge 0$, $\overline{x}$ is the average of the pixel intensity values on $\overline{b}$, and m is called the *length* of the contour. This definition is different from conventional contour definitions [43] - [45] in condition (9.19), which is introduced to prevent mixing of the upper and lower edge brims into one contour. The added constraint in (9.19) essentially breaks longer contour into segments, which leads to a problem similar to the straight-line-fitting problem in [29], with the exception that the straight-line-segments in [29] are replaced by constant-intensity curve-segments here.

The extraction algorithm for contours satisfying conditions (9.18) and (9.19) works as follows. Starting from a pixel $(i^0, j^0, x^s_{i^0,j^0}) \in \mathcal{B}_T(\mathcal{X}^s)$, in its eight neighboring pixels, search for pixels in $\mathcal{B}_T(\mathcal{X}^s)$, among which we choose a pixel that gives the minimum maximum-variation; if this maximum-variation is lower than T_c, we take this pixel as the next pixel $(i^1, j^1, x^s_{i^1,j^1})$ on the contour. We then move the search center to $(i^1, j^1, x^s_{i^1,j^1})$, delete $(i^0, j^0, x^s_{i^0,j^0})$ from $\mathcal{B}_T(\mathcal{X}^s)$, and repeat the above procedure again. The process stops when there is no pixel that is in $\mathcal{B}_T(\mathcal{X}^s)$ having a maximum-variation lower than T_c and is in the neighborhood of the current search center. This is the basic form of the algorithm.

Note that taking an arbitrary pixel in $\mathcal{B}_T(\mathcal{X}^s)$ to begin a contour tracing would only get a portion of this contour using the basic algorithm. To circumvent this problem, a flag is introduced so that when in the first search step there is more than one neighboring pixels in $\mathcal{B}_T(\mathcal{X}^s)$ giving a maximum-variation lower than T_c, this flag is set to 1; otherwise, it is set to 0. When this flag is 1, the algorithm first traces to one end of the contour and then traces back to the other end.

Due to the iterative nature of the algorithm for generating the stressed image, it is possible that a few pixels are lost on some edge brim contours for a given threshold T_c. Some of the missing pixels may be recovered by a predictive process based on the information of the generated contours [41]. Having determined all contours with the above algorithm, those contours of length smaller than a certain threshold T_ℓ can be deleted. Since strong edges are represented by two contours, namely, lower and upper edge brims, this pairing property can be used to further reduce spurious contours. More precisely, let us define the d-neighborhood of a pixel $(i, j, x^s_{i,j})$ as $\{(i, j, \cdot) : \max(|i - i^k|, |j - j^k|) \leq d\}$, where $d \geq 0$. We say that a pixel in $\mathcal{B}_T(\mathcal{X}^s)$ is paired in distance d if its d-neighborhood contains at least one pixel on another contour in $\mathcal{B}_T(\mathcal{X}^s)$, and that a contour in $\mathcal{B}_T(\mathcal{X}^s)$ is paired in distance d if it has at least T_ℓ pixels paired in distance d on another contour. Those contours not paired in distance d are then deleted from $\mathcal{B}_T(\mathcal{X}^s)$. We will denote the final set of pixels on the remaining contours by $\mathcal{B}_T(\mathcal{X}^s)$ again, and will refer to it as the collection of extracted strong edge brims.

4.2 The Three-Component Decomposition

Let us now define an image $\mathcal{P} = \{(i, j, p_{i,j}),\ i, j = 0, 1, ..., M - 1\}$, solely from $\mathcal{B}_T(\mathcal{X}^s)$, such that the difference image $\mathcal{X} \ominus \mathcal{P} \equiv \{(i, j, x_{i,j} - p_{i,j})\}$ has no strong edges represented by $\mathcal{B}_T(\mathcal{X}^s)$. The image $\mathcal{P}$ is generated with the minimum information principle introduced in Section 2 (9.3) and is called the *primary image* (or *primary component*) of the original image $\mathcal{X}$, since the strong edge information contained in image $\mathcal{P}$ is most important to the HVS.

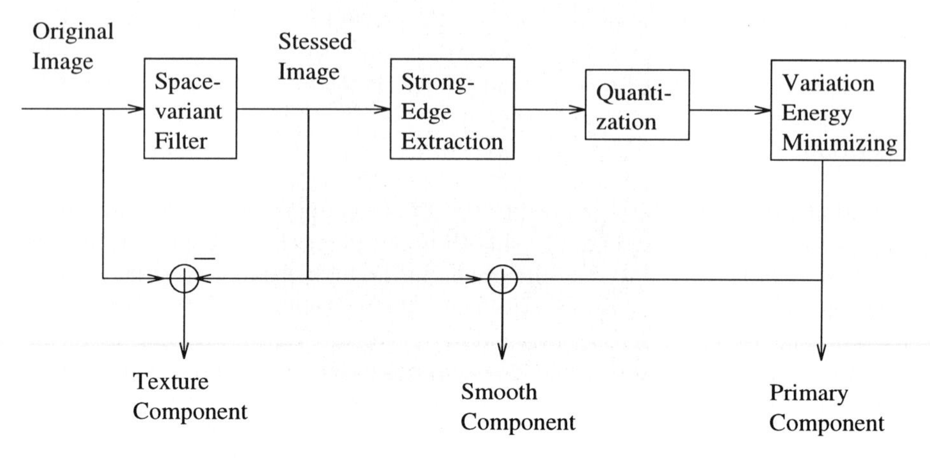

Figure 7 Block diagram for generating the three-component image model.

The stressed image $\mathcal{X}^s$, from which the set $\mathcal{B}_T(\mathcal{X}^s)$ is derived, can be considered as the output of a space-variant filter with input $\mathcal{X}$ [31]. This filter is basically low-pass except at locations of strong edges. Thus the stressed image $\mathcal{X}^s$ contains the low-frequency (smooth) component, and the strong edge information of the image $\mathcal{X}$. Therefore, the difference image $\mathcal{X} \ominus \mathcal{X}^s$ consists of the high-frequency component, but without the strong edge information. We define $\mathcal{T} \equiv \mathcal{X} \ominus \mathcal{X}^s = \{(i,j,t_{i,j})\} = \{(i,j,x_{i,j} - x^s_{i,j})\}$ and $\mathcal{S} \equiv \mathcal{X}^s \ominus \mathcal{P}$ $= \{(i,j,s_{i,j})\} = \{(i,j,x^s_{i,j} - p_{i,j})\}$ as the texture and smooth components, respectively. Thus, the image $\mathcal{X}$ can be expressed as

$$\mathcal{X} \equiv \mathcal{T} \oplus \mathcal{S} \oplus \mathcal{P} \equiv \{(i,j,t_{i,j} + s_{i,j} + p_{i,j})\}. \tag{9.20}$$

In the sequel, this model will be referred to as the *three-component image model.* The generation of a three-component image model is summarized in the block diagram shown in Fig. 7.

Now let us look at an example of a three-component image model. The test image is of the size 256 × 256 and is shown in Fig. 8 with a circular strong edge in the middle and textures and smooth areas in the rest of the image. Two typical edge segments along the circular strong edge are shown in Fig. 9. Note that the width of the edge is not limited to one pixel in contrast to the restrictions in [27], [28].

For simplicity, it is assumed, in the generation of the stressed image, that $\lambda^2_{i,j} = \lambda^3_{i,j} \equiv \lambda_{i,j}$, for all (i,j); thus, $\lambda_{i,j}/\lambda^1_{i,j}$ is the only independent parameter, and instead of updating $\lambda^2_{i,j}$, $\lambda^3_{i,j}$, $\lambda^1_{i,j}$ is updated with the formula, $\lambda^1_{i,j} =$

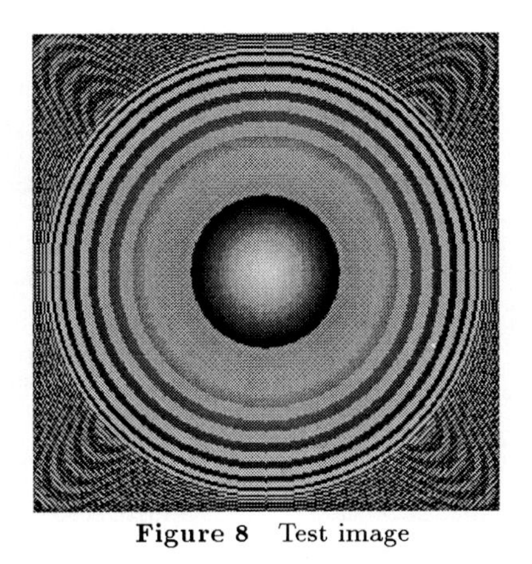

Figure 8 Test image

$\gamma(C^r_{i,j} + C^c_{i,j})$, where γ is a constant. The resultant stressed image is shown in Fig. 10 (a). Edge brim contours are extracted with $T = 512$, $T_c = 32$, $T_\ell = 16$ and $d = 3$. The resulting two contours are shown in Fig. 10 (b).

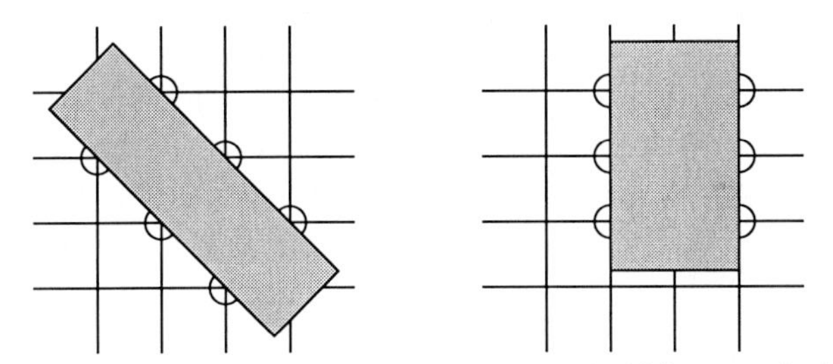

Figure 9 Two typical strong edge segments, one of width 1.414 pixel and one of width 2 pixels, in the test image.

The primary image associated with the extracted strong edge is shown in Fig. 11 (a). The other two components, namely, the smooth and texture components, are shown in Fig. 11 (b) and (c). Notice that the smooth and texture components are differences of two images, and thus may have negative intensity values. Constants are added to these two components to bring their intensity values into the displayable range. As expected, the primary image contains the strong edge information, the smooth component provides the background slow intensity variations and the texture component contains all the textures. The edge extracting part of the above example is compared in [31] with the Laplacian-Gaussian operator scheme [27]. The Laplacian-Gaussian operator scheme gives location error, and, more importantly, fails to provide the correct intensity variation of the extracted edge.

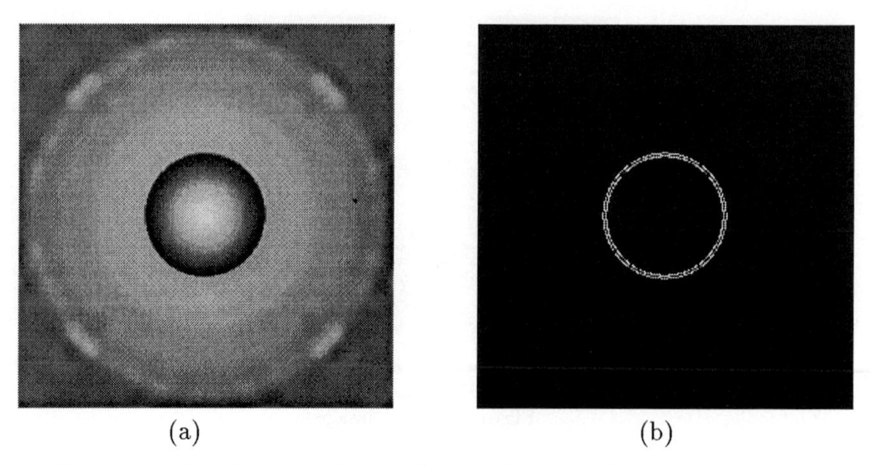

(a) (b)

Figure 10 (a) The stressed image, (b) the picture of edge brim contours.

Some traces of bright pixels can be noticed around the location of the circular edge in the texture image. The reason for this phenomenon is basically the orientation anisotropy of the EMM problem since there are only two directions for the bars and the discrete structure of the problem. This orientation anisotropy can be seen most clearly around edge segments of an orientation slightly tilted away from the vertical or the horizontal directions. Since the primary image gives a perceptually good copy of the edge, these traces of brighter pixels can be treated as textures.

Starting from the next section, we concentrate on the applications of the proposed three-component image model in image compression.

5 CODING OF PRIMARY COMPONENT

In coding the primary component, the objective is to reconstruct the location and intensity of edge brim contours with little or no perceptual distortion. To this end, the sequence of contour pixel locations, $\overline{c} \equiv \{(i^0, j^0), (i^1, j^1), \ldots, (i^{m-1}, j^{m-1})\}$, is encoded by using chain coding [45], [43], [46] and [47]. When the well-known Freeman code [45] (also known as the 1-ring code [46]) is used, the sequence is losslessly encoded. To achieve lower bit rates, the N-ring code with $N > 1$ can be used. This procedure results in lossy coding of the contour locations. An example of the 2-ring code is given in [58].

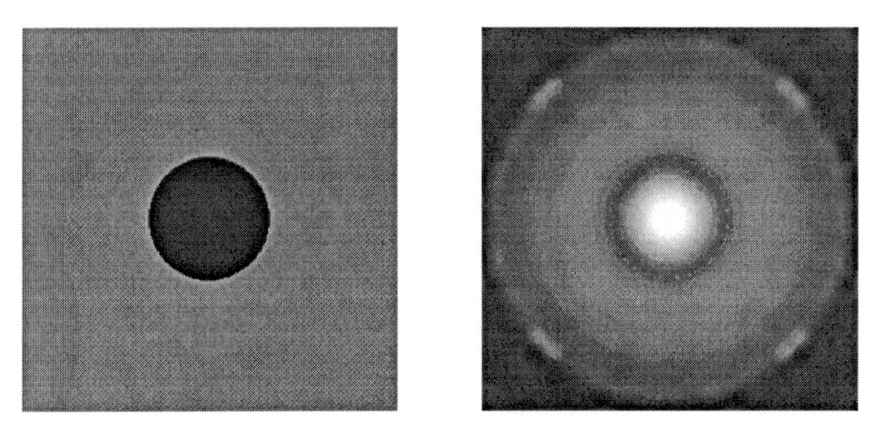

(a) (b)

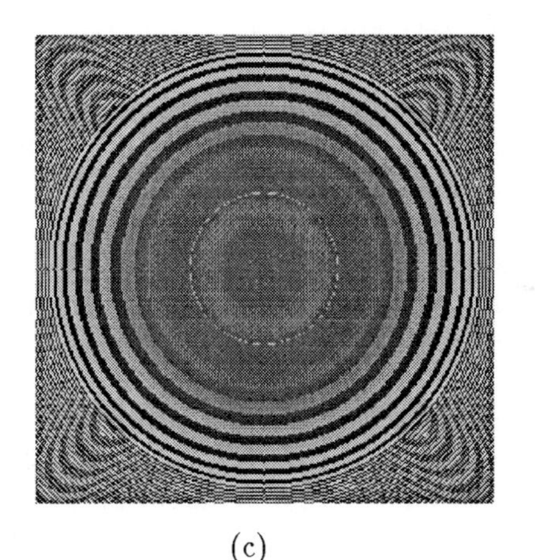

(c)

Figure 11 (a) The primary image, (b) the smooth component, (c) the texture component of the test image.

Let $\{s^1, s^2, s^3, \ldots\}$ denote the sequence of indices on the N-ring obtained in successive steps of the chain coding operation. Because the contours are generally smooth, it is more efficient to code the differences between the neighboring indices than to code them directly. Therefore, upon defining $r^i = (s^i - s^{i-1}) \bmod 8N$, $i = 2, 3, \ldots$, a code sequence $\{(i^0, j^0), s^1, r^2, r^3, \ldots\}$, is generated where (i^0, j^0) is called the starting location, s^1 the first direction, and $r^2, r^3, \ldots$, the relative directions. To encode the contour intensity values, the sequence $\{x_{i^0,j^0}, x_{i^1,j^1}, \ldots, x_{i^{m-1},j^{m-1}}\}$ is quantized to a constant sequence $\{\overline{x}, \overline{x}, \ldots, \overline{x}\}$, where $\overline{x} = [m^{-1} \sum_{\ell=0}^{m-1} x_{i^\ell,j^\ell}]$ and $[x]$ denote the integer closest to

Configuration	image	1-ring	2-ring	3-ring
A	512 × 512 Lenna	0.048	0.038	0.032
B	512 × 512 Lenna	0.017	0.013	0.011
A	256 × 256 Lenna	0.020	0.017	0.014

Table 1 Bit rate (bits/pixel) for encoding the primary components.

x. Simulation results show that this quantization gives rise to little perceptual degradation on edges as compared with the case where the intensity values are losslessly coded. The sequence $\{\overline{x}, m, (i^0, j^0), s^1, r^2, r^3, \ldots\}$ is used to represent each strong edge brim contour. While $\overline{x}, m$ and (i^0, j^0) are encoded by a fixed number of bits, s^1 and $\{r^i\}$ are encoded separately by means of arithmetic coding [48], [49].

Two examples of primary components are provided in the following along with their companion components for the 512 × 512 Lenna. Two different sets of parameters are used: The A-configuration with $N_\nu = 20$, $T = 64$, $T_c = 32$, $T_\ell = 8$ and $d = 3$ and the B-configuration with $N_\nu = 10$, $T = 64$, $T_c = 32$, $T_\ell = 4$ and $d = 3$. Here, N_ν is the number of iterations used to generate the stressed image.

The original image, the stressed image, the strong edge brim contours and the resulting three components are shown in Figs. 12 and 13 for the two examples. Again, since the smooth and texture components can have negative values, for display purposes, appropriate constants are added to these components to render all pixel values non-negative. In both cases, the primary component is obtained from the quantized strong edge information and the corresponding quantization distortion is included in the smooth component [31]. These distortions can be noticed near the strong edges in Figs. 12(e) and 13(e). Note that the A-configuration results in more strong edges than the B-configuration due to its larger N_ν. Generally, larger values of N_ν lead to extracting a larger number of strong edges [41]. Additional 3-component decomposition examples can be found in [41] and [58]. The average rate for encoding the primary component is denoted by r_p bits/pixel and tabulated in Table 1.

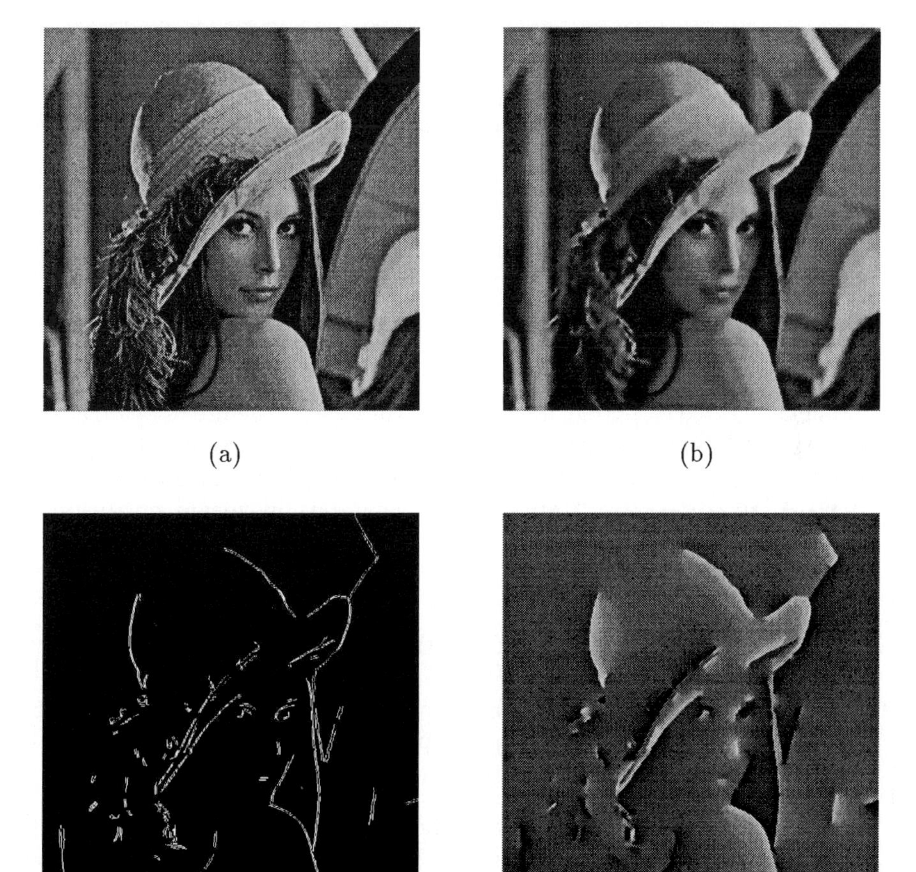

(a) (b)

(c) (d)

Figure 12 (a) Lenna image, (b) stressed image, (c) strong edge brim contours, (d) primary component (1-ring coding).

6 CODING OF SMOOTH AND TEXTURE COMPONENTS

Let us now consider the coding of the smooth and texture components. The basic approach is to encode these components using waveform coding techniques. Two different entropy-coded image coding techniques are used; they are adaptive (2-D) DCT coding and 2-D subband coding.

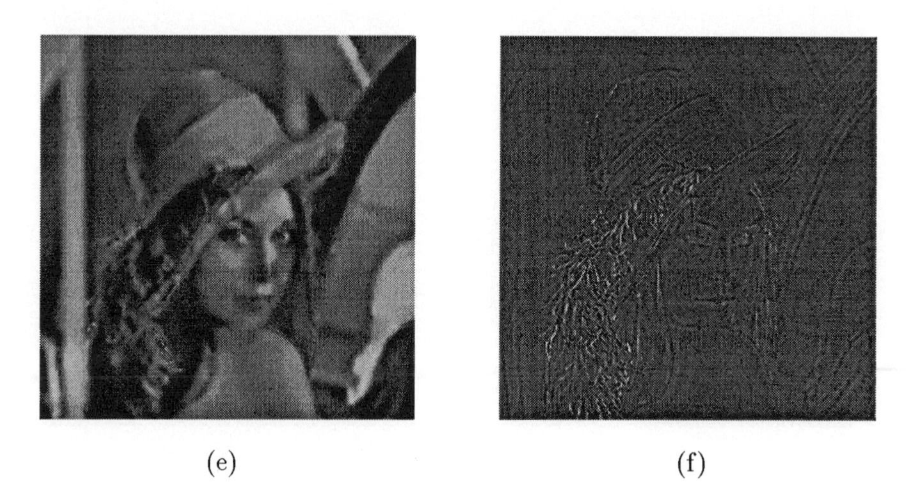

(e) (f)

Figure 12 *(continued)*
(e) smooth component and (f) texture component associated with A-configuration.

6.1 Adaptive DCT Coding

Let the smooth and texture components of the original image $X = \{x_{i,j}\}$ be denoted by $S = \{s_{i,j}\}$ and $T = \{t_{i,j}\}$, respectively, where $i, j = 0, 1, \ldots, M - 1$. Let S and T be segmented into $L \times L$ blocks $\mathbf{s}_{m,n}$ and $\mathbf{t}_{m,n}$, $m, n =$

(a) (b)

Figure 13 (a) Lenna image, (b) stressed image.

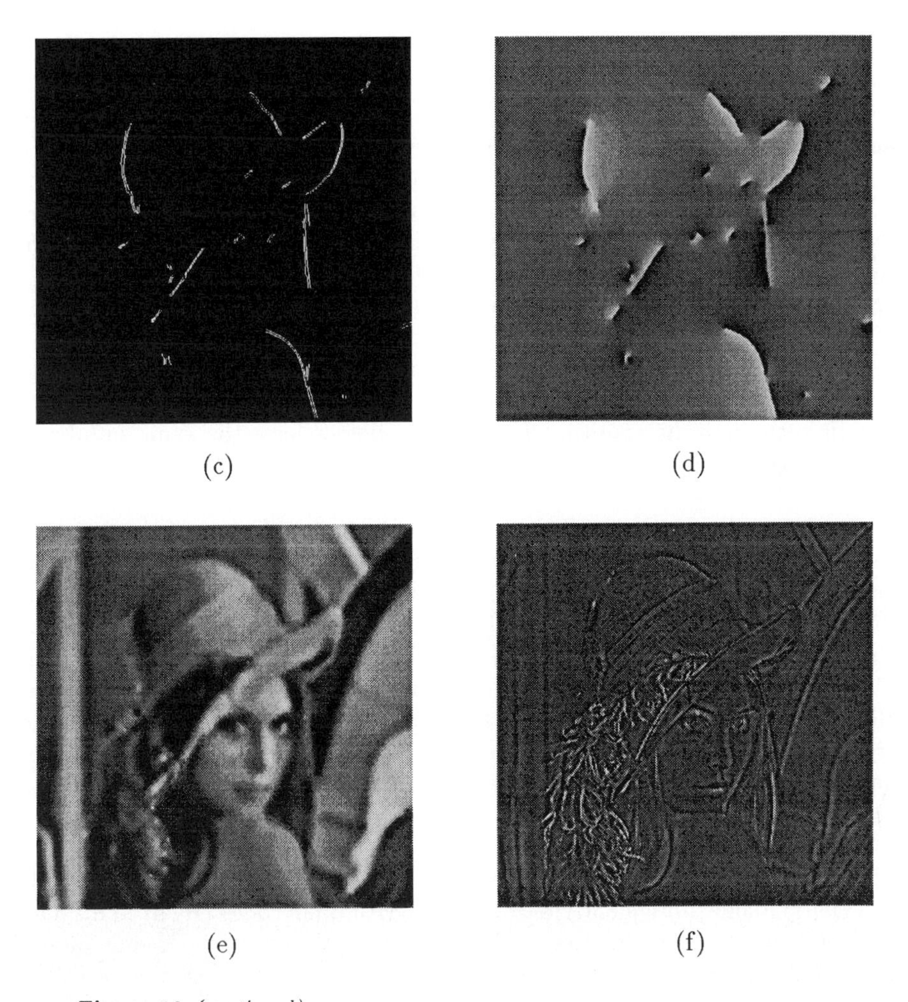

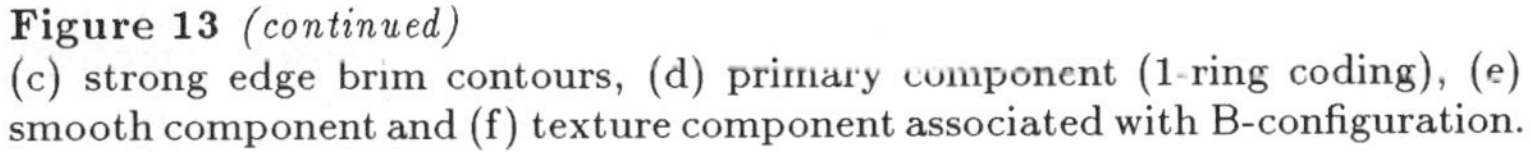

Figure 13 *(continued)*
(c) strong edge brim contours, (d) primary component (1-ring coding), (e) smooth component and (f) texture component associated with B-configuration.

$0, 1, \ldots, (M/L) - 1$, respectively, and denote the block of 2-D DCT coefficients of $\mathbf{s}_{m,n}$ and $\mathbf{t}_{m,n}$ by $\overline{\mathbf{s}}_{m,n}$ and $\overline{\mathbf{t}}_{m,n}$, respectively.

The structure of the adaptive DCT coding scheme used here is similar to that in [2]. After computing the 2-D DCT of $\mathbf{u}_{m,n} \equiv (\mathbf{s}_{m,n} + \mathbf{t}_{m,n})$, the blocks $\overline{\mathbf{u}}_{m,n} \equiv (\overline{\mathbf{s}}_{m,n} + \overline{\mathbf{t}}_{m,n})$ are classified into one of a finite number of classes. In [2], the classification is done based on the ac energies of the blocks. As indicated in

[50], the classification scheme in [2] neglects the frequency distribution of the ac energy which should be considered for a more efficient coding. Here, due to the three-component model, the low and high frequency energies of a block are already separated and are contained in $\overline{\mathbf{s}}_{m,n}$ and $\overline{\mathbf{t}}_{m,n}$, respectively. Thus, a more efficient classification is possible by using the ac energies of $\overline{\mathbf{s}}_{m,n}$ and $\overline{\mathbf{t}}_{m,n}$. The classification procedure used here consists of two stages. In the first stage, the transform blocks are classified into one of K_1 classes by comparing the ac energies of $\overline{\mathbf{s}}_{m,n}$ against T_k^1, $k = 0, 1, \ldots, K_1 - 2$; the thresholds $\{T_k^1\}$ are chosen such that each of the resulting K_1 classes contains the same number of blocks. In the second stage, each resulting class of the first stage, say k, is further divided into K_2 classes by comparing the ac energies of $\overline{\mathbf{t}}_{m,n}$ against another set of thresholds, $T_{k,\ell}^2$, $\ell = 0, 1, \ldots, K_2 - 2$. These thresholds are also chosen such that the resulting $K = K_1 K_2$ classes have the same number of blocks in them.[5]

In a study similar to that of [51], the approximate distribution of the 2-D DCT coefficients of the different classes are determined by comparing their empirical distributions against the so-called generalized Gaussian distribution [52] using the Kolmogorov-Smirnov test [51]. This is done using a database consisting of a large number of different images. Based on this study, it was concluded that the $(0, 0)$th, $(0, 1)$th and $(1, 0)$th coefficients have an almost Gaussian distribution; all other coefficients (except those that have very small variances and hence are subsequently assigned very small bit rates) are best approximated by a Laplacian distribution. This observation holds for the coefficients in all classes in the case where $K = 4$.

The 2-D DCT coefficients are quantized with uniform-threshold quantizers (UTQs) [52] and subsequently encoded using Huffman codes (HCs). This leads to a performance close to that of optimal entropy-constrained scalar quantization [13]. Let us denote the variance-normalized mean squared error (MSE) associated with the UTQ-HC pair operating at rate r bits/sample for the Gaussian and Laplacian distributions by $d_G(r)$ and $d_L(r)$, respectively. If the variance and the coding rate associated with the (u, v)th coefficient in the kth class

[5] The constraint that each class should contain the same number of blocks is not a requirement; it just makes the system simpler to implement. Improved performance can be expected if this constraint is removed.

are denoted, respectively, by $\sigma_k^2(u,v)$ and $r_{u,v}^k$, the overall MSE is given by[6]

$$D_{st} = \frac{1}{L^2}\{\sigma^2(0,0)d_G(r_{0,0}) + \frac{1}{K}\sum_{k=0}^{K-1}[\sigma_k^2(0,1)d_G(r_{0,1}^k) + \sigma_k^2(1,0)d_G(r_{1,0}^k) + \sum_{(u,v)\neq(0,0),(0,1),(1,0)} \sigma_k^2(u,v)d_L(r_{u,v}^k)]\}, \quad (9.21)$$

where $\sigma^2(0,0)$ and $r_{0,0}$ are the variance and the coding rate, respectively, for the dc coefficient which is encoded in the same manner regardless of the class it belongs to. The average bit rate for encoding the smooth and texture components, r_{st}, is given by

$$r_{st} = r_o + \frac{1}{L^2}\{r_{0,0} + \frac{1}{K}\sum_{k=0}^{K-1}\sum_{u,v\neq(0,0)} r_{u,v}^k\}, \text{ bits/pixel}, \quad (9.22)$$

where r_o is the bit rate for coding the overhead information.

At this point, it remains to determine the optimal allocation of bit rates among the 2-D DCT coefficients of the different classes. This bit allocation which is the solution of the following constrained minimization problem,

$$\min_{\{r_o, r_{0,0}, r_{u,v}^k\}:\ r_{st}\leq r_d} D_{st}, \quad (9.23)$$

where r_d is the design average bit rate, can be obtained efficiently using the steepest descent algorithm described in [53]. To reduce the overhead information for the bit allocation maps and to have a simple system, $r_{0,0}$ and $\{r_{u,v}^k\}$ are limited to be an integer multiple of 0.1 bits; they are also constrained to be less than 5.0 bits for the Laplacian distribution and 8.0 bits for the Gaussian distribution.

To reconstruct the DCT coefficients at the receiver side, estimates of $\{\sigma_k^2(u,v)\}$ are required ($\sigma^2(0,0)$ is transmitted directly). If the variances of quantization errors of the DCT coefficients, denoted by $\epsilon_k^2(u,v)$, are known, then $\{\sigma_k^2(u,v)\}$ can be estimated from the allocated bit rates $\{r_{u,v}^k\}$, since, for $(u,v)\neq(0,0)$,

$$\epsilon_k^2(u,v) = \begin{cases} \sigma_k^2(u,v)d_G(r_{u,v}^k), & (u,v)=(0,1) \text{ or } (1,0), \\ \sigma_k^2(u,v)d_L(r_{u,v}^k), & \text{otherwise}. \end{cases} \quad (9.24)$$

[6]Throughout this paper, the 2-D DCT used is as defined in [18]. This transformation is unitary and hence the MSE in the spatial domain is the same as that in the transform domain.

As mentioned before, the performance of the UTQ-HC pairs is very close to that of optimal entropy-constrained scalar quantizers. These quantizers, in turn, have a performance which for high bit rates is about 0.255 bits/sample above the rate-distortion function [52], [54]. Approximating the rate-distortion function by the Shannon Lower Bound [55] and using this observation, we have

$$d_G(r) \approx \frac{\pi e}{6} 2^{-2r}, \qquad d_L(r) \approx \frac{e^2}{6} 2^{-2r}. \tag{9.25}$$

Substituting (9.25) into (9.21) to solve (9.23), one can easily show that when the bit rates are optimally chosen, the quantization error of the DCT coefficients with a positive bit rate have equal variances. In the ideal case if the distribution of the coefficients is exactly what has been assumed, the knowledge of this common quantization error variance, called the normalization factor and denoted by c, at the receiver, can be used to compute the variances $\{\sigma_k^2(u,v)\}$ using the bit maps (9.24). In the system actually implemented, the average of those $\epsilon_k^2(u,v)$ corresponding to coefficients with $r_{u,v}^k > 0$, $(u,v) \neq (0,0)$ is used as the value of c.

The overhead information that needs to be sent consists of the classification information ($\log_2 K$ bits per block), the mean and variance of the dc coefficient (32 bits for each), the normalization factor c (32 bits) and the bit maps[7] $(2K\lceil \log_2 L \rceil + \lceil \log_2 B \rceil (1 + \sum_{k=0}^{K-1}(u_k v_k - 1)))$. For an $M \times M$ image, this amounts to $r_o = (\log_2 K)/L^2 + (96 + 2K\lceil \log_2 L \rceil + \lceil \log_2 B \rceil (1 + \sum_{k=0}^{K-1}(u_k v_k - 1)))/M^2$, bits/pixel: for example, for encoding the 512×512 Lenna at 0.5 bits/pixel with $L = 16$, $r_o = 0.017$ bits/pixel, $r_p = 0.048$ bits/pixel with 1-ring coding and the A-configuration leaving 0.435 bits/pixel for r_{st}.

6.2 Subband Coding

The entropy-coded 2-D subband coding (SBC) system used here for encoding the sum of the smooth and texture components (hereafter called the smooth-texture component) is identical to that of [13]. The smooth-texture component is analyzed into 16 subband images which are then encoded separately. In the decoder, a replica of the smooth-texture component is synthesized from

[7] The integers u_k and v_k are defined to be the smallest integers such that $u \geq u_k$ or $v \geq v_k$ implies $r_{u,v}^k = 0$. Thus, for the kth class, we need to transmit $2\lceil \log_2 L \rceil$ bits for u_k and v_k, and $\lceil \log_2 B \rceil (u_k v_k - 1)$ bits to specify $r_{u,v}^k$, $u < u_k$, $v < v_k$, $(u,v) \neq (0,0)$, where $\lceil x \rceil$ denotes the smallest integer larger than x and B is the number of possible different values of $r_{u,v}^k$. Since $r_{0,0}$ is the same for all classes, the overall number of bits for the bitmaps is $2K\lceil \log_2 L \rceil + \lceil \log_2 B \rceil (1 + \sum_{k=0}^{K-1}(u_k v_k - 1))$.

the decoded subband images. The analysis and synthesis are performed using 2-D separable quadrature mirror filter (QMF) banks. The reader is referred to [10] for details. All subbands except the lowest frequency subband (LFS) are encoded by means of appropriately designed UTQ-HC pairs; the LFS is encoded by means of a non-adaptive 2-D DCT encoding system in which the DCT coefficients are encoded also by UTQ-HC pairs. The overall system design and the bit allocation procedure are described in [13].

Perhaps we should mention that in the context of SBC, it suffices to have a two-component model which separates the primary component from the smooth-texture component. This is because in a SBC system, the input image is separated into subband images based on their frequency contents and therefore the system has the ability to extract the smooth and texture components from the smooth-texture input. In the adaptive DCT coding system introduced in the last subsection, however, the three-component model is needed for an efficient block classification and the corresponding bit allocations.

7 IMAGE CODING SYSTEMS

In this section let us look at the overall structure of the different image coding systems developed based on the three-component model.

7.1 The ADCT-based scheme

The block diagram of the ADCT-based scheme is illustrated in Fig. 14. The primary component is encoded using the contour coding scheme of Section 5. The sum of the smooth and texture components is encoded using the ADCT coding scheme of Subsection 6.1. At the decoder, a replica of the original image is obtained by adding the reconstructed versions of the primary and smooth-texture components. This scheme is referred to as the 3C-ADCT-HC scheme, where 3C stands for the three-component model and HC for Huffman coding of the quantized DCT coefficients.

7.2 The SBC-based scheme

The block diagram of the SBC-based scheme is illustrated in Fig. 15. The only difference with 3C-ADCT-HC is in the coding of the smooth and texture

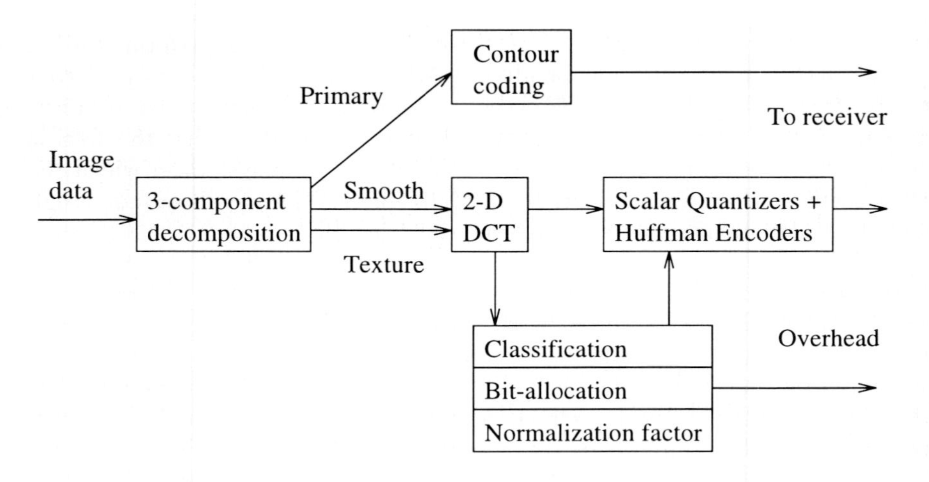

Figure 14 Block diagram of the ADCT-based image coding scheme using the three-component model.

components which in this case are encoded using the SBC coding scheme of Subsection 6.2. This scheme will be denoted by 3C-SBC-HC.

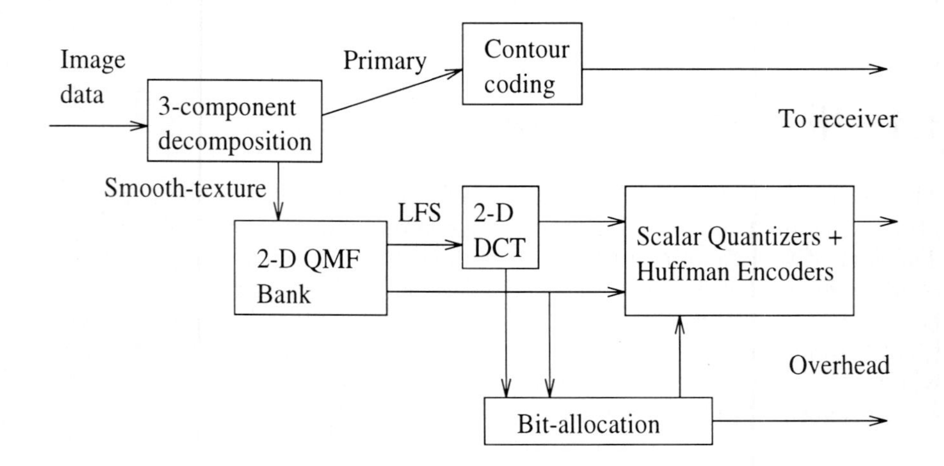

Figure 15 Block diagram of the SBC-based image coding scheme using the three-component model.

7.3 Other related schemes

For comparison purposes, we also studied the performance of an ADCT-based scheme like that of Subsection 6.1 which operates on the original image directly and hence does not utilize the three-component model; in this scheme the classification method is exactly like that of [2]. This scheme, referred to as 1C-ADCT-HC, is essentially an entropy-coded version of the system in [2]. The other differences between 1C-ADCT-HC and the scheme in [2] are in the bit assignment algorithm, the assumption on the distribution of the transform coefficients, and the estimation of $\{\sigma_k^2(u,v)\}$ at the receiver.

Also, we considered the SBC-based coding scheme of [13] (called System B in [13]) directly applied to the image (thus ignoring the three-component model) and refer to it as 1C-SBC-HC. Finally, the Huffman coded version of JPEG is referred as JPEG-HC. All JPEG simulation results are obtained using a software package described in [5].

As an alternative to the Huffman coding, we have also examined arithmetic coding of the quantized DCT coefficients in all DCT-based schemes considered. The arithmetic coded counterpart of JPEG-HC, 1C-ADCT-HC and 3C-ADCT-HC are denoted, respectively, as JPEG-AC, 1C-ADCT-AC and 3C-ADCT-AC. For the theory behind arithmetic coding, its operation and advantages the reader is referred to [48], [49].

7.4 Simulation results

The performance of the above mentioned schemes are summarized for the 512×512 Lenna in Table 2 and in Fig. 16. Additional sets of simulation results can be found in [41] and [58]. For these simulations, in the 1C-ADCT-HC(AC) and 3C-ADCT-HC(AC) schemes the block size used is 16×16; JPEG-HC(AC) uses an 8×8 block size. In 1C-SBC-HC(AC) and 3C-SBC-HC(AC), the lowest frequency subband is encoded using a nonadaptive DCT encoder with block size 4×4 as in [13]. In the 3C-ADCT-HC(AC) schemes, $K_1 = K_2 = 2$; in the 1C-ADCT-HC(AC) schemes, $K = 4$.

In Table 2, the PSNR results associated with the three-component model are those obtained based on 1-ring codes; both A- and B-configuration results are included. In some cases, 2- or 3- ring codes yielded slightly better PSNRs, but in no case did we observe a noticeable perceptual difference. At low rates, the B-configuration results in slightly better PSNR.

Encoding Scheme	Configura.	Design Bit Rate (bpp)		
		0.125	0.25	0.5
JPEG-HC		24.91 (0.148)	31.25 (0.283)	34.69 (0.513)
1C-ADCT-HC		30.10 (0.126)	32.92 (0.247)	35.91 (0.485)
3C-ADCT-HC	A	29.50 (0.128)	33.14 (0.252)	36.22 (0.498)
	B	30.11 (0.127)	33.29 (0.250)	36.25 (0.492)
1C-SBC-HC		29.77 (0.125)	32.55 (0.250)	35.67 (0.500)
3C-SBC-HC	A	29.46 (0.130)	32.46 (0.254)	35.61 (0.504)
	B	29.50 (0.127)	32.46 (0.250)	35.73 (0.496)
JPEG-AC		28.45 (0.128)	32.08 (0.265)	34.95 (0.491)
1C-ADCT-AC		30.10 (0.123)	32.92 (0.241)	35.91 (0.471)
3C-ADCT-AC	A	29.52 (0.125)	33.14 (0.247)	36.23 (0.487)
	B	30.11 (0.123)	33.29 (0.244)	36.25 (0.479)

Table 2 PSNR (in dB) performance of various encoding schemes for 512 x 512 Lenna at different design bit rates. Numbers in parentheses indicate actual bit rates.

The reconstructed images for the results in Table 2 are omitted here due to space limit, but can be found in [41], [58]. The subjective and the objective performance improvements of 1C(3C)-ADCT-HC(AC) and 1C(3C)-SBC-HC over JPEG-HC(AC) are quite significant, especially at low bit rates. In comparing the subjective performances of 1C-ADCT-HC(AC) and 3C-ADCT-HC(AC), it became evident that 3C-ADCT-HC performs better near the strong edges; the difference is quite visible at low bit rates [41], [58]. An additional interesting observation is that even though 3C-ADCT-HC is designed to provide good perceptual quality and not to maximize PSNR, it, in fact, resulted in a PSNR larger than that of 1C-ADCT-HC – a system designed to maximize the PSNR. Additional studies have revealed that the PSNR performance improvement of 3C-ADCT-HC(AC) over 1C-ADCT-HC(AC) (for the same block size) are primarily due to the more efficient 2-stage block classification used in 3C-ADCT-HC(AC) and not because of the separate encoding of the primary component [41], [58]. However, we should mention that the separate encoding of the primary component is responsible for the improved perceptual quality of 3C-ADCT-HC near the strong edges.

Additionally, we made comparisons in Fig. 16 to the constant block distortion ADCT (CBD-ADCT) of [4], the 2-D adaptive entropy coded SBC (2-D ECSBC)

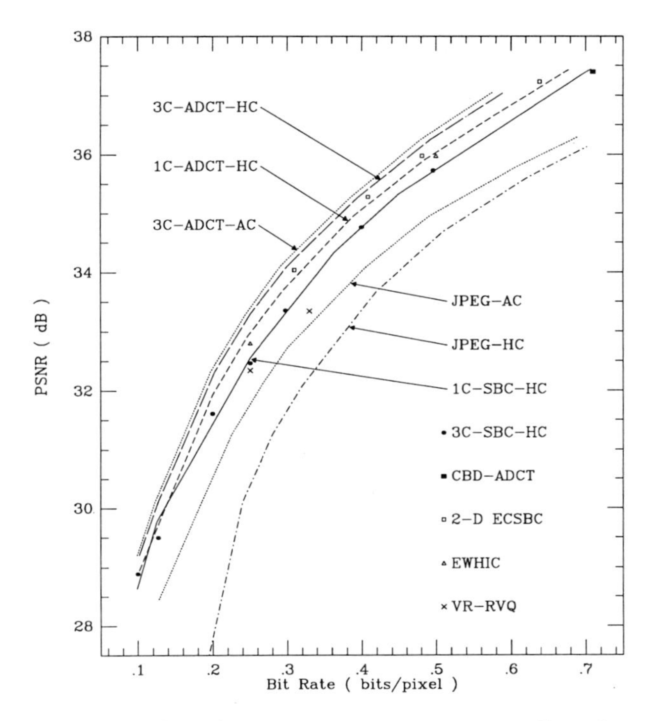

Figure 16 PSNR (in dB) versus bit rate for various encoding schemes; 512 × 512 Lenna.

of [14] (System C with arithmetic coding in [14]), the embedded wavelet hierarchical image coder (EWHIC) of [56] and the variable-rate residual vector quantizer (VR-RVQ) of [57].

It was observed that despite the superior strong edge reconstruction of 3C-ADCT-HC, at very low bit rates, 0.125 bits/pixel, 1C-ADCT-HC(AC) produces less blockiness in smooth areas than does 3C-ADCT-HC(AC), simply because the portion of the overall bit rate spent on encoding the strong edge contours becomes significant in these cases. This problem can be alleviated by using a perceptual weighting of distortion in conjunction with the three-component image model as described in Section 8.

In comparing 1C-SBC-HC against 3C-SBC-HC, the significant reduction of "ringing" near the strong edges in 3C-SBC-HC was observed; these ringing effects, which are well-known in subband coding systems, were quite visible in the 1C-SBC-HC scheme especially at low bit rates [41], [58]. The PSNR improvement of 3C-SBC-HC over 1C-SBC-HC is almost negligible, referring Table 2.

8 PERCEPTUAL WEIGHTING OF DISTORTIONS

In this section, in order to achieve high perceptual quality at low bit rates, a method is developed, which integrates the three-component ideas with the contrast sensitivity [34]-[36] of the HVS. The contrast sensitivity approach is applied to the ADCT-based schemes of Subsection 7.1 only; similar ideas have been used in a subband coding framework in [59].

To determine the contrast sensitivity of the HVS, a test stimulus of the form of a grating pattern (vertical bars with sinusoidal-intensity scanlines) is presented to an observer [34]. The sine wave has mean $\overline{m}$, amplitude a and frequency f, defined as the reciprocal of the angle, subtended at the observer's eye, of one complete cycle, in cycles/degree (c/deg). It is shown that for a given spatial frequency f, the visibility threshold of the pattern depends primarily on $a/\overline{m}$, and not separately on a and $\overline{m}$. Upon defining $a/\overline{m}$ as the *contrast* of the grating, the *threshold contrast* is defined as the contrast below which the grating pattern is barely detectable. The reciprocal of the threshold contrast at frequency f is called the *contrast sensitivity* at f and denoted by $m(f)$. The contrast sensitivity curve was measured and presented in [34]. We have approximated this curve by the parametric expression $m(f) \approx a \times b^{(\log(\frac{f}{c}))^d}$. The optimum values of the parameters (in the sense of minimizing the squared fitting error for $f > 2$ c/deg) are determined to be: $a = 621.31$, $b = 0.14$, $c = 1.73$ and $d = 1.83$ [41]. In the sequel, we abuse notation slightly and use $m(f)$ to denote the least-squares fit to the empirical contrast sensitivity curve.

Now let us use the contrast sensitivity of the HVS for a perceptual tuning of the ADCT-based image coding scheme of Subsection 7.1. Consider a generic block of 2-D DCT coefficients, $\{d(u,v);\, u,v = 0,1,\ldots,L-1\}$. Let the error in quantizing $d(u,v)$ be denoted by $e(u,v)$. Then the resulting error in the spatial domain is given by [18]

$$\frac{2}{L}\alpha(u)\alpha(v)e(u,v)\cos\frac{\pi u(2i+1)}{2L}\cos\frac{\pi v(2j+1)}{2L}, \quad i,j = 0,1,\ldots,L-1, \quad (9.26)$$

where $\alpha(0) = 1/\sqrt{2}$ and $\alpha(u) = 1,\ u \neq 0$. In the special case where $d(u,v) = 0$, for all $(u,v) \neq (0,0)$, and $e(u,v) = 0$, for all $(u,v) \neq (0,v_1)$, where $v_1 \neq 0$, the corresponding block in the spatial domain is given by

$$\frac{1}{L}(d(0,0)+\sqrt{2}e(0,v_1)\cos\frac{\pi v_1(2j+1)}{2L}), \quad i,j=0,1,\ldots,L-1, \tag{9.27}$$

which is exactly the form of the sinusoidal-intensity grating used as the test stimulus for determining the contrast sensitivities in [34]. Therefore, the error $e(0, v_1)$ will be essentially undetectable by a human observer if [8]

$$e(0,v_1) \leq \frac{d(0,0)+c_1L}{\sqrt{2}m(f(v_1))}, \tag{9.28}$$

where $f(v_1)$ is the spatial frequency of the waveform $\cos\frac{\pi v_1(2j+1)}{2L}$ measured in c/deg, and the constant c_1 is the luminance measurement of the block when $d(u,v)=0$ for all (u,v).

The above over-simplified case can be used to provide a guideline for incorporating the contrast sensitivity of the HVS into the ADCT-based image coding system. Specifically, we argue that for good perceptual quality the quantization error $e(u,v)$ should be proportional to $(d(0,0)+c_1L)/m(f(u,v))$, for $(u,v)\neq(0,0)$, where $f(u,v)$ is the spatial frequency in c/deg of the waveform associated with the (u,v)th DCT coefficient. Obviously, such a condition necessitates a different quantization rule for each block. To circumvent this problem, we require, instead, that (i) $e(u,v)$ be proportional to $\hat{d}(0,0)+c_1L$, where $\hat{d}(0,0)$ is the quantized version of $d(0,0)$, and (ii) the variance of $e(u,v)$ be inversely proportional to $m^2(f(u,v))$. In the following, a scheme is designed to achieve these two requirements.

Denote the (m,n)th block of 2-D DCT[9] coefficients by $\mathbf{d}_{m,n}$, $m,n=0,1,\ldots,(M/L)-1$, where $\mathbf{d}_{m,n}\equiv\{d_{m,n}(u,v);\ u,v=0,1,\ldots,L-1\}$. We assume that the classification is performed as described in Subsection 6.1. The dc coefficients $d_{m,n}(0,0)$ are quantized independently of the classification and the perceptual weighting described below; their quantized versions are denoted by $\hat{d}_{m,n}(0,0)$. To account for the perceptual role of the block luminance, the other 2-D DCT coefficients, $d_{m,n}(u,v)$, $(u,v)\neq(0,0)$, are modified as follows:

$$d'_{m,n}(u,v)=\frac{d_{m,n}(u,v)}{\hat{d}_{m,n}(0,0)+c_1L}, \quad m,n=0,1,\ldots,(M/L)-1. \tag{9.29}$$

---

[8] Note that according to the definition of 2-D DCT in [18], $d(0,0)/L$ is the mean intensity of the block.

[9] The 2-D DCT could be performed either on the image itself or on its smooth-texture component.

The variances of the modified 2-D DCT coefficients are denoted by ${\sigma'}_k^2(u,v)$, $(u,v) \neq (0,0)$, where $k = 0,1,\ldots,K-1$, is the classification index. The quantization is performed on $d'_{m,n}(u,v)$; the quantized version of $d_{m,n}(u,v)$ is obtained from that of $d'_{m,n}(u,v)$ by multiplying back by the scale factor, $(\hat{d}_{m,n}(0,0) + c_1 L)$. We denote the quantization errors for $d'_{m,n}(u,v)$ and $d_{m,n}(u,v)$ by $e'_{m,n}(u,v)$ and $e_{m,n}(u,v)$, respectively.

To ensure that the quantization error variance for the (u,v)th coefficient is inversely proportional to $m^2(f(u,v))$, let us introduce the weighted variances ${\sigma'}_k^2(u,v)\times\, m^2(f(u,v))$. These weighted variances are subsequently used for bit allocation among all 2-D DCT coefficients of the K classes except for the dc coefficient; the number of bits used for the dc coefficient is as given before the perceptual weighting. Let us use ${\epsilon'}_k^2(u,v)$ to denote the variance of $e'_{m,n}(u,v)$ when the (m,n)th block is in class k. Then, using arguments based on the Shannon Lower Bound similar to those given in Subsection 6.1, we can conclude that $m^2(f(u,v)){\epsilon'}_k^2(u,v)$ is *constant*, which implies that ${\epsilon'}_k^2(u,v)$ is inversely proportional to $m^2(f(u,v))$ — exactly what we had set out to achieve.

The ADCT coding part of both the 1C-ADCT-HC and 3C-ADCT-HC schemes were modified according to the above mentioned perceptual weighting and obtained simulations at different bit rates. The reconstructed images (512×512 Lenna) of 3C-ADCT-HC at bit rates 0.25 and 0.125 bpp are shown in Fig. 17. In these simulations, we have used $f_k(u,v) = 2.5\sqrt{u^2+v^2}$ and $c_1 = 512$, [41]. The choice of c_1 is determined by experimentation.

The results of perceptually-weighted 3C-ADCT-HC were compared with those without perceptual-weighting [41], [58]. It was clearly seen that the perceptually-weighted scheme offers a better subjective performance; the improvement is quite visible at low bit rates. However, for the perceptually-weighted 1C-ADCT-HC results the reverse conclusion was observed; it suffered from a visible degradation near the strong edges. This phenomenon can be explained as follows. When compared with the straight 1C-ADCT-HC scheme, its perceptually-weighted version results in allocating more bits to the low frequency coefficients and fewer bits to the high frequency coefficients. Therefore, for texture regions, the quantization error generally increases but this has little perceptual effect. In smooth regions, the image is better reproduced and in particular the blockiness is reduced due to the extra bits received for low frequency coefficients. However, in those blocks dominated by strong edges, the perceptual weighting has a detrimental effect: The high frequency components (which are generally strong due to the sharp transitions of the strong edges) are more coarsely quantized and the resulting quantization errors are spread over the entire block in

Figure 17 3C-ADCT-HC with perceptual weighting at design bit rates (a) 0.25 and (b) 0.125 bpp.

the spatial domain resulting in a visible degradation of the uniform regions on both sides of the strong edge. In contrast to 1C-ADCT-HC, the strong edges in 3C-ADCT-HC are extracted and encoded separately; the perceptual weighting can thus be applied solely to the smooth and texture components without damaging the integrity of the strong edges.

9 VIDEO CODING USING THE THREE-COMPONENT IMAGE MODEL

Until now, only continuous-tone still image decomposition has been addressed. The techniques and schemes developed so far can be applied directly to components of color images and to the intra-coded frames in motion video sequences [35].

We have generalized the concept of the three-component image model to a complex-valued 1-D case for an efficient representation of planar curves [60]. Likewise, the concept of the three-component image model can be generalized to a 3-D case for video coding and processing; this approach is currently not being pursued due to a lack of documented psychovisual understanding of human perception of motion videos. Instead, we are pursuing an approach based on the digital image warping techniques [61], [62].

In this new approach, the motion video frames are subsampled in time domain with the maximum number of skipped frames constrained by the system delay requirement. The actual number of skipped frames after a sampled frame is determined based on the video content, along with other system considerations such as output buffer fullness; for slow motion video segments the number of skipped frames is larger, while for fast motion segment, it is smaller. The primary components of the sampled frames are then extracted using the scheme of Section 4. The extracted strong edges in the two consecutive sampled frames are paired to one another to specify affine transforms [61], [62] between these two frames. A prediction of the second frame in the two frames is created from the reconstructed first frame with the affine transforms[10]. The prediction error may or may not be coded depending on the motion content of the video. All the skipped frames are reconstructed (using zero bits) through warping from the reconstructed first frame, or from both reconstructed frames if the prediction error is coded. Here, the primary components of the three-component image models provide perceptually meaningful features for the specification of warping [61]. Work on this new approach is in an early stage at the moment of writing this chapter.

10 CONCLUSIONS

We have discussed some important properties of the HVS and attempted to describe them mathematically. In particular, strong edges and areas of smooth intensity variation are carefully studied and their effect as well as their interaction in formation of perception are described in terms of simple minimization problems. An algorithm, based on a space-variant low-pass filtering operation, is developed for generating the so-called stressed image from which the strong edges can be extracted. This has led to a three-component image model based on the (i) strong edges, (ii) areas of smooth intensity variation and (iii) textures – the three components which apparently play different roles in the formation of perception.

We have developed a framework for image coding in which a combination of waveform coding and second generation coding techniques is used to achieve high perceptual quality at low bit rates. The coding schemes revolve around a perceptually motivated 3-component model by means of which the image is decomposed into three components. The primary component, which contains

[10] The very first frame of a video segment is always intra-coded; more intra-coded frames are needed for videos with fast motions and scene cuts.

the strong edge information, is encoded separately with little or no perceptual distortion. The remaining two components (smooth and texture) are encoded by entropy-coded ADCT or entropy-coded SBC. While the novelty of the proposed schemes reside primarily in the use of the three-component model and the separate encoding of its constituent components, there are some new elements in the adaptive DCT coding scheme, such as the classification procedure and the estimation of coefficient variances in the receiver, that have contributed to its good objective performance.

We have also shown that the well-understood contrast sensitivity of the HVS can be used for the perceptual tuning of the ADCT-based encoding scheme. We have developed a method for weighting the 2-D DCT coefficients based on the contrast sensitivity. When this perceptual tuning is used in conjunction with the 3-component model, the subjective performance of the ADCT-based scheme is further improved. Use of perceptual weighting without the 3-component model leads to a visible degradation of the strong edges and hence is not desirable.

The techniques and schemes developed in the three-component image model have been generalized to a complex-valued 1-D case for an efficient representation of planar curves, and can be generalized to a 3-D case for video coding and processing. An approach of using the three-component image model in combination with the digital image warping techniques in video coding is adopted, where the primary components of the three-component image models provide perceptually meaningful features for the specification of warping.

REFERENCES

[1] A. Habibi and P. A. Wintz, "Image coding by linear transformation and block quantization," *IEEE Trans. Commun. Tech.*, vol. COM-19, pp. 50-62, Feb. 1971.

[2] W. H. Chen and C. H. Smith, "Adaptive coding of monochrome and color images ," *IEEE Trans. Commun.*, vol. COM-25, pp. 1285-1292, Nov. 1977.

[3] W. H. Chen and W. K. Pratt, "Scene adaptive coder," *IEEE Trans. Commun.*, vol. COM-32, pp. 225-232, Mar. 1984.

[4] W. A. Pearlman, "Adaptive cosine transform image coding with constant block distortion," *IEEE Trans. Commun.*, vol. COM-38, pp. 698-703, May 1990.

[5] G. K. Wallace, "The JPEG still picture compression standard," *Commun. ACM*, vol. 34, pp. 30-44, Apr. 1991.

[6] M. Liou, "Overview of the px64 kbit/s video coding standard," *Commun. ACM*, vol. 34, pp. 60-63, Apr. 1991.

[7] D. LeGall, "MPEG: a video compression standard for multimedia applications," *Commun. ACM*, vol. 34, pp. 47-58, Apr. 1991.

[8] ITU-T Draft Recommendation H.262, "Generic coding of moving pictures and associated audio," 1993.

[9] ITU-T Draft Recommendation H.263, "Video coding for narrow telecommunication channel at $<$ 64 kbit/s," 1995.

[10] J. W. Woods and S. D. O'Neil, "Subband coding of images," *IEEE Trans. Acoust. Speech Signal Processing*, vol. ASSP-34, pp. 1105-1115, Nov. 1986.

[11] H. Gharavi and A. Tabatabai, "Sub-band coding of monochrome and color images," *IEEE Trans. Circuits and Systems*, vol. 35, pp. 207-214, Feb. 1988.

[12] P. H. Westerink, D. E. Boekee, J. Biemond and J. W. Woods, "Subband coding of images using vector quantization," *IEEE Trans. Commun.*, vol. 36, pp. 713-719, Jun. 1988.

[13] N. Tanabe and N. Farvardin, "Subband image coding using entropy-coded quantization over noisy channels," *IEEE Journal of Selected Areas in Communications,* vol. 10, pp. 926-943, Jun. 1992.

[14] Y. H. Kim and J. W. Modestino, "Adaptive entropy coded subband coding of images," *IEEE Trans. Image Processing*, vol. 1, pp. 31-48, Jan. 1992.

[15] M. Antonini, M. Barlaud, P. Mathieu and I. Daubechies, "Image coding using wavelet transform," *IEEE Trans. Image Processing*, vol. 1, pp. 205-220, Apr. 1992.

[16] Y. Linde, A. Buzo and R. M. Gray, "An algorithm for vector quantizer design," *IEEE Trans. Commun.*, vol. COM-28, pp. 84-95, Jan. 1980.

[17] N. M. Nasrabadi and R. A. King, "Image coding using vector quantization: a review," *IEEE Trans. Commun.*, vol. COM-36, pp. 957-971, Aug. 1988.

[18] N. S. Jayant and P. Noll, *Digital coding of waveforms, principles and applications to speech and video*, Englewood Cliff, NJ: Prentice-Hall, 1984.

[19] A. Gersho and R. M. Gray, *Vector quantization and signal compression*, Kluwer Academic Publishers, Boston, 1992.

[20] B. Ramamurthi and A. Gersho, "Classified vector quantization of images," *IEEE Trans. Commun.*, vol. COM-34, pp. 1105-1115, Nov. 1986.

[21] D. Chen and A. Bovik, "Visual pattern image recognition," *IEEE Trans. Commun.*, vol. 38, pp. 2137-2146, Dec. 1990.

[22] A. N. Netravali and B. Prasada, "Adaptive quantization of picture signals using spatial masking," *Proc. IEEE*, vol. 65, pp. 536-548, 1977.

[23] J. O. Limb and C. B. Rubinstein, "On the design of quantizers for DPCM coders: a functional relationship between visibility, probability and masking," *IEEE Trans. Commun.*, vol. COM-26, pp. 573-578, 1978.

[24] R. J. Safranek and J. D. Johnston, "A perceptually tuned sub-band image coder with image dependent quantization and post-quantization data compression," *Proc. IEEE ICASSP*, pp. 1945-1948, May 1989.

[25] M. Kunt, A. Ikonomopoulos, and M. Kocher, "Second-generation image-coding techniques," *Proceedings of the IEEE*, vol. 73, No. 4, pp. 549-574, Apr. 1985.

[26] M. Kunt, M. Benard, and R. Leonardi, "Recent results in high-compression image coding," *IEEE Trans. Circuit and Systems*, vol. CAS-34, No. 11, pp. 1306-1336, Nov. 1987.

[27] S. Carlsson, "Sketched based coding of grey level images," *Signal Processing*, vol. 15, No. 1, pp. 57-83, Jul. 1988.

[28] S. Carlsson, C. Reillo, and L. H. Zetterberg, "Sketched based representation of grey value and motion information," in *From Pixels to Features* by J. C. Simon (ed.) Elsevier Science Publishers B.V. (North-Holland), 1989.

[29] J. K. Yan and D. J. Sakrison, "Encoding of images based on a two-component source model," *IEEE Trans. on Communications*, vol. COM-25, No. 11, pp. 1315-1322, Nov. 1977.

[30] D. Marr and E. Hildreth, "Theory of edge detection," *Proc. R. Soc. Lond.* B 207, pp. 187-217, 1980.

[31] X. Ran and N. Farvardin, "A perceptually motivated three-component image model - Part I: Description of the model" *IEEE Trans. Image Processing*, vol.4, pp. 401-415, Apr. 1995.

[32] H. Helmholtz, *Treatise on Physiological Optics*, edited by J. Southall, vol. III, *The Perceptions of Vision*, the Optical Society of America, Menasha, Wisconsin: George Bonta Publishing Company, 1925.

[33] T. Cornsweet, *Visual Perception*, New York and London: Academic Press, 1970.

[34] F. W. Campbell and J. G. Robson, "Application of Fourier analysis to the visibility of gratings," *J. Physiol.*, 197, pp. 551-566, 1968.

[35] A. N. Netravali and B. G. Haskell, *Digital pictures, representation and compression*, Plenum Press, New York, 1988.

[36] J. L. Mannos and D. J. Sakrison, "The effects of a visual fidelity criterion on the encoding of images," *IEEE Trans. Inform. Theory*, vol. IT-20, pp. 525-536, Jul. 1974.

[37] W. Grimson, "Surface consistency constraints in vision," *Computer Vision, Graphics, and Image Processing* 24, pp. 28-51, 1983.

[38] D. G. Luenberger, *Linear and Nonlinear Programming*, Menlo Park, CA: Addison-Wesley Publishing Company, 1984.

[39] W. Hackbusch, *Multi-Grid Methods and Applications*, Berlin, Heidelberg, New York and Tokyo: Springer-Verlag, 1985.

[40] D. M. Young, *Iterative Solution of Large Linear Systems*, New York and London: Academic Press, 1971.

[41] X. Ran, "A three-component image model for human visual perception and its application in image coding and processing," Ph.D. dissertation, University of Maryland, College Park, MD. Aug. 1992.

[42] P. Chou and N. Pagano, *Elasticity, Tensor, Dyadic, and Engineering Approaches*, Princeton, NJ: D. Van Nostrand Company, 1967.

[43] T. S. Huang, "Coding of two-tone images," *IEEE Trans. on Communications*, vol. COM-25, No. 11, pp. 1406-1424, Nov. 1977.

[44] D. N. Graham, "Image transmission by two-dimensional contour coding," *Proceedings of IEEE*, vol. 55, No. 3, pp. 336-346, Mar. 1967.

[45] H. Freeman, "On the encoding of arbitrary geometric configuration," *IRE Trans. Electron. Comput.*, EC-10, pp. 260-268, Jun. 1961.

[46] D. L. Neuhoff and K. G. Castor, "A rate and distortion analysis of chain codes for line drawings," *IEEE Trans. Inform. Theory*, vol. IT-31, pp. 53-67, Jan. 1985.

[47] M. Eden and M. Kocher, "On the performance of a contour coding algorithm in the context of image coding. Part I: Contour segment coding," *Signal Processing*, vol. 8, pp. 381-386, 1985.

[48] J. J. Rissanen, "Generalized Kraft inequality and arithmetic coding," *IBM J. Res. Develop.*, pp. 198-203, May 1976.

[49] I. H. Witten, R. M. Neal and J. G. Cleary, "Arithmetic coding for data compression," *Commun. ACM,* vol. 30, pp. 520-540, Jun. 1987.

[50] O. R. Mitchell and A. Tabatabai, "Adaptive transform image coding for human analysis," *Proc. ICC,* pp. 23.2.1-23.2.5, 1979.

[51] R. C. Reininger and J. D. Gibson, "Distributions of the two-dimensional DCT coefficients for images," *IEEE Trans. Commun.*, vol. COM-31, pp. 835-839, Jun. 1983.

[52] N. Farvardin and J. W. Modestino, "Optimum quantizer performance for a class of non-Gaussian memoryless sources," *IEEE Trans. Inform. Theory*, vol. IT-30, pp. 485-497, May 1984.

[53] A. V. Trushkin, "Optimal bit allocation algorithm for quantizing a random vector," *Problems of Inform. Transmission,* vol. 17, pp. 156-161, 1981.

[54] H. Gish and J. N. Pierce, "Asymptotically efficient quantizing," *IEEE Trans. Inform. Theory,* vol. IT-14, pp. 676-683, Sep. 1968.

[55] T. Berger, *Rate distortion theory*, Englewood Cliff, NJ: Prentice-Hall, 1971.

[56] J. M. Shapiro,"An embedded wavelet hierarchical image coder," in *Proc. IEEE ICASSP*, pp. IV.657–IV.660, March 1992.

[57] F. Kessentini, M. J. T. Smith and C. F. Barnes, "Image coding with variable rate RVQ," in *Proc. IEEE ICASSP*, pp. III.369-III.372, March 1992.

[58] X. Ran and N. Farvardin, "A perceptually-motivated three-component image model - Part II: Applications to image compression," *IEEE Trans. Image Processing*, vol.4, pp. 430-447, Apr. 1995.

[59] M. G. Perkins and T. Lookabaugh, "A psychophysically justified bit allocation algorithm for subband image coding systems," *Proc. IEEE ICASSP*, pp. 1815-1818, May 1989.

[60] X. Ran and N. Farvardin, "On planar curve representation," *Proc. of IEEE ICIP*, pp. I-676-I-680, Nov. 1994.

[61] T. Beier and S. Neely, "Feature-based image metamorphosis," *Computer Graphics*, 26, 2, pp. 35-42, Jul. 1992.

[62] G. Wolberg, *Digital image warping*, IEEE Computer Society Press, Los Alamitos, CA, 1990.

10

EXTENDED SIGNAL-THEORETIC TECHNIQUES FOR VERY LOW BIT-RATE VIDEO CODING

Haibo Li and Robert Forchheimer

Image Coding Group,
Department of Electrical Engineering,
Linköping University,
S-58183, Linköping,
Sweden

ABSTRACT

The goal of this chapter is not to focus on known signal-theoretic techniques but to extend their use more than is normally done in conventional image coding. The methods described are research oriented but are fully concievable considering the fast pace of the hardware development for signal processing.

1 INTRODUCTION

Video coding at very low bit-rates is motivated by its potential applications for *videophones, multimedia electronic mail, remote sensing, electronic newspapers, interactive multimedia databases, multimedia videotex, video games, interactive computer imagery, multimedia annotation, surveillance, telemedicine and communication aids for hearing impaired people.* Due to practical limitations, the main problem of introducing these applications lies in how to compress a huge amount of visual information into a very low bit-rate stream for transmission or storage purposes.

Traditional and mature techniques for low bit-rate video coding are mainly based on Signal Theory. A well-developed video coding scheme is the *hybrid coding scheme. Hybrid* originally meant the combination of transform coding with predictive coding so that the attractive features of both schemes can be utilized together [5][11]. After efforts made over a period of several years, the hybrid coding scheme has evolved into its current version, *the Motion Compensated DCT scheme (MC-DCT)* as shown in Figure 1. This scheme today has governed the efforts on video compression standards, ranging from terrestrial broadcasting of HDTV, and digital video, such as MPEG I, II into the videophone standards H.26p. Although there are current efforts to use this scheme to achieve a very low bit-rate transmission of video, e.g. through the existing telephone network, the decoded image quality is far from the level which is acceptable to the viewers. This is due to the inherent drawbacks of this scheme. Some drawbacks which are identified are

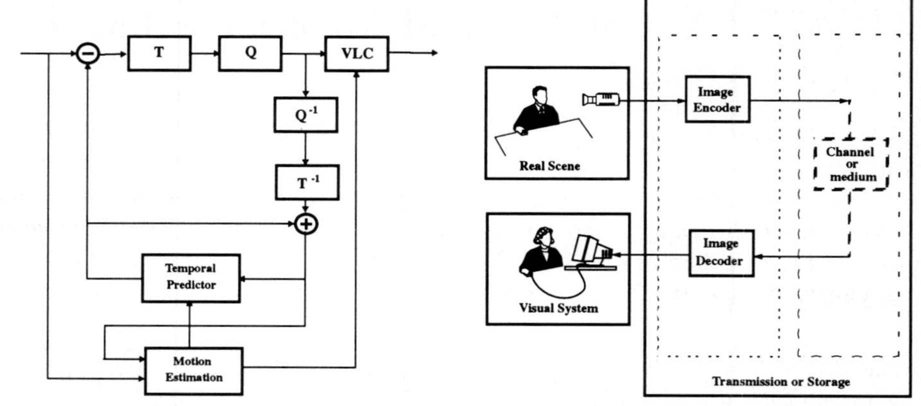

Figure 1 The hybrid image coding scheme.

Figure 2 The general image coding system.

- *The properties of the Human Visual System (HVS) are not made full use of.*

- *Motion information is not efficiently extracted, represented and utilized.*

- *Except for one previous frame, the transmitted information is discarded, which is a serious waste for some applications which need very low bit-rate.*

- *Implicit constraints from the application itself are not well taken advantage of , e.g. signal-independent basis functions are used.*

- *The coding architecture is not suitable to accommodate future functions, e.g. scalability, interactivity, or lossy transmission.*

For more details the reader is referred to Chapter 2.

The efforts on low bit-rate video coding to be described here come from the authors own research activities and are based on extensions of traditional signal theoretic methods. These efforts are made around the complete image coding system shown in Figure 2. This system unifies the transmission and storage systems. In the general coding system, two important aspects are the *scene* and the *human observer*. Our investigations are based on the understanding of the captured scene contents and the human visual system. Therefore, we choose the following three *coding methodologies* as our cornerstone of low bit-rate video coding:

- *Perception-based Coding*
- *Fractal-based Coding*
- *Model-based Coding*

In **perception-based coding**, we mainly study how to imitate the human visual system when designing image coders. Our own work in this area started with the study of subband coding. Today, wavelet coding and the extraction and employment of motion information are examples of perception-based coding.

In **fractal-based coding**, fractal-theory is used to exploit the self-similar redundancy existing in image volumes. A lot of effort is spent on the building of self-similar models for image volumes and the combination of fractal coding with traditional coding techniques.

Different from waveform-based coding where an image sequence is treated as a 3-D signal waveform exploiting the inherent statistical or deterministic properties, **model-based image coding** views the input image as a 2-D projection of a 3-D real physical scene. The coding is performed by first modelling the 3-D scene, and then extracting model parameters at the encoder, finally synthesizing the image at the decoder by using the extracted and quantized parameters. Depending on the amount of a priori information assumed, different schemes have been proposed such as: *Semantic-based Facial Image Coding* and *Object-Oriented Image Coding*.

Based on these three methodologies, the following research directions will be addressed:

- *role of vision in image coding*
- *the estimation and employment of motion information*
- *signal-dependent coding*

- *re-use of the transmitted information*

This chapter mainly introduces our approaches to low bit-rate video coding. However, some related works done by other groups around the world are also included. This chapter is organized as follows: after the introduction part, the estimation and employment of motion information is addressed in Section 1. Section 2 discusses how to design video coders based on the properties of the human visual system and how to assess the quality of compressed images at low bit-rates. In Section 3, we argue that it is very useful to store the transmitted information to be used by future frames, especially for some applications with repeated scenes. How to store and employ the transmitted information in both waveform-based coding and model-based coding is addressed. For videophone-type applications, we suggest in Section 4 that signal-dependent basis functions should be employed instead of the commonly used fixed-basis function. How to construct the signal-dependent basis functions and how to update the basis functions is discussed.

2 MOTION INFORMATION

Efficient *extraction*, *representation* and *employment* of motion information will play a very important role in low bit-rate video coding. The extraction of motion information from video sequences has been extensively addressed in Chapter 6. As a complement, we mainly deal with the *representation* and *employment* of motion information in this section.

In the employment of motion compensation techniques, besides *motion compensated residual information*, *motion information* also needs to be transmitted. Therefore in very low bit-rate video coding, motion information has to be given a serious treatment. We should choose such motion estimation algorithms that can provide both reliable and compact motion vectors. An example is introduced in Section 2.1.

Since a coding operation normally involves not only prediction, but also interpolation, or compression of motion information, the reliability of motion fields becomes particularly important. We may worry about how to identify the reliability of motion vectors, how to represent the reliability and how to employ it. These problems are addressed in Section 2.2.

The last problem is how much motion information is required and how to allocate bits among motion information and residual information. This problem is discussed in Section 2.3.

2.1 Computation and transmission of a motion vector field

Due to its simplicity and efficiency in reducing the residual error, the *block-matching algorithm* (BMA) based on a full search strategy is now the most often used motion estimation algorithm. In this algorithm, the optimal motion vector for each block (normally with a size of 16×16) is determined by choosing a minimum matching cost among points within a match window. This strategy has two significant drawbacks: 1) the decision of motion vector for the current block is independent of that for previous blocks. 2) The computed matching costs at different points within a matching window become all but one useless after taking a minimum. We know that the computation of these costs is done with a great deal of effort. The independency of motion vector decision results in unreliable motion vectors for blocks in uniform areas and the estimated motion vectors are more or less independent, which is not suitable for lossless coding of motion vectors (e.g. DPCM coding).

In [25], a low-entropy motion estimation algorithm developed by the authors is described. This algorithm is a modified version of the blocking-matching algorithm (BMA). The main difference lies in how to decide motion vectors based on the same matching results. The decision here is no longer independent and local but interdependent and global. The basic idea behind this algorithm is that motion estimation should minimize the transmission costs of the residual information (assuming a coder structure according to Figure 1) and the motion information simultaneously.

This can be achieved by defining a cost function which contains two factors: the residual factor R_k and the motion factor M_k.

$$\Delta_k = R_k + \alpha M_k \qquad (10.1)$$

where k is the index of blocks and α is a tradeoff between the residual and motion information.

The motion factor M_k is defined as

$$M_k(i,j) = |\mathsf{v}_k - \mathsf{v}_{k-1}| \tag{10.2}$$

where v_k and v_{k-1} are motion vectors of the current block and previous block, respectively. They are also a function of i and j. M_k measures the distance between the motion vector of the current block and that of the previous block. Minimizing M_k makes the adjacent motion vectors more consistent, which will benefit DPCM coding of motion vectors.

In a general block-matching algorithm, the motion vector of the current block is obtained by

$$\mathsf{v}_k = \min_i R_k(i) \tag{10.3}$$

where i are possible motion vectors within a search window.

Obviously, the decision of the optimal motion vectors is locally made. However, in the new motion estimation algorithm, the decision is made in a global sense. The global cost function can be generated as

$$C_k = C_{k-1} + \Delta_k \tag{10.4}$$

where C_k is the accumulated cost from block 0 to block k. We define that $C_0 = R_0$.

If there is a total of N blocks, then the required motion vectors should form a shortest path p from block 0 to block $N-1$ which minimizes the total cost

$$p = \min_i C_N(i) \tag{10.5}$$

where i again represents possible motion vectors within a search window.

With the shortest path p, motion vectors can be obtained as

$$\mathsf{v}_k = p(k) \tag{10.6}$$

Now the problem is how to find the shortest path. From (10.5) (10.4) and (10.1), the accumulated cost function can be written as

$$C(k)(i) = R_k(i) + \min_j\{C_{k-1}(j) + \alpha M_k(i,j)\} \tag{10.7}$$

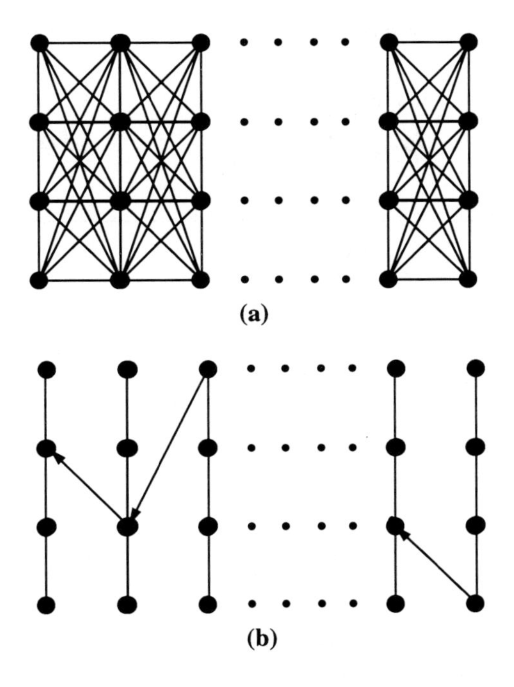

Figure 3 Shortest path for motion vectors.

This expression allows us to employ the powerful *Dynamic Programming* technique to efficiently compute the shortest path. The basic strategy to estimate motion vectors is shown in Figure 3. Choosing a suitable tradeoff α we will then achieve a low-entropy motion estimation.

A computer simulation was performed on the test sequence Claire [25]. Experimental results show that motion compensation operation is a powerful tool to handle temporal redundancies and that there is not a significant difference between the reconstructed image using different motion vectors. However, there is a large difference in the motion vector fields estimated from different methods. This can be verified from Figure 4. The motion field provided by the new scheme is quite compact. The quantitative results are listed in Table 1. From this table, 1.3-2 bits can be saved for each motion vector if the new motion estimation method is employed.

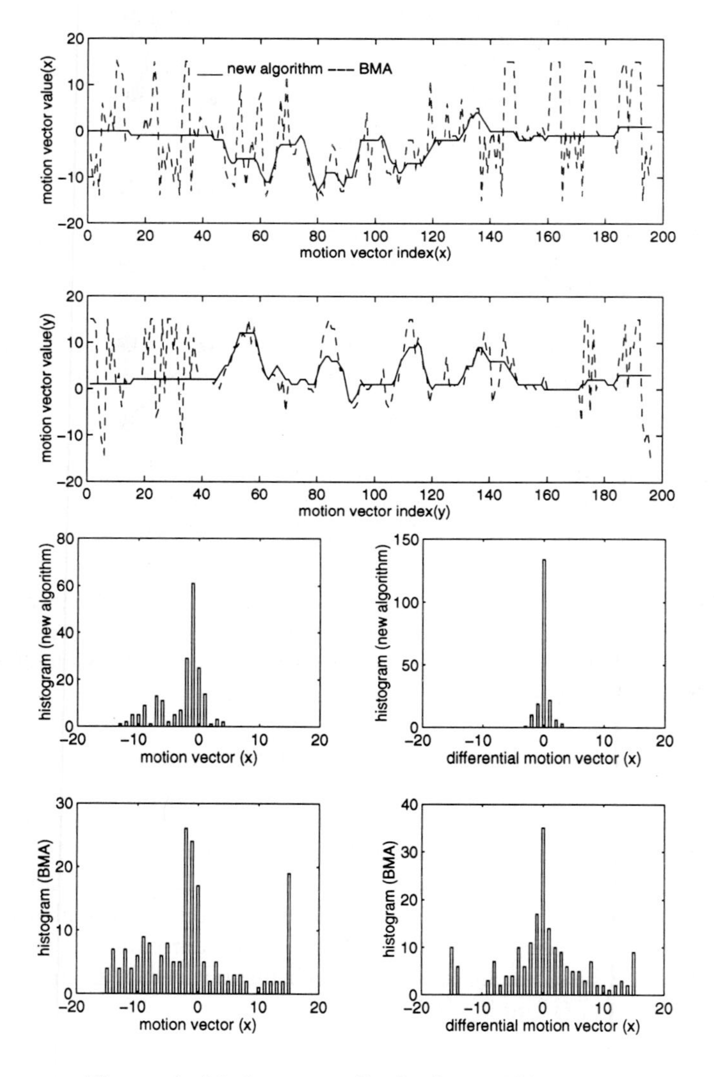

Figure 4 Motion vector distribution and histogram.

		Residual (MSE)	Entropy
BMA	T=0	2.797	3.937
	T>3	2.936	3.212
New Method		3.481	1.928

2.2 Can we trust the estimated motion information?

Since a coding operation normally involves not only prediction, but also interpolation, or compression of motion information, motion fields that are to be used for image coding must be faithful estimates of true motion. This problem can be handled by use of more sophisticated motion models. Even with these models, however, motion estimation will have problems in certain areas, especially where spatial luminance changes are so small that accurate motion estimates are difficult or impossible to achieve. Therefore, it is better to provide the estimated motion vectors with certainty indexes which can be used to guide further processing.

Several different ways to measure motion certainty have been proposed [26]. A typical definition is [15]

$$C = \frac{k}{l}$$

where k and l are estimated directly from the match-values, $E(x, y)$,

$$E(x, y) = k(x^2 + y^2) + l.$$

Motion certainty measures can be used for the following purposes

Motion Certainty Controlled Pre-filtering

Motion compensation has to be applied in pre-filtering to avoid blurring. Motion detection has been used in practical systems to allow effective filtering of stationary areas while blurring of moving areas is prevented. Motion compensated pre-filtering was introduced in [15] and allows for much lower cutoff frequencies without blurring than straight (non-motion compensated) pre-filtering. In the same way as motion detection was used to reduce the amount of filtering in the case of straight temporal filtering, a certainty measure can be used when doing motion compensated pre-filtering to indicate that the motion is not reliable at this point, a signal that the bandwidth of the temporal filter should be increased. In Figure 5 we see the benefit of letting a certainty measure control the amount of temporal filtering.

Figure 5 Cut-outs from the test image sequence 'Salesman'. Top: original; The two images below have been temporally filtered so that on average the same amount of noise has been removed, their motion compensated prediction error have the same entropy. Left: conventional motion compensation and fixed temporal filter; Lower right: Bandwidth of the filter controlled by the certainty of the motion estimation.

Motion Certainty Weighted 3D Motion Parameter Estimation

If a motion field can be modeled in a mathematical form, then the motion field can be very efficiently represented. In the employment of a model-based motion representation, we need to estimate the parameters of the model. For example, under orthographic projection, a planar surface undergoing 3D translation and linear deformation gives rise to a 2D affine transformation in the image plane. This is described as

$$\begin{aligned} u(x,y) &= a_1 + a_2x + a_3y \\ v(x,y) &= a_4 + a_5x + a_6y \end{aligned} \tag{10.8}$$

where (u, v) is the motion field.

Using vector notation this can be rewritten as follows

$$\mathsf{u}(\mathsf{x}) = \mathsf{X}(\mathsf{x})\mathsf{a} \tag{10.9}$$

where a denotes the motion parameter vector $[a_1, a_2, ..., a_6]^T$, $\mathsf{u}(\mathsf{x}) = [u(x,y), v(x,y)]^T$ and

$$\mathsf{X}(\mathsf{x}) = \begin{bmatrix} 1 & x & y & 0 & 0 & 0 \\ 0 & 0 & 0 & 1 & x & y \end{bmatrix}$$

If the motion field $\mathsf{u}(\mathsf{x})$ is available, a common way of estimating the affine parameters a is to minimize an error measure function such as

$$\mathsf{E}(\mathsf{a}) = |\mathsf{u}(\mathsf{x}) - \mathsf{X}(\mathsf{x})\mathsf{a}| \tag{10.10}$$

If the obtained motion field is accurate, an LS solution of the affine parameters can be obtained. The problem is that the estimated motion field is normally noisy. Several bad estimates may completely destroy the estimation of the affine parameters. However, if a motion certainty can be provided with each motion vector, then motion certainty can be used to weight the estimation:

$$\mathsf{E}(\mathsf{a}) = \mathsf{W}(\mathsf{x})|\mathsf{u}(\mathsf{x}) - \mathsf{X}(\mathsf{x})\mathsf{a}| \tag{10.11}$$

where $\mathsf{W}(\mathsf{x})$ are weights specified by the motion certainty. In this way, the estimation of motion parameter becomes very robust.

Motion Certainty Aided Image Coding

When the motion certainty is low less consideration should be taken to old information. One way of doing this would be to lower the prediction coefficient(s) within the predictor, or letting the motion certainty control an adaptive loop-filtering.

A more advanced approach that can lead to a substantial increase in coding efficiency is to let the motion certainty control the time between the updates

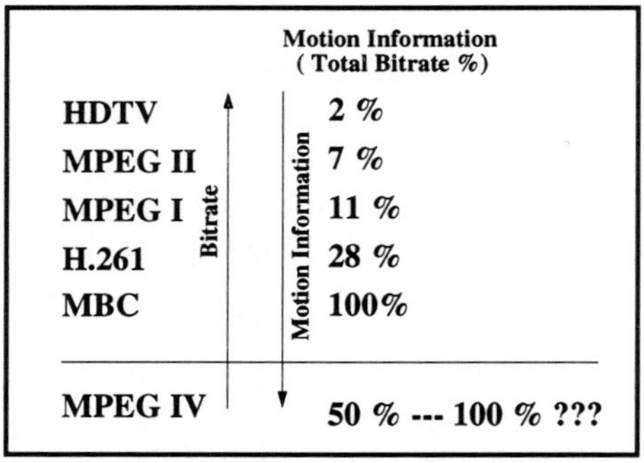

	Motion Information (Total Bitrate %)
HDTV	2 %
MPEG II	7 %
MPEG I	11 %
H.261	28 %
MBC	100%
MPEG IV	50 % --- 100 % ???

Bitrate ↑ Motion Information ↓

Figure 6 Relation between the percentage of bits used for motion vector and total bit-rate.

of the image information. One way of looking at the scheme is to compare it with a scheme which classifies the image into stationary background and moving foreground, where the background is coded once and for all and only the foreground is continuously updated. The motion certainty controlled system would consider all areas that can be satisfactorily described by motion compensation as a sort of background that is coded with image information and motion parameters only once (or, rather, until they disappear out of the picture or are covered by new image information). This is a more complex way of doing image coding than the conventional hybrid scheme. Conventional hybrid schemes strive to achieve the same effect by the use of 'not coded' blocks, i. e. blocks that are reconstructed with only the motion compensated prediction and no update image information.

2.3 How much motion information is required

The percentage of bits used for motion information with respect to the total number of bits depends on several aspects, for example, *the employed coding scheme*, *scene content*, *the way to represent motion*, and *the total bit-rate*. Roughly speaking, more motion information will be utilized as the total bit-rate decreases. An approximate relation between the percentage of bits for motion information and the bit-rate is shown in Figure 6. This figure seems to give us a hint that at least 80% total bit-rate should be used for motion information in very low bit-rate video compression.

In traditional approaches, motion description, estimation and motion field encoding are performed on a block by block basis. Such motion information processing is tailored to translational motion only and is inappropriate to handle complex motion. Thus interframe variances caused by complex motion can not be compensated. A large amount of bits has to be used to update these interframe variances. This results in a bit allocation strategy in which the majority of available bits are devoted to update the residual information and only a small remaining part of bits are used to transmit motion information. Such a bit allocation is suitable for high- or medium bit-rate video transmission because there is a relatively large body of available bits. However, at low bit-rate, especially, in the very low bit-rate case, there are not enough bits to update the interframe variances. We have to balance the bit allocation between the residual information and motion information. Normally, interframe information can be more efficiently represented through motion information than through representation by residual information. To obtain maximum benefit, the limited bits should be spent more on motion information. In this situation, a dense motion field, e.g. one motion vector per pixel, is appreciated. If the dense motion field is perfect, then there is no prediction error to be transmitted. In this way, 100% of the bits are used for motion information. Even if the motion field is not perfect, a smooth motion compensated prediction can be obtained as long as the motion field is relatively smooth. In this case, only some minor parts which are perceptually important need to be updated. Therefore, in very low bit-rate video compression, we should adopt a bit allocation in which the majority of available bits is used for motion information instead of coding residual information.

Due to the large data set generated by a dense motion field, we can not afford to transmit it to the receiver. Therefore, it is necessary to compress the dense motion field. Obviously, lossless coding methods are not powerful enough to compress dense motion fields, so we have to adopt lossy compression methods. Experiments show that motion field compression has to be performed with high precision if the gain of the improved motion estimates is to be retained [34]. Therefore, we must carefully choose the lossy compression methods for motion vector encoding. Normal waveform coding methods such as DCT and VQ have been tested for this purpose. Although these methods can provide high coding gains for image intensity coding, how to make them more suitable for motion field coding needs further investigation. Another approach to compress dense motion fields is the employment of **model-based motion coding**. In model-based motion coding, a dense motion field is fitted into a parametric model, e.g. a global or local affine motion model. Thus, the motion field allows estimation, interpretation and transmission in the parametric domain. As a major drawback, parametric estimators fail in the case of complex motion which disobeys the parametric model. Accordingly, the motion field is split into several smaller

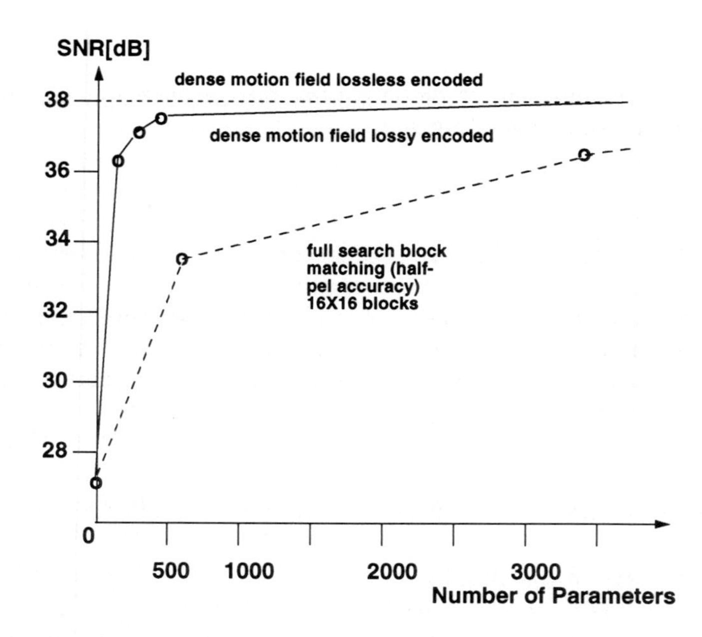

Figure 7 The SNR of the motion compensated prediction versus the number of selected basis functions. (from [34])

areas each obeying the parametric motion constraint. However, this will bring into play a risk of introducing artificial object boundaries.

A good solution is to use the second generation image coding methods to compress dense motion fields. C. Stiller used a contour/content coding method to achieve a compact, lossy motion field coding [34]. In his method, a dense motion field is segmented into several areas with homogeneous motion. The contours of the areas are approximated by a set of polygons and quadratic splines defined by a set of vertices. The contents of the areas are coded using polynomial basis functions. Figure 7 depicts the SNR of the motion compensated prediction versus the number of selected basis functions. This figure shows a significant advantage of dense motion field followed by a lossy coding over the traditional mode: *sparse motion field with lossless coding.*

Recently, Gisladottir and Orchard introduced a motion-only video compression scheme [10]. In their scheme, besides transmission of motion vectors, the residual information is encoded as motion rather than intensity. They claim that image sequences can be more efficiently represented through their motion field than the hybrid representation used in standard video coding algorithms.

Therefore, the employment of dense motion field in very low bit-rate image coding is worth being studied further.

3 VISUAL PERCEPTION

Visual perception plays a key role in second generation image coding [16]. Properties of the human visual systems (HVS) and their relation to video coding have been discussed in Chapter 1. In this section, we will discuss how to design image coders based on some properties of the HVS.

3.1 Eye-movement and image coding

The spatial acuity of the visual system decreases as retinal image velocity increases due to the temporal integration time (limited temporal bandwidth) of the visual processing. This is reflected in the spatio-temporal modulation transfer function (MTF) of the human visual system shown in Figure 8(a). This property of the HVS was used in the design of early motion video coders [33]. However, this property was wrongly applied, e.g. substantial reduction in image spatial resolution was allowed in regions of motion. However, recent experiments show that human observers may be highly sensitive to spatial blur even at high image velocities. This means that fast image motion does not necessarily permit reduction of image spatial resolution. This can be explained by the tracking of moving objects through eye-movements. It is the eye-movement that compensates the motion of objects. Now let us examine the effect of eye-movements on the MTF of the human visual system.

Spatio-temporal variations of image intensity based on the velocity field can be described as

$$I(\mathsf{x}, t) = I(\mathsf{x} - \int \mathsf{v} dt, t-1) \tag{10.12}$$

where $I(\mathsf{x}, t)$ is the image intensity at spatial point x and v is the velocity field.

Eye movements can be modeled as shown in Figure 9. The image $I(\mathsf{x}, t)$ is compensated by eye movements before reaching the retina.

$$I_r(\mathsf{x}, t) = \mathcal{K}(I(\mathsf{x}, t-1)) \tag{10.13}$$

where $I_r(\mathsf{x}, t)$ is the retina image and $\mathcal{K}$ is the motion compensated operator performed by eye movements. Eye movements shift all velocities in the image by the eye velocity $\mathsf{v^e}$,

$$I_r(\mathsf{x}, t) = I(\mathsf{x} + \int \mathsf{v^e} dt, t - 1) \quad (10.14)$$

The difference between the image velocity and eye velocity constitutes the velocity of the image on the retina:

$$\mathsf{v}^r(\mathsf{x}, t) = \mathsf{v}(\mathsf{x}, t) - \mathsf{v}^e(t) \quad (10.15)$$

where the eye velocity is $\mathsf{v}^e(t) = [v_x^e, v_y^e]^T$

The image on the retina is

$$I_r(\mathsf{x}, t) = I(\mathsf{x} - \int \mathsf{v}(\mathsf{x}, \mathsf{t}) - \mathsf{v^e}(\mathsf{t}) dt, t - 1) \quad (10.16)$$

If translation of the eye relative to the screen is assumed, the frequency response of the human visual system can be easily translated into the screen coordinate system using

$$\begin{cases} \omega_x &= \omega_x^* \\ \omega_y &= \omega_y^* \\ \omega_t &= \omega_t^* - \omega_x^* v_x^e - \omega_y^* v_y^e \end{cases} \quad (10.17)$$

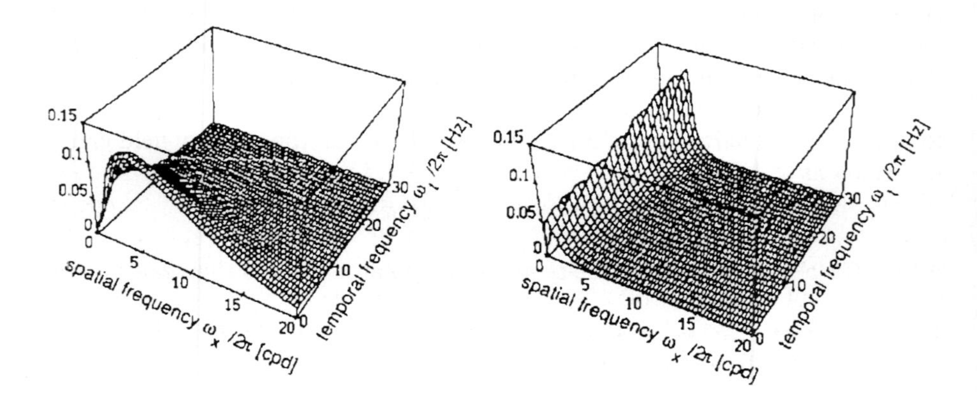

Figure 8 The spatio-temporal modulation transfer function. (a) without considering the effect of eye movements, (b) with considering the effect of eye movements. (from [9])

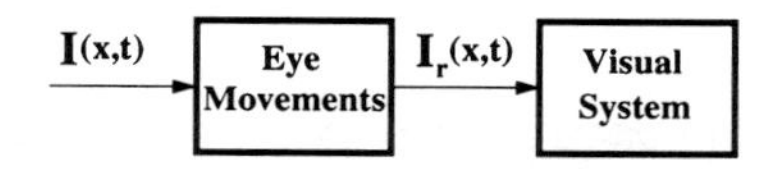

Figure 9 The model of eye movements. (from [4])

Using this transformation, the spatio-temporal MTF for smooth-pursuit eye movements of $v_x^e = 10\ deg/sec$ is shown in Figure 8(b). From this figure, we find that the effective spatio-temporal frequency response of the human visual system is dramatically altered by eye movements. Through eye movements very high temporal frequencies can be perceived at high spatial frequencies. Therefore, for the visual system, spatial acuity is only a function of retinal velocity not image velocity itself. Therefore a more correct description should be: the spatial acuity of the human visual system decreases as the velocity of the image traveling across the retina increases.

In the visual perception process, smooth-pursuit eye movements accomplish two tasks [4]: (1) *to maintain the object of interest in the area of highest spatial acuity of the visual field* and (2) *to minimize the velocity slip of the image across the retina by matching eye velocity to image velocity.* When working with image coding it is important to make use of the characteristics of such eye-movements.

The visual system has a maximum spatial acuity on the region-of-interest. To maintain the maximum spatial acuity tracking region-of-interest is necessary through eye-movements. This provides us a hint that if we could determine regions-of-interest then more bits will be used for these important regions. An illusion is shown in Figure 10. From this figure, we can see that although it is reconstructed in a spatially-variable quality, the image still gives us an impressive of high visual quality. Potential applications are very low bit-rate coding schemes and several image transmission applications where a spatially-constant reconstruction quality is not necessary, e.g. head-and-shoulder scenes, or remote surveillance. Allowing coarser compression for non-relevant components of the signal can save bit-rate. Alternatively, this can enhance local reconstruction quality for relevant components for a given bit-rate. Obviously, this obeys the basic image coding principle that we only transmit the parts which are important for visual perception. In practice we will be faced with three problems: (a) *how to select region-of-interest*; (b) *how to design a suitable motion compensator*; and (c) *how to achieve selective compression.* These three problems will be addressed in the following.

A.1 How to choose regions-of-interest

Figure 10 An example of an image with spatially-variable quality.

To determine the region-of-interest (ROI), we have at least two different ways: *Active* and *Passive* ways. In active ways, the region-of-interest is detected through following the trace of eyes by means of some physical instruments, e.g. eye-movement detectors. In addition, active ways are often related to the interactive ways. For example, the region-of-interest is determined directly by the human observers. In passive ways, the region-of-interest is determined by image analysis techniques. As is known, human eyes use a lot of visual cues, such as, depth, color, shape, motion to guide their focus. Therefore, the task of the image analysis is to identify the region-of-interest using these visual cues. The following example shows how to employ motion cue to select ROI [24].

The main moving object in typical videophone scenes is the head-and-shoulder of the speaker. Facial area is the most active and attractive part. The main visual information comes from this part. For a typical videophone scene a reasonable assumption is that the moving speaker stands in front of a stationary background. Besides the exposed background, only the head and shoulder of the speaker are visible. In most such cases people are constantly moving. Especially when involved in video transmission. Due to the narrow view-angle of the camera, we have to fidget and adjust our body position, nod our heads, look around, etc., so that we are in a suitable position within the picture frame. Motion information, therefore, becomes a principal cue to distinguish the speaker from the stationary background. Most segmentation algorithms only utilize the apparent motion feature. It can be observed that the head seldom undergoes the same motion as the shoulder! This is because the head is flexibly, not firmly attached to the shoulder.

Based on the above observation, the reasonable assumption is made that the projections of the motions of head and shoulder onto the image plane can be approximated by two different affine motions. Although the affine motion does not accurately model the actual 2D motion, this does not matter because our objective here is simply to localize the interest region, the facial area of the speaker, not to estimate the egomotion of the speaker.

Assume that the image velocity at point $\mathbf{x} = (x, y)$ *is* $(u(x, y), v(x, y))$, *then it can be modeled as an affine motion*

$$\mathbf{u} = \mathbf{X}\mathbf{a} \tag{10.18}$$

where $\mathbf{u} = [u, v]^t$, $\mathbf{a}$ *denotes the coefficients of the affine mapping* $[a_1, a_2, ..., a_6]^t$ *and* $\mathbf{X}$ *is a position matrix*

$$\mathbf{X} = \begin{bmatrix} x & y & 1 & 0 & 0 & 0 \\ 0 & 0 & 0 & x & y & 1 \end{bmatrix} = \begin{bmatrix} \mathbf{x}_1 \\ \mathbf{x}_2 \end{bmatrix} \tag{10.19}$$

Using (10.18), the affine mapping from the image velocity field can be calculated.

When the image velocity field is not available, the affine mapping can be directly computed from the image intensity. This can be cited in [24].

Motion segmentation of the facial area is based on the observation that the head and shoulder of the speaker undergo different motions. Assume that the motion of the head is specified by the parameter vector $\mathbf{a}_h$ *and that of the shoulder by* $\mathbf{a}_s$.

An attributive mark d_i *for the moving point* i *is assigned as follows.*

$$d_i = \begin{cases} 0 & i \in \textit{head} \quad \textit{part} \\ 1 & i \in \textit{shoulder} \quad \textit{part} \end{cases} \tag{10.20}$$

Motion segmentation consists of the following three steps:

- *step 1. compute affine mappings using some initial segmentation:*

$\mathbf{a}_h$ *can be obtained by minimizing the following error measure*

$$E(\mathbf{a}_h) = \sum_i (1 - d_i) w_i (\mathbf{u} - \mathbf{X}\mathbf{a}_h)^2 \tag{10.21}$$

where w_i *is a weight whose initial value is set to be* 1.

Similarly, $\mathbf{a}_s$ *can be obtained by*

$$E(\mathbf{a}_s) = \sum_i d_i w_i (\mathbf{u} - \mathbf{X}\mathbf{a}_s)^2 \tag{10.22}$$

- *step 2. revise the attribute:*

The model error can be calculated using the obtained $\mathbf{a}_h$ *and* $\mathbf{a}_s$,

$$e_h = (\mathbf{u} - \mathbf{X}\mathbf{a}_h)^2 \tag{10.23}$$

$$e_s = (\mathbf{u} - \mathbf{X}\mathbf{a}_s)^2 \tag{10.24}$$

d_i *is revised according to*

$$d_i = \begin{cases} 0 & e_h < e_s \\ 1 & e_h \geq e_s \end{cases} \tag{10.25}$$

Weight is given as

$$w_i = \psi(min(e_h, e_s)) \tag{10.26}$$

where $\psi(.)$ *is the Tukey's bi-weight function*

$$\psi(x) = \begin{cases} [1 - (\frac{x}{t})^2]^2, & |x| < t; \\ 0, & otherwise. \end{cases} \tag{10.27}$$

where t *is called the 'rejection point'.*

- *step 3. remove isolated attributes.*

The whole process is iterated until the attributive marks become stable.

Simulation results are shown from the test image sequence 'Miss America'. The two successive frames used are shown in Figure 11(a)(b), from which it can be found that miss America is nodding her head and moving her shoulder at the same time. We employ the Horn's algorithm to compute optical flow field between these two frames. Points with apparent motions are marked as Figure

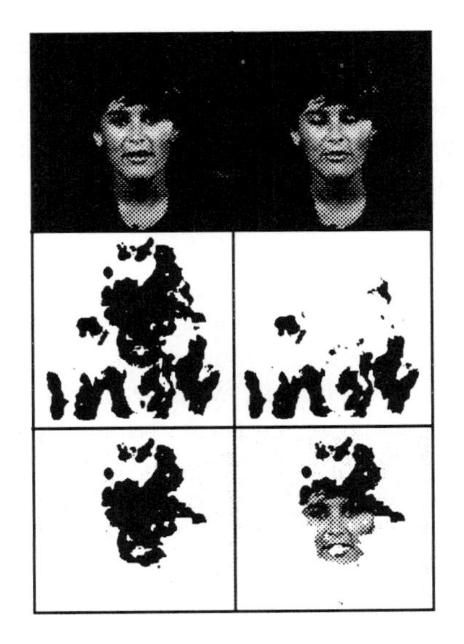

Figure 11 Miss America image sequence. (a)(b) two successive frames of Miss America, (c) points with apparent motions, (d) separated shoulder, (e) segmented head, (f) extracted facial area.

11(c). From this it is seen that both the head and shoulder are in motion. The developed algorithm is then used to separate the facial area. The separated shoulder is shown in Figure 11(d) and the extracted facial area from Figure 11(c) is shown in Figure 11(e). This can be further verified from Figure 11(f) which shows the original image 'gated by' the extracted facial area pixels.

Using this algorithm the image can be split into different layers. This will benefit region-of-interest coding as different priorities can be allocated to different layers.

A.2. How to design a suitable motion compensator

Since the spatial acuity of the observer depends solely on retinal velocity rather than absolute image velocity, we should measure the retinal velocity of the scene crossing eyes. To measure the retinal velocity, we have to know the eye movements. In practice, it is difficult to specify eye movements. Smooth-pursuit eye movements depends on many factors, e.g. predictability of object motion, object size, background illuminance, prior expectations of motion, image velocity and image acceleration [4]. These factors make eye movements unpredictable.

If we example the function of eye movements, we will find that eye movements always try to fix the scene of interest on the retina, or to minimize the difference between the new incoming image and the percepted image, so as to achieve visual perception. In this sense, we can imitate this function by the motion compensation technique.

Motion compensation is not a new concept. It had been used in video coding since 1964. Today, motion compensation has become a powerful tool to remove temporal redundancy. However, the widely used motion compensation does not minimize what the eyes do. Eye movements try to align the input visual information to the received one, that is, eye movements directly perform motion compensated operation on the input frame while in the conventional image coding concept the motion compensated operation is imposed on the previous frame. Although this sounds very similar, there is a great deal of difference between them. Given two successive frames I_{i-1} and I_i, this can be explained as follows.

For the human observer, the frame I_i is aligned to I_{i-1} as I^*_{i-1},

$$I^*_{i-1} = \mathcal{K}(I_i) \tag{10.28}$$

where $\mathcal{K}(.)$ is the motion compensation generated by eye movements.

Then the difference signal

$$d = I^*_{i-1} - I_{i-1}$$

is used for subsequent visual perception. Therefore, the useful visual information is contained in the difference signal d. It is unnecessary to keep the input frame I_i as faithfully as possible. Instead, we should do our best to transmit the difference signal d.

In conventional video coding schemes, motion compensation is performed in a different way (as shown in Figure 12(a)): namely to align the previous frame I_{i-1} to the input frame I_i as I^*_i,

$$I^*_i = \mathcal{H}(I_{i-1}) \tag{10.29}$$

where $\mathcal{H}(.)$ is the motion compensation.

The residual signal $r_o = I_i - I^*_i$ is sent to the receiver as updating information. However, the residual information r_o is normally not the same as the difference signal d required by visual perception but is the updating information required

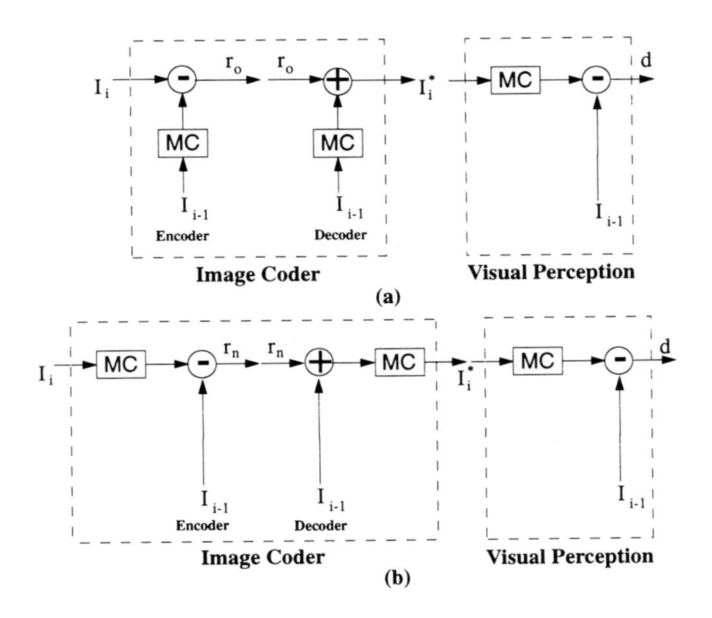

Figure 12 Motion compensated operation in conventional scheme (a), in new scheme (b).

for the reconstruction of the input frame I_i. This is verified by the simulation of the effect of eye movements on the reconstructed input frame.

At the receiver, the reconstructed frame is $\hat{I}_i = I_i^* + r_o$. Through eye movements, this frame is aligned into

$$\hat{I}_{i-1}^* = \mathcal{K}(\hat{I}_i) = \mathcal{K}(I_i^* + r_o) = (\mathcal{K}o\mathcal{H})(I_{i-1}) + \mathcal{K}(r_o) \tag{10.30}$$

The obtained difference signal

$$d = \hat{I}_{i-1}^* - I_{i-1} = (\mathcal{K}o\mathcal{H})(I_{i-1}) - I_{i-1} + \mathcal{K}(r_o) \tag{10.31}$$

Due to the fact that $\mathcal{K}(.)$ and $\mathcal{H}(.)$ are two independent motion compensated operations, normally $(\mathcal{K}o\mathcal{H})(I) \neq I$. Therefore, d and r_o do not correspond to the same information. This implies that not right thing is sent to the receiver.

To imitate the function of eye movements, motion compensated operation should be done the following way (as shown in Figure 12(b)): the input frame is aligned to the previous frame and the residual signal r_n between them is then sent to the receiver.

$$r_n = I_{i-1}^* - I_{i-1} \tag{10.32}$$

where $I_{i-1}^* = \mathcal{F}(I_i)$ and $\mathcal{F}$ is the forward motion compensation which aligns the input frame to the previous frame.

At the receiver, the input frame can be reconstructed by a backward motion compensated operation:

$$\hat{I}_i = \mathcal{B}(I_{i-1}^*) = \mathcal{B}(I_{i-1} + r_n) \tag{10.33}$$

where $\mathcal{B}$ is a backward motion compensation. To recover the input frame the backward motion compensation must satisfy

$$(\mathcal{F}o\mathcal{B})(I) = I \tag{10.34}$$

When the reconstructed image is observed, the input signal to the retina via eye movement for this frame is

$$I_{i-1}^* = \mathcal{K}(\hat{I}_i) = (\mathcal{K}o\mathcal{B})(I_{i-1} + r_n).$$

If the forward motion compensation is made to be very close to the motion compensation performed by eye movement $\mathcal{K} \approx \mathcal{F}$, then $(\mathcal{K}o\mathcal{B})(I) \approx I$ according to (10.34).

The difference signal

$$d = (\mathcal{K}o\mathcal{B})(I_{i-1} + r_n) - I_{i-1} \approx I_{i-1} + r_n - I_{i-1} = r_n.$$

This reflects that what is transmitted is just what the visual system needs.

According to this principle, a new motion-compensated hybrid coding scheme is suggested as shown in Figure 13. In comparison to the traditional hybrid coding scheme shown in Figure 1, computational complexity is added. However, the additional complexity helps in forward and backward image decoding, which is useful in multimedia applications.

A.3 How to achieve selective compression

To achieve selective compression of an image sequence based on a priori selection of regions-of-interest, we need to carefully choose suitable image coding schemes to compress the residual information, otherwise, false contours may

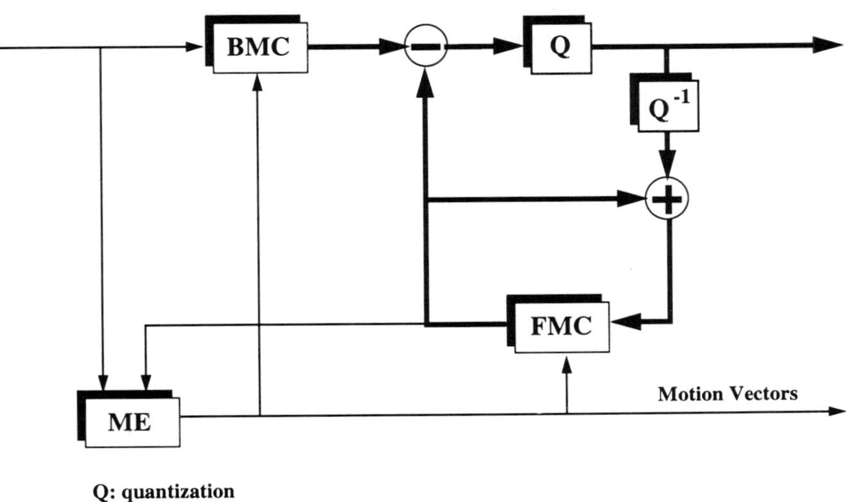

(a) Encoder

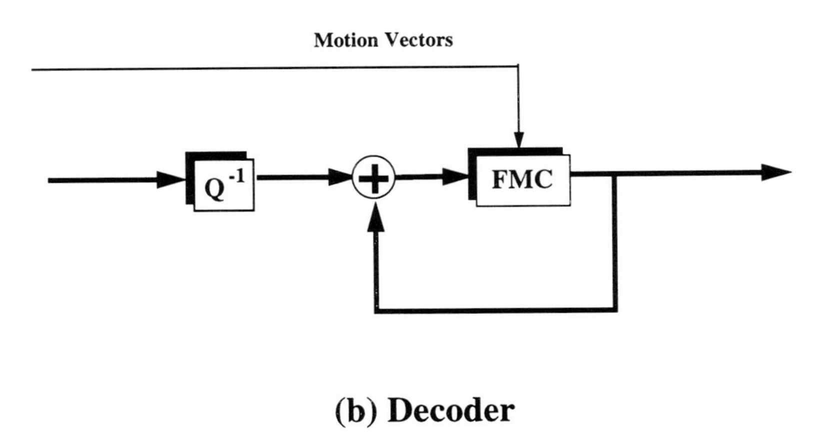

(b) Decoder

Figure 13 A new motion compensated hybrid coding scheme.

appear in between the regions-of-interest. The false contours may pull our focusing from the vital regions, which will be perceptually annoying. It seems that the hierarchical image coding scheme is suitable for this purpose. With hierarchical coding, selection can be made in either spatial domain or frequency domain or both. Preliminary experiments on this subject have appeared in literature. The interested reader is referred to [17] [29].

3.2 Subjective quality metrics

In image coding, an important issue is how to measure the quality of compressed images. The most commonly used quantity for measuring the distance between a given grey-valued image I and an approximation of the image $\hat{I}$ is the so called SNR-value, the L^2-metric.

$$MSE = E(I - \hat{I})^2 \tag{10.35}$$

However, the SNR criterion is far different from subjective assessment made by human observers. This becomes especially significant when the SNR is used to assess the image quality at a very high compression ratio (or very low bit-rate image coding). In very low bit-rate image coding, the coding distortions normally manifest no longer as quantization distortion but rather as geometric distortion. Therefore, the SNR is not competent for quality measure due to the fact that it is only based on differences at each pixel; it does not take into consideration spatial variation. Therefore, even small translations of an image may give rise to a fairly large distance. We need to search for more suitable distance measures which are consistent with subjective assessments.

With the emerge of fractal coding the use of image similarity becomes particularly important. In order to measure self-similarity it is necessary to have a distance-function for images. In the already famous book [2] on fractals by M. Barnsley, the author uses two other metrics for measuring distances between images. For black and white images Barnsley uses the Haussdorff metric and for grey-valued images of equal mass, Barnsley uses the Kantorovich metric which he calls the Hutchinson metric.

A nice thing with the Haussdorff metric is that there exists very efficient algorithms for its computation. Certainly, the Haussdorff metric can be used for grey-valued images after some binary operations. For example, a greyscale image is transferred into a binary image first by means of some operators, e.g.

edge operator, or threshold operator. Although the Kantorovich metric can directly operate on grey-valued images, heavy computational loads with this metric must be solved first. Before we discuss the computation of this metric, let us look at the definition of the Kantorovich metric for grey-valued images.

Let K denote the rectangular set $\{(i,j) : 1 \leq i \leq N, 1 \leq j \leq M\}$, and let $P = \{p(i,j) : (i,j) \in K, p(i,j) \geq 0\}$ and $Q = \{q(i,j) : (i,j) \in K, q(i,j) \geq 0\}$ represent two rectangular images of the same size and having the same total grey value.

We define a *transportation plan* as follows: A transportation plan between P and Q is a 4-dimensional non-negative function $t(i,j,x,y)$ such that the following conditions are satisfied

$$\sum_{x,y} t(i,j,x,y) = p(i,j), \quad (i,j) \in K \tag{10.36}$$

$$\sum_{i,j} t(i,j,x,y) = q(i,j), \quad (x,y) \in K \tag{10.37}$$

We denote the set of all transportation plans by $T(P,Q)$.

Next let $d(i,j,x,y)$ denote a distance-function on the set K on which P and Q are defined. Normally, the distance function takes the Manhattan metric as

$$d(i,j,x,y) = |i-x| + |j-y| \tag{10.38}$$

The Kantorovich metric $D(P,Q)$ between P and Q is simply defined as the minimum of the transportation cost where the minimum is taken over all possible transportation plans,

$$D(P,Q) = min\{\sum_{i,j,x,y} t(i,j,x,y)d(i,j,x,y), \quad t \in T(P,Q)\} \tag{10.39}$$

Thus, the computation of the Kantorovich metric means that one has to solve a so called balanced transportation problem.

Figure 14 (a) the original Lenna image, (b) the fractal-decoded Lenna image.

From (10.39), we may understand why the Kantorovich metric will take a very long time to compute. In Jacquin's thesis [12] the author also refers to the Kantorovich metric, but does not use it because "it is extremely difficult to compute". The first to use the Kantorovich metric as a distance function between two-dimensional grey-valued images was probably M. Werman [42]. But in Werman's thesis one makes the observation that the computational complexity for computing the Kantorovich metric is so large that it hardly can be used for practical applications.

Using a primal-dual algorithm, T. Kaijser achieves a reasonable computation time [13]. If two images are fairly similar and not larger than 256×256 then the Kantorovich distance can be computed in approximately 2 hours on a SUN4/690 computer. For images consisting of 1000-2000 pixels (which fractal images and motion estimation based on block-matching algorithms quite often do) the Kantorovich distance can be computed within a few seconds.

One of the interesting aspects of the Kantorovich metric is that the computation process also creates a transportation plan T between the two images consisting of a large set of 5-vectors: $T = \{(i_n, j_n, x_n, y_n, m_n), n = 1, 2, .., N\}$ (the transportation image). Figure 15 illustrates the transportation image (without the masses) obtained when computing the Kantorovich distance between two images of size 256×256 namely the original Lenna (Figure 14(a)) and a fractal-coded Lenna (Figure 14(b)). One application for this transportation image could be to use it as a replacement for difference images in conventional hybrid coding schemes. To the authors' knowledge, no such experiments have so far been reported. Thus, the usefulness of the Kantorovich metric outside the area of fractal coding still needs to be verified.

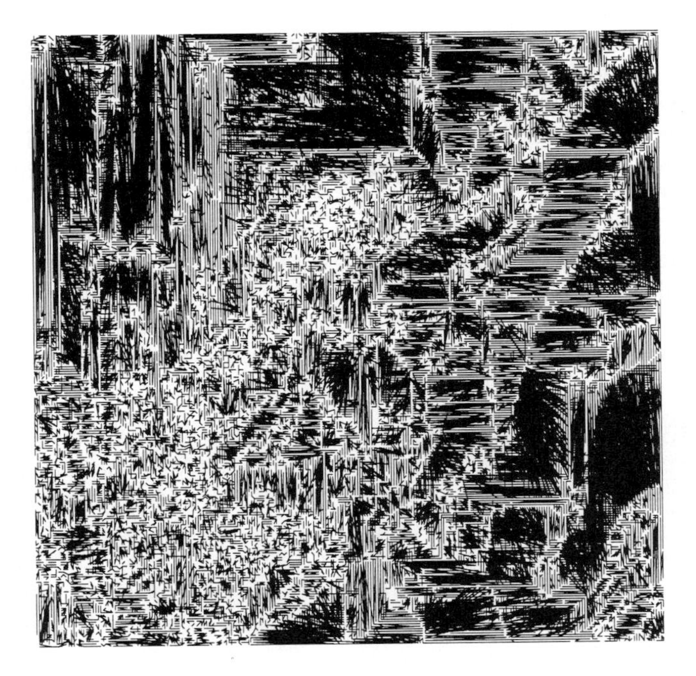

Figure 15 The transportation image obtained from Figure 14.

4 SIGNAL-DEPENDENT CODING

Almost all transform coding schemes are employing transforms with fixed basis functions, such as DCT. Up until today, the statistically optimum, signal-dependent transform, Karhunen-Loeve Transform(KLT) is still believed to be impractical! The reasons are twofold: First, it is necessary to estimate the data covariance matrix, a time-consuming task. Second, in a transmission system employing this type of coding, the basis functions also need to be sent to the receiver along with the coded data due to their signal-dependent nature. In contrast, DCT has become very popular. This is because DCT is a fixed transform with a fast algorithm and DCT is believed to have the best performance of all data-independent transforms and to be very close to that of the KLT. However, it should be pointed out that the latter conclusion is obtained under a *questionable* assumption. In this assumption, image data are modeled by a first-order Markov process with a high interpixel correlation. Obviously, this assumption is not valid for realistic images and it is only approximately obeyed for local patches. This is why the size of the encoding block patches takes the 8×8 pixel format.

Unlike DCT, the basis functions used in KLT are adaptively generated from the signal to be coded. Therefore one can expect that a signal can be more compactly represented by KLT than DCT. Due to the employment of signal-dependent basis functions, a larger size of blocks can be utilized in the KLT coding. In [22], blocks with the size 64×64 are used. The generated basis functions from the input image is shown in Figure 16. Experimental results are shown in Figure 17, from which we can see that a satisfactory image reconstruction can be obtained with as few as 8 basis functions (or eigenblocks). It is not likely that blocks with the size 64×64 are adopted in DCT coding. Therefore, employing blocks with large size will significantly improve transform coding efficiency, and normally the improvement is enough to compensate the cost of the transmission of basis functions.

Figure 16 The basis functions for the image 'Claire' used in KLT.

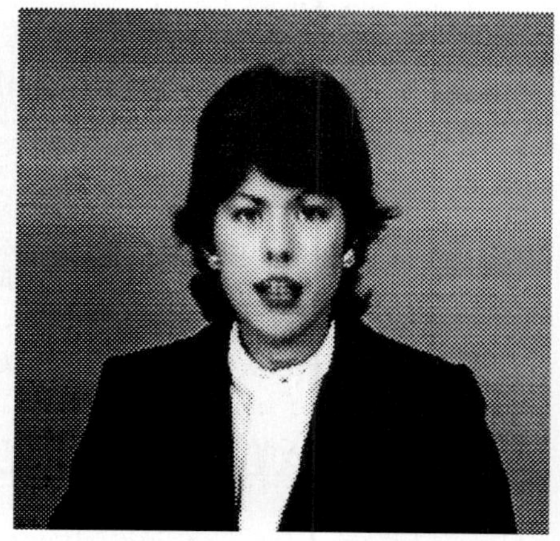

Figure 17 The synthesized 'Claire' image with only 8 eigenblocks.

As an extreme, we can take blocks as large as the whole frame. That is, an $N \times N$ full frame is taken as a coding unit, which is equivalent to viewing an image as a point in N^2 dimensional space. A typical value of N is 256, that is, we have to manipulate images in a $65,536$ dimensional space. In principle, the efficiency of transform coding can be maximally obtained by taking a frame as a block. However, we will face the problem of a huge number of images. If each sample is quantized with 8 bits, then we have $8N^2 = 524,288$ bits/image. That means there are $2^{524288} \approx 10^{158000}$ possible images. It is impossible to do coding operations on such a huge number of images. Fortunately, the number of combination is much larger than that of actual images. Although it is impossible to know the exact number of actual images, we can, however, obtain a rough estimate: *up to now, the number of all images which have been perceived by human beings is only about* 10^{19} *[18]*[1]. Therefore, most of the combinations correspond only to random patterns.

In practice, pictorial data contain significant structure, which make them differ from randomness. Even when considering the structure property, there is still a huge number of possible images. However, for an image sequence, each frame is typically a projection of a 3-D scene leading to a lot of temporal correlation. In many very low bit-rate video coding situations various constraints exist such as a relatively fixed scene, limited motion or repetition of the same scenes. Therefore, for a short time interval, image sequence only evolves in a subspace of the huge space. We argue that, to utilize this, it is wiser to employ a set of signal-dependent basis coding methods instead of a set of fixed-basis coding methods. This illustrates that the image sequence can be compactly described in a low dimensional subspace.

In fact, this conclusion has been verified in the computer vision community. For example, Breuel [3] has proven that only $O(\alpha^2)$ 2-D aspects are needed to cover the entire 3-D viewing sphere with a 2-D matching error bounded by α. For the face scene, only 30 faces with different views may be required to cover all viewing directions and only five to cover 180 degree of horizontal views [37]. For well aligned facial images, a total of 50 aspects is enough to represent any facial image [14]. Moreover, it may be possible to reduce the number of aspects even further. For instance, Ullman [38] and Poggio [32] have both argued that 3-D recognition can be accomplished using a linear combination of as few as four or five 2-D views.

This fact reduces the number of image frames needed to be stored and manipulated to a reasonable level. The following problem is, since the image sequence subspace can be described using low dimensional vectors, how to characterize the low-dimensional space. A solution is to introduce a statistically optimal basis vectors extraction method by means of Karhunen-Loeve Transform techniques [8].

Assume that the images to be encoded have been collected into an ensemble of images $\{B^{(n)}\}_{n=1}^{p}$, where each $B^{(n)}$ is an N^2 dimensional vector, normally a frame, and p is the total number of images.

[1]This number is estimated in this way: Assume that 30 frames per second are percepted by human eyes. If the average life-span is 70 years and 8 hours per day are spent on sleep, then about 10^8 frames are received by one person during his life. Furthermore, if the total number of human beings who lived on the earth is 10^{11}, then, the total number of images which have been perceived is approximately 10^{19}.

In order to increase efficiency, we first compute the ensemble average vector

$$\bar{B} = \frac{1}{p}\sum_{n=1}^{p} B^{(n)} \tag{10.40}$$

Then we remove the ensemble average vector from each image

$$C^{(n)} = B^{(n)} - \bar{B} \tag{10.41}$$

A covariance matrix is built from $C^{(n)}$,

$$\mathbf{D} = \frac{1}{p}\sum_{n=1}^{p} C^{(n)} C^{(n)T} \tag{10.42}$$

$\mathbf{D}$ is an $N^2 \times N^2$ matrix.

The eigenvectors $\{\phi^{(k)}\}_{k=1}^{p}$ of the ensemble averaged covariance matrix $\mathbf{D}$

$$\mathbf{D\Phi} = \mathbf{\Phi}\lambda \tag{10.43}$$

with $\lambda_1 > \lambda_2 > \lambda_3 > ...$, need to be computed.

Due to its huge dimension, it is difficult to calculate eigenvectors from the covariance matrix $\mathbf{D}$. However, we should notice that only p frames are available in the N^2 dimensional vector space. $\mathbf{D}$ is a function of p or less linearly independent vectors. The rank of $\mathbf{D}$ should therefore be p or less

(10.42) can be written as

$$\mathbf{D} = \frac{1}{p}\mathbf{U}\mathbf{U}^T, \tag{10.44}$$

where $\mathbf{U}$ is a $N^2 \times p$ matrix called *a sample matrix*

$$\mathbf{U} = [C^{(1)}...C^{(p)}]. \tag{10.45}$$

Instead of using the $N^2 \times N^2$ matrix $\mathbf{D}$, let us calculate the eigenvalues and eigenvectors of the following matrix

$$\mathbf{S} = \frac{1}{p}\mathbf{U}^T\mathbf{U} \tag{10.46}$$

where $\mathbf{S}$ is an $p \times p$ matrix. The computation of eigenvectors and eigenvalues is here greatly reduced, e.g. if 64 images whose sizes are 256×256 are used, the dimension of $\mathbf{S}$ is only 64×64 while that of $\mathbf{D}$ is $65,536 \times 65,536$!

The eigen-equation of $\mathbf{S}$ is

$$\mathbf{S}\Psi = \Psi\Lambda \tag{10.47}$$

where Ψ and Λ are the eigenvectors and eigenvalues of the matrix $\mathbf{S}$.

Multiplying $\mathbf{U}$ into (10.47) from the left side, we obtain

$$\mathbf{U}\mathbf{S}\Psi = \mathbf{U}\Psi\Lambda$$

$$\mathbf{D}(\mathbf{U}\Psi) = (\mathbf{U}\Psi)\Lambda \tag{10.48}$$

Thus, $(\mathbf{U}\Psi)$ and Λ are the p eigenvectors and eigenvalues of $\mathbf{D}$. The other $(N^2 - p)$ eigenvalues are all zero and their eigenvectors are indefinite. The matrix $(\mathbf{U}\Psi)$ represents orthogonal but not orthonormal vectors. To obtain orthonormal vectors $\phi^{(i)}$, a normalization is needed

$$\phi^{(i)} = \frac{\mathbf{U}\psi_i}{(p\lambda_i)^{1/2}}. \tag{10.49}$$

Then each image can be exactly reconstructed from these eigenvectors using

$$B^{(n)} = \sum_{k=1}^{p} a_{kn}\phi^{(k)} + \bar{B} \tag{10.50}$$

In order to compress the image sequence, we keep only q eigenvectors, that is

$$B^{(n)} \approx \sum_{k=1}^{q} a_{kn}\phi^{(k)} + \bar{B} \tag{10.51}$$

Then each image can be represented by a fixed set of eigenvectors $\{\phi^{(k)}\}_{k=1}^{q}$ and a sequence of coefficients a_{kn} corresponding to each element of the fixed set. Due to the eigenvectors corresponding to images, they can be called *eigen-images*. Using eigenimages, any input image of this image sequence can be represented by an optimal linear combination of these eigenimages. In comparison with bilinear prediction used in MPEG, this method can be viewed as an optimal multiframe prediction. A preliminary experiment was performed on the image sequence 'Claire'. The extracted eigenimages are shown in Figure 18. Using 16 eigenimages, the synthesized image sequence is illustrated in Figure 19. When the same eigenimages are available at both the transmitter and receiver, each input image can be represented well with only 16 parameters!

Based on this idea, it is possible to develop a full-frame image sequence coding scheme [20]. A basic schematic diagram of such a scheme is shown in Figure 20. In this scheme there are three main functional modules: *frame storage*, *image transformation & representation*, and *information updating*. This scheme is different from the conventional schemes in four ways 1) *the full-frame image is taken as a coding unit instead of the usual block-patches*; 2) *the coding scheme is signal-dependent*; 3) *multiframes or transformed frames are used to make the prediction of the current frame*; 4) *the majority of the bit stream are used to update the basis functions.*

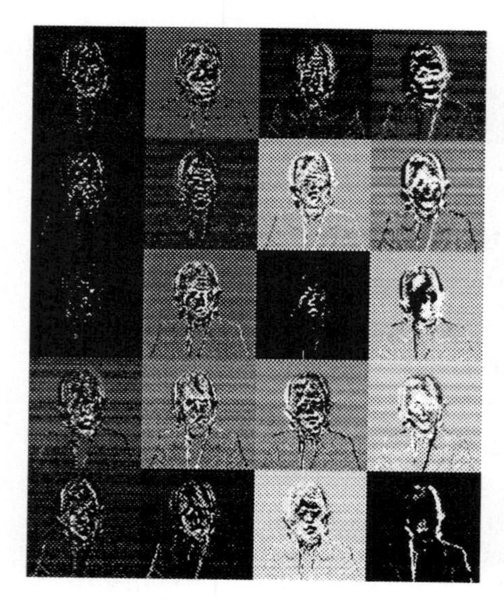

Figure 18 The extracted eigenimages from the image sequence 'Claire' by KLT.

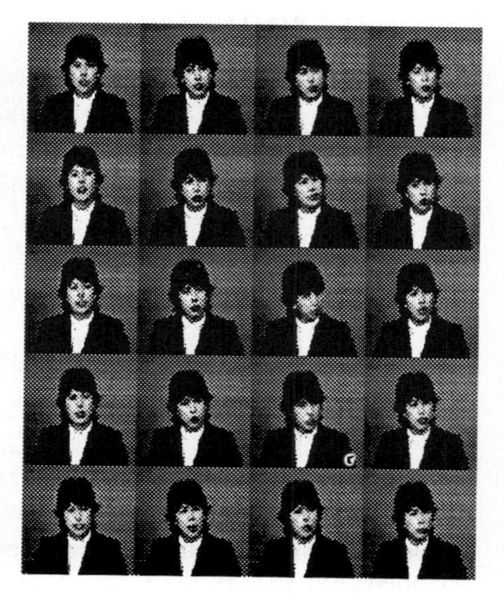

Figure 19 Every fourth frame of the reconstructed image sequence 'Claire' with 16 eigenimages.

It is worth pointing out that interframe motion and illumination changes will cause temporally varying statistics in the image sequence. The interframe changes can be modelled as transformations, such as small rotations, small scale changes, and changes in the illumination position or brightness, which are approximately linear over small variations in the transformation parameters. In contrast, transformations such as global translation, large rotation, or large scale change will generate shape changes in the image set at projective discontinuities corresponding to rapid albedo changes, high surface curvatures, and contours. The interframe motion will destroy the generalization of the eigenimages. To avoid this, we have to compensate for the interframe motion. A technique, namely motion compensated alignment[20], can be used for this purpose. In this technique, input images are aligned to a reference frame. In this way, interframe variations due to large interframe motion are greatly reduced [18].

5 SPATIO-TEMPORAL INFORMATION

5.1 Introduction

A significant feature of current video compression schemes is simple prediction based on the previous frame. This originated in the early limitation of storage capacity and computing ability at the time when only a single frame could be stored in the frame memory and simple prediction was made using this frame. Three significant drawbacks of the current use of a single frame storage are:

· *only a single frame used for prediction*

· *simple frame operation*

· *continuous frame refresh*

A single frame is insufficient to represent the related segments of the image sequence. This is because some parts visible from some viewing angles may be occluded from others, making it difficult to use a single frame to represent the scene from different viewing directions. This becomes particularly significant in the case of covered/uncovered background situations.

When making prediction, the existing video compression schemes employ simple operations, e.g. translational shift, on the stored frame image. As translational shift is not a good approximation of interframe motion, making it difficult to obtain a highly efficient motion compensated prediction.

When only a previous frame is stored in the frame storage, no information is stored over a longer period. Potentially, new information is transmitted for each image and it may be that much of this information is redundant, having been previously transmitted. Therefore, in some situations, image information should be stored over a period of many image frames and be re-used when possible. For example, in videophone scenes, if the head appears in the same position as in some previous image frame, it needs not be re-transmitted. Obviously,

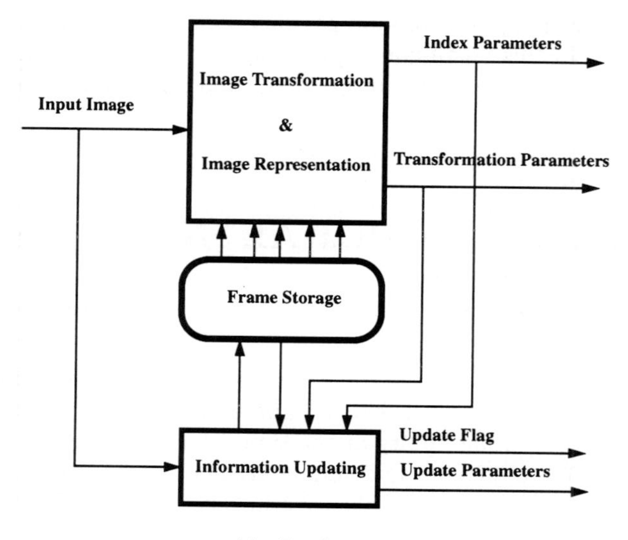

(a) Encoder

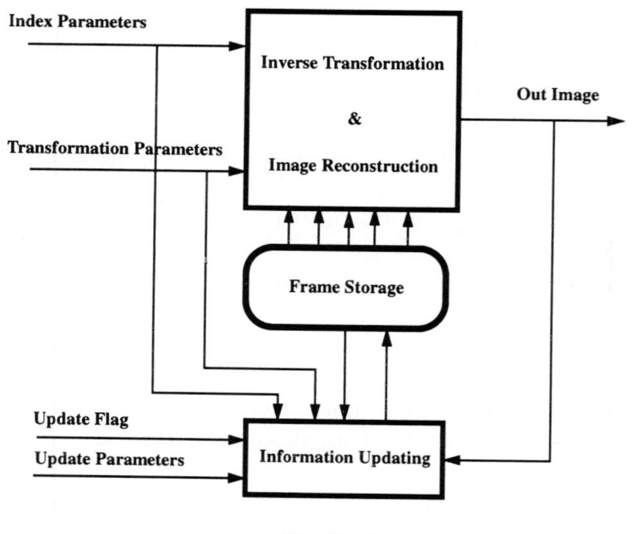

(b) Decoder

Figure 20 Schematic diagram of full-frame image sequence encoding and decoding.

the more frames that are stored, the less information needs to be transmitted. This strategy has been verified implicitly by the MPEG standards, where two frames are used to make bidirectional interpolation.

In comparison with a single frame prediction, a simple but very powerful strategy is to keep the transmitted spatio-temporal information for future use instead of simply discarding it as in conventional hybrid coding schemes! This is motivated by the following aspects: (1) Two commonly occurring events in movies and video communication sequences are repetitive or cyclical activity and transient actions [39]. If we could hold recurring visual elements such as cyclically revealed information and actors that persist through several images, then a great reduction in bit-rate can be achieved when the same or similar scene reappears. (2) The existence of the spatio-temporal information facilitates interactive and content-based operations on the video sequence such as browsing, searching by content and categorizing. (3) The spatio-temporal information can be used to reduce image noise, or to enhance spatial and temporal resolution.

When using the transmitted spatio-temporal information to aid low bit-rate video coding, we should consider the following three important issues: spatio-temporal information storage, spatio-temporal information operation, and spatio-temporal information updating. The spatio-temporal information can be stored by taking either blocks, frames, scenes or objects as basic storage entities. Normally, in waveform-based coding, blocks or frames are utilized as storage entities while in model-based image coding scenes or objects are the storage entities.

5.2 The employment of spatio-temporal information in waveform-based coding

We will now discuss how to employ spatio-temporal information to enhance waveform-based video coding, e.g. several international coding standards such as H26P and MPEG. The basic idea is to use a cache memory to save the transmitted spatio-temporal information taking either blocks or frames as basic storage entries.

In the scheme described in [39], the entries are blocks in the cache memory. The additional memory in the coding process is used to allow prediction bases to be saved for more than a single frame period. A library of elements of the frame is built, which persists until changes in the longer term statistics of the video sequence renders them obsolete. Vector Quantization formalism is used to organize and search the library. For example, a vector quantizer divides

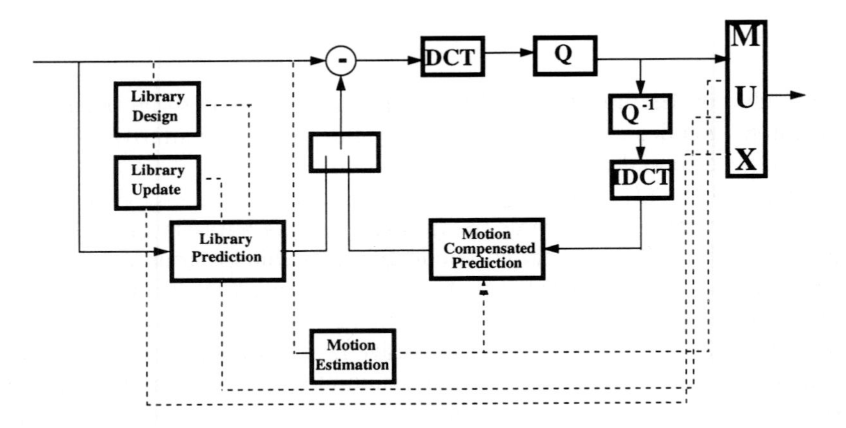

Figure 21 Motion compensated DCT coder with a cache. (from [39])

its input space into a set of clusters, by grouping together vectors (blocks) with similar characteristics, minimizing the overall distortion. The famous LBG algorithm is used for designing the library. Once the library used to encode a given frame is designed, the new library entries must be transmitted as side information that will be used by the decoder. The following well-known mechanisms for the organization of computer memories [35] can be borrowed to manage the library: *the least recently used (LRU) strategy*, and *the least frequently used strategy (LFU)*, *the first-in/first-out strategy (FIFO)*, and *the working set model(WSM)*.

The coding scheme of motion compensated DCT combined with a cache is shown in Figure 21. Each input frame is segmented into a set of square blocks, which are then processed to minimize both temporal and spatial correlation. Two different prediction structures are used for temporal processing: library- (or cache-) based predict and motion compensated prediction. The DCT is used for encoding the prediction error signal. The DCT coefficients are scalarly quantized and entropy coded. Besides the normal terms needed to be transmitted, the new cache entries which are encoded as regular image blocks, and the index of the blocks from the cache need to be transmitted.

An experimental coding scheme equipped with a motion compensated predictor and a library-based predictor has been tested. Experimental results show that motion compensation is in general chosen for the prediction of blocks without motion or under translational motion (like the background), and the library prediction handles the more complex cases of revealed objects and nontranslational motion. Therefore library-based coding gives a much faster response to scene changes. Experimental results reported in [39] show that the peak-gain provided by the library-based encoder is about 5 dB over a pure motion compensated encoder although the overall gain is about 0.5 dB.

The major problem of this scheme lies in the complexity of codebook design. In addition, since information is stored by taking blocks as entities, it is difficult to align the input blocks to the stored blocks. This will significantly reduce the efficiency of library-based prediction. To overcome these drawbacks, it seems that it is wiser to take image frames as the storage entities. Specifically, we can use the past reconstructed frames to predict blocks of the current frame instead of using a previous frame only. In the previous-frame prediction mode, although most blocks can be found in the previous frame there is still a certain number of blocks that can be found in past frames. Normally, these blocks are perceptually important ones. Obviously, if we can make good predictions from these frames, a significant saving of bits can be achieved. In addition, according to our experiments it seems that using past 10 frames is sufficient to predict the current frame. This is important because otherwise a larger frame memory and a more extensive motion estimation would have been needed. However, we should point out that the results are obtained from a videophone scene which normally contains no scene transition. If there are other activities, e.g. different actions, another 10 or several 10 frames are required.

Since the frame is used as stored entines, a current block can be predicted from the shifted-form. In this way, a more efficient prediction is achieved. Of course, more complex predictions like the transformations being done in fractal coding are possible. Experimental results shown in Figure 22 (a) and (b) illustrate that with prediction by past frames the number of blocks needed to be transmitted is greatly reduced.

Due to the introduction of multiple frames, motion estimation may face some problems. e.g. the search scope, multiframe search and so on. However, making use of the interframe correlation, we can design a frame-by-frame index technique to avoid these difficulties. In addition, since the indexed blocks lie in the area of interest, it is unnecessary to store a whole frame, only parts of the frames need be stored. This will reduce the memory requirement.

5.3 The employment of spatio-temporal information in model-based coding

Due to its extremely high compression efficiency, model-based image coding, including object-oriented coding and semantic-based coding, has been exten-

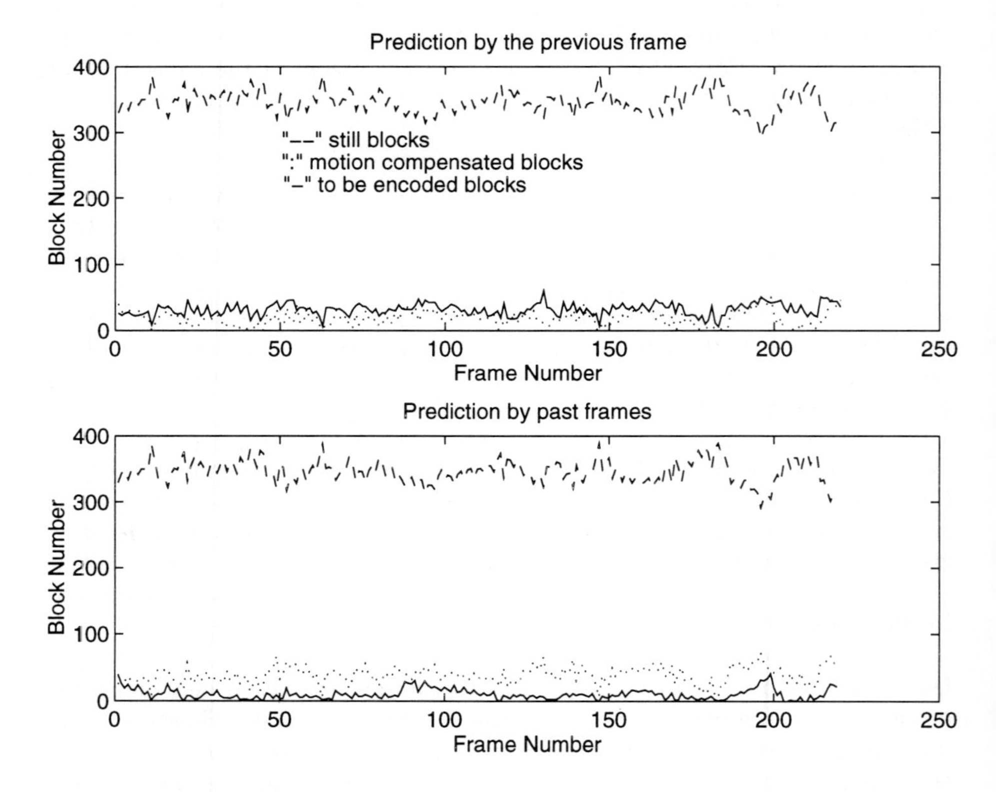

Figure 22 The distribution of different types of blocks over time: (a) prediction is made by using a previous frame, and (b) prediction by using past frames.

sively investigated and developed [1][6][7][19][27][30][41]. Although in model-based coding the input images have been operated on a structural level, there is not a systematic framework to re-use the transmitted information. For example, in a typical object-oriented coder, the transmitted information, that is, the residual information about model-failure areas, is discarded via a frame-by-frame refresh strategy. This strategy originates from the hybrid coding scheme. Similarly, in semantic-based coding, current discussion is now only over how to update the texture information [28][31]. No attempt is made to store the coming new texture information for future usage. In this way, when information is repeated, we have to re-transmit it! This greatly decreases image transmission efficiency.

Therefore, at very low bit-rate video transmission, a key technique is how to re-use the transmitted information. There are a few scattered discussions about the re-use of transmitted information in the literature e.g. [40][43], however, the ways of re-using the information are still at a lower level: block-level. We need a systematic framework to study the re-use of sources. Generally speaking, we should give up the employment of a frame storage. We should, instead, employ a scene storage which does not necessarily have the same format as the input image. The input image is viewed as just a window over scene. By integrating image information exposed in a series of windows, the scene can be reconstructed for future use. Only in this way can the transmitted information be employed with high efficiency. In practice, there are two types of scene reconstruction depending on the description of scenes: *background reconstruction* and *object reconstruction*.

In many applications, a background can be modeled as a still or slowly moving 2-D planar object. 2-D affine motion, either global or local, is enough to model its motion. Through estimating affine motion parameters, the background shown in a series of windows can be registered and then reconstructed. An example is illustrated in Figure 23. Equipped with a background storage, the efficiency of object-oriented coding will be significantly increased, especially for the background. However, for objects in the scene, the storage of their textures is more difficult. To register textures, we first have to recognize the objects, which is unnecessary in the background reconstruction. Therefore, we require a lot of related information, such as orientation, shape, depth, about the objects. Object construction will add considerable complexity to the coding process of normal image coding schemes and is not adopted there. However, object construction is an important topic in future semantic image coding. For example, in semantic facial image coding, a human face is recognized in advance and a 3-D wireframe is used to provide a potential way to organize textures. Each input frame just corresponds to a particular view, which contributes a

Figure 23 A reconstructed scene from a series of image frames.

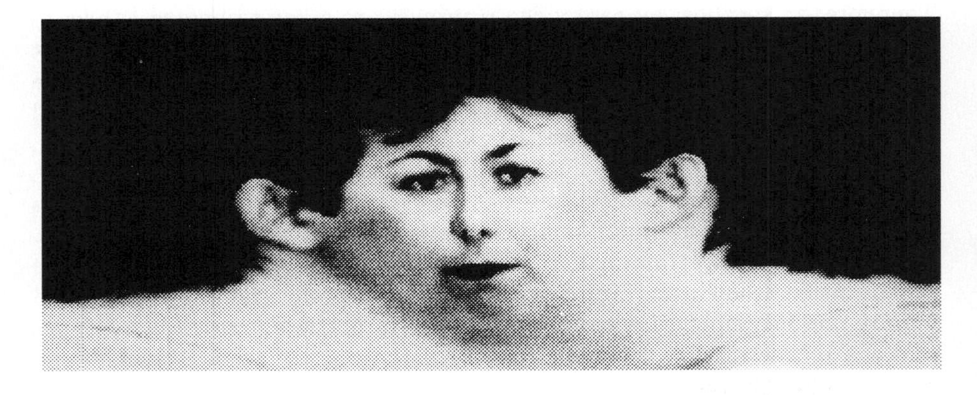

Figure 24 A pantoscopic facial texture image. (from [36])

part of the pantoscopic texture. The successive frame provides the revealed information. Through transmission, the fresh information is registered in the pantoscopic texture through a 3-D motion estimation [21][23] and is then used to update the pantoscopic textures. In this way, the pantoscopic texture can be gradually reconstructed. After a series of views, a complete pantoscopic facial texture can be obtained. A pantoscopic texture is very useful for facial image synthesis and browsing. An example of a pantoscopic facial texture image for facial image coding is shown in Figure 24.

With the rapid development of computer graphics, more and more objects or parts of them can be rendered with greater reality. Background and object reconstruction will depend more and more on the graphics techniques. In other applications, where narrow communication channels may cause other problems, e.g. communication delay, the usage of the transmitted information or the known information becomes particularly important. A promising approach is to mix live video and computer generated imagery with the respective geometries registered together. If the background is known in advance or is completely virtual, then the scene obtained from the camera can be artificially synthesized by means of Virtual Reality (VR) technology at the receiver. Therefore, we can prolong that in the future image coding decoders which will use more and more artificial images.

6 ACKNOWLEDGMENTS

The authors would like to acknowledge the contribution from the other members of the Image Coding Group. Special thanks are due to Astrid Lundmark for her contribution on Section 2, Thomas Kaijser for his contribution on Section 3.2, and Niclas Wadströmer for making Figure 14 and 15.

REFERENCES

[1] K. Aizawa, H. Harashima, and T. Saito, Model-Based Analysis Synthesis Image Coding for a Person's Face, Image Communication, Vol. 1, No. 2, pp. 139-152, 1989.

[2] Michael Barnsley. Fractals Everywhere. Academic Press, 1988

[3] T. Breuel, Indexing for Recognition from a Large Model Base. MIT AI Memo 1108, 1990.

[4] M. Eckert and G. Buchsbaum, The Significance of Eye Movements and Image Acceleration for Coding Television Image Sequences. In Digital Images and Human Vision, ed. A. Watson, MIT Press 1993.

[5] R. Forchheimer, Differential Transform Coding – A New Hybrid Coding Scheme, in Proc. Picture Coding Symp. (PCS-81), Montreal, Canada, June. 1981, pp. 15-16.

[6] R. Forchheimer, O. Fahlander, and T. Kronander, Low Bit-Rate Coding Through Animation, Picture Coding Symposium'83, Davis, CA, pp. 113-114, March, 1983.

[7] R. Forchheimer and T. Kronander, Image Coding-From Waveforms to Animation, IEEE Trans. on ASSP, Vol. 37, No. 12, pp. 2008-2023, December, 1989.

[8] K. Fukunaga, Introduction to Statistical Pattern Recognition, Academic Press, 1990.

[9] B. Girod, Motion Compensation: Visual Aspects, Accuracy, and Fundamental Limits. In Motion Analysis and Image Sequence Processing, ed. M. Sezan and R. Lagendijk, 1993.

[10] J. Gisladottir and M. Orchard, Motion-only Video Compression. In Proceedings of International Conf. Image Processing, 1994.

[11] A. Habibi, Hybrid Coding of Pictorial Data, IEEE Trans. on Communications, COM-22, No. 5, May 1974, pp. 614-624.

[12] A. Jacquim, A Fractal Theory of Iterated Markov Operators with Applications to Digital Image Coding. Ph.D. Thesis, Georgia Institute of Technology, August, 1989.

[13] T. Kaijser, On the Kantorovich Metric for Images, Proc. SSAB, 1993.

[14] M. Kirby and L. Sirovich, Application of the Karhunen-Loeve Procedure for the Characterization of Human Faces, IEEE Trans. Pattern Anal. Machine Intell. Vol. 13, No. 10, 1991.

[15] T. Kronander, Some Aspects of Perception Based Image Coding, Ph.D. dissertation 203, Dept. Elec. Eng., Linköping Univ., Jan. 1989.

[16] M. Kunt, A. Ikonomopoulos and M. Kocher, Second Generation Image Coding Techniques, Proc. IEEE, Vol. 73, No. 4, pp.795-812, 1985.

[17] O. Kwon and R. Chellappa, Region-based Subband Image Coding Scheme. In Proceedings of International Conf. Image Processing, 1994.

[18] H. Li, Low Bitrate Image Sequence Coding, Ph.D. dissertation 318, Dept. Elec. Eng., Linköping University, Nov. 1993.

[19] H. Li, A. Lundmark and R. Forchheimer, Image Sequence Coding at Very Low Bitrates: A Review, IEEE Trans. Image Processing, Vol. 3, No. 5, pp.589-609, 1994.

[20] H. Li and R. Forchheimer, Full-frame Image Sequence Coding, Technical Report LiTH-ISY-R-1777, 1993.

[21] H. Li, P. Rovivainen, and R. Forchheimer, 3-D Motion Estimation in Model-Based Facial Image Coding, IEEE Trans. PAMI, Vol. 15 No. 6, pp.545-555, 1993

[22] H. Li and R. Forchheimer, KLT-Based Image Coding, in Proceediings of SSAB on Image Analysis, Mar., 1993.

[23] H. Li and R. Forchheimer, Two View Facial Motion Estimation, IEEE Trans. CSVT, Vol.4 No.3, pp.276-287, June, 1994.

[24] H Li, Segmentation of the Facial Area for Videophone Applications", Electronics Letters, Vol. 28, No. 20, pp.1915-1917, Sep. 1992.

[25] H. Li and R. Forchheimer, Low Entropy Motion Computation. In Proceedings of Very-Low Bitrate Video Coding, Tokyo, 1995.

[26] A. Lundmark, Using Certainty of Motion Estimation in Image Coding, Licentiate Thesis (under preparation), Linköping University, 1995

[27] H. G. Musmann, M. Hötter and J. Ostermann, Object-Oriented Analysis-Synthesis Coding of Moving Images, Image Communication, Vol. 1, No. 2, pp.117-138, Oct., 1989.

[28] Y. Nakaya, K. Aizawa, and H. Harashima, Texture Updating Methods in Model-Based Coding of Facial Images, Picture Coding Symposium '90, Boston, 1990.

[29] E. Nguyen, C. Labit and J. Odobez, A ROI Approach for Hybrid Image Sequence Coding. In Proceedings of International Conf. Image Processing, 1994.

[30] D. Pearson, Developments in Model-Based Video Coding, Proc. IEEE, Vol. 83, No. 6, pp.892-906, June 1995.

[31] D. Pearson, Texture Mapping in Model-based Image Coding, Image Communication, Vol. 2, No. 4, pp.377-396, 1990.

[32] T. Poggio, and S. Edelman, A Network That Learns to Recognize Three Dimensional Objects. Nature Vol. 3433, No. 6255, 1990.

[33] Image Transmission Techniques, ed. W. Pratt. Academic Press, 1979.

[34] C. Stiller, Object-oriented Video Coding Employing Dense Motion Fields. In Proceedings of ICASSP, Vol. 5, 1994.

[35] H. Stone, High-Performance Computer Architecture, Addison-Wesley, 1987

[36] Y. Suenaga and Y. Watanabe, A Method for the Synchronized Acquisition of Cylindrical Range and Color Data, IAPR Workshop on Machine Vision Applications, Nov. 1990, Tokyo.

[37] M. Turk, and A. Pentland, Eigenfaces for Recognition, J.Cognit Neurosci. Vol. 3, No. 1, 1991.

[38] S. Ullman, and R. Basri, Recognition by Linear Combinations of Models, IEEE Trans. Pattern Anal. Machine Intell. Vol. 13, No. 10, 1991.

[39] N. Vasconcelos and A. Lippman, Library-based Image Coding. In Proceedings of ICASSP, Vol. 5, 1994.

[40] J. Wang and E. Adelson, Representing Moving Images with Layers, IEEE Trans. Image Processing, Vol. 3, No. 5, pp.625-638, Sep. 1994.

[41] W. Welsh, Model-Based Coding of Videophone Images, Electronics & Communication Engineering Journal, Vol. 3, No. 1, pp. 29-36, Feb, 1991.

[42] M. Werman, S. Peleg and A. Rosenfeld, A Distance Metric for Multidimensional Histograms, CVGIP 32, 1985.

[43] M. Wollborn, Object-oriented Analysis-synthesis Coding Using Prototype Prediction for Color Update. In Proc. Picture Coding Symp. (PCS-93), Mar. 1993. Lausanne, Switzerland,

Index